Wharton on managing emerg
 technologies

WHARTON
on
MANAGING
EMERGING
TECHNOLOGIES

WHARTON

on

MANAGING EMERGING TECHNOLOGIES

Edited by

GEORGE S. DAY

and

PAUL J.H. SCHOEMAKER

with

ROBERT E. GUNTHER

John Wiley & Sons, Inc.

New York • Chichester • Weinheim • Brisbane • Singapore • Toronto

Published by John Wiley and Sons, Inc.
Published simultaneously in Canada.

This publication is designed to provide accurate and authoritative information in
regard to the subject matter covered. It is sold with the understanding that the
publisher is not engaged in rendering legal, accounting, or other professional services.
If legal advice or other expert assistance is required, the services of a competent
professional person should be sought.

$ 3495

Library of Congress Cataloging-in-Publication Data:

Day, George S.
    Wharton on managing emerging technologies / George S. Day, Paul J.H.
Schoemaker, Robert E. Gunther.
        p.   cm.
    Includes bibliographical references and index.
    ISBN 0-471-36121-6 (cloth : alk. paper)
    1. Technology—Management.   2. Technological innovations—Management.   I.
Schoemaker, Paul J.H.   II. Gunther, Robert E., 1960–   III. Title.

    T49.5 D395 2000
    658.4′062—dc21

                                                                    99-056131

Printed in the United States of America.

10   9   8   7   6   5   4   3   2   1

2001 5330
♭4273645l

PREFACE

THE CHALLENGE

Although Terry Fadem is officially director of corporate new business development at DuPont, in reality he has dropped off the organization chart. He now works in a vortex of relentless experiments to gain advantage from the company's 18,000 or so patents. His division cannot be considered a business unit in the traditional sense, nor can his job be considered management. He behaves much more like a venture capitalist.

After its restructuring in the 1990s, DuPont charged a core group to develop new markets from inventions such as a light-emitting material that could replace a light bulb. "Their job was to build new businesses any way they could," Fadem said. "We don't start with teams of people reporting to us, but over time, we develop a business plan, an organization, and so on." Even this approach wasn't fast enough, however, so the company began to assemble an array of venture teams. "We pulled together people on a team and told them their job was to find an opportunity the company could pursue, do a business plan, and if the company approved we would fund it."

DuPont's core businesses in nylon, fluorocarbons, and other areas, continue to operate within a formal organization. But life for Fadem and managers at the frontiers is very different. How do they manage numerous emerging technologies that create new opportunities or disrupt established businesses? How do they navigate the complexity, uncertainty and rapid change inherent in this untamed territory? How do they remain a part of a larger established firm without being swallowed or slowed by it? Where do they find the knowledge and talent to manage emerging technologies?

When a pair of researchers from the Wharton Emerging Technologies Program asked Fadem where he gets his insights on managing emerging technologies, he responded, "I try not to read the thinking of the day. My job is to build new businesses and I have to look for the edge all the time. If everyone is doing it, it may be too late for us. The only way to gain a competitive advantage is to go beyond that."

Emerging technologies, based on advances in information technology, biotechnology, and other scientific disciplines, are the future of some industries and will transform many others. These technologies are creating and restructuring industries at an unprecedented rate, making traditional practices obsolete, and creating a need for new best practices, core competencies, and competitive strategies. Established firms often have no choice but to become players in external technologies that could redefine their future. Yet established firms sometimes face the greatest challenges from within because managing emerging technologies requires a very different set of skills, frameworks, and strategies than managing existing technologies. How do managers master these new approaches? Like Fadem, they look for the edge.

LOOKING FOR THE EDGE

The Wharton Emerging Technologies Management Research Program was established in 1994 (before the web was a household word) to address these challenges. A diverse team of faculty began working with senior executives such as Fadem to understand the unique challenges of managing emerging technologies and develop strategies for success. This work was based on a growing awareness from our work in industry that emerging technologies would be important to future management success. Since then, the explosion of the Internet and maturation of biotechnology have put emerging technologies front and center in management thinking.

This initiative represents one of the first and broadest attempts to bring an element of rigor to a field that has been characterized by simplistic prescriptions, hype, and occasional hyperbole. Our research program marries the experience of practice with emerging tools, frameworks, and research by professors, consultants, and thoughtful practitioners. Some of these tools have been in our hands for decades (such as scenario planning and real options analysis) but are just beginning to find their way into popular use. The potential of other insights in this book, such as perspectives from evolutionary biology and network strategies, have yet to be widely recognized.

This book builds on an important research vein exploring the management of technology-based innovation. Some notable writings on this subject include Richard Foster's *Innovation: The Attacker's Advantage* (Summit Books, 1986), Geoffrey Moore's *Crossing the Chasm* (Harper, 1991), James Brian Quinn's *Innovation Explosion* (Free Press, 1997), and more recently Clay Christensen's *The Innovator's Dilemma* (Harvard Business School Press, 1997). While all of these works offer important insights into the management of emerging technologies, the goal of this book is to provide more than a single framework or perspective. This project draws together a wide range of authors from diverse disciplines who have focused their attention on emerging technologies. The management challenges presented by emerging technologies cut across the entire organization, calling on disciplines including finance, managing alliances, human resources, entrepreneurship, network theory, marketing, strategy, and many others. All these perspectives are represented in this book.

The present volume offers a rigorous distillation of our collective wisdom about the study of emerging technologies to date. Among the questions we address: How do managers need to change their approaches to financial analysis, marketing, competitive strategy, and internal organization? How do they avoid common traps and work with partners? How do they anticipate the public policy issues that will shape the emerging markets? How can they manage the organizational issues that make or break emerging technology initiatives? These discussions help prepare managers in established and new firms to compete, survive, and succeed.

Emerging Knowledge

This book examines the management challenges posed by emerging technologies at the point where scientific research reveals a technological possibility and goes all the way to the commercialization of the technology into lead markets. This is roughly the zone in the graph on page viii between the intersections labeled "competing modalities" and "competing applications." We do not deal extensively with the management of R&D, nor with the later-stage commercialization of a proven technology. We approach the issue from a top management perspective and do not deal with the intricacies of individual technologies. We've chosen this zone because it presents the greatest challenges for management.

The ability to master emerging technologies is essential to survival in a growing number of industries. This promises to be even more true in the

Figure P.1
How Emerging Knowledge Evolves

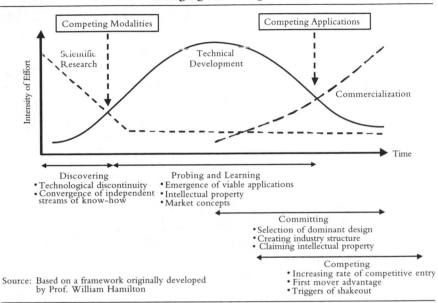

Source: Based on a framework originally developed
by Prof. William Hamilton

future. We are just at the beginning of a wave of breakthroughs that will
equal or surpass those of the past decade. A quick glance at the laboratories
in industry and science reveals technologies such as microelectromechanical
motors and sensors, made-to-order materials constructed an atomic layer at
a time, personalized medicine through genetically tailored treatments, organic electronics, and iris scanning for personal identification. Many others
promise to make life interesting and challenging for managers in the future.

Snow melts first at the edge. As emerging technologies on the edge begin
to melt down the established structures of many industries, we can learn the
most by looking at the periphery. This is where we can best understand
how to compete in industries being transformed by emerging technologies.
Traditional ways of thinking, analyzing, and organizing are no longer adequate. The different approaches that emerge at the periphery will eventually become part of mainstream management.

We have spent several years exploring the melting snows through a fortunate synergy of thoughtful practitioners trying to learn from their experiences and each other along with enquiring academics who are devising
and testing new management theories and methods by studying winners

and losers. These discussions with more than 20 leading companies have helped to shape a set of priority issues (see box) that formed the foundation for the development of this book. Since there is no established model or paradigm for managing emerging technologies—but at best a diverse array of perspectives and approaches—our book seeks to balance a plurality of views around the common core management challenges of playing the rather different game of emerging technologies better.

Priority Issues of Emerging Technologies Management

1. *Evaluation of Emerging Technologies.* How do firms make decisions and commitments in the face of the extreme uncertainty of an emerging technology?

2. *Designing and Managing Alliances.* How can established and entrepreneurial firms ally themselves to capitalize on their core strengths to achieve mutual benefits? How does the Alliance Lifecycle evolve?

3. *Strategies for Participating in Emerging Technologies.* How do firms decide whether to aggressively commit to a new strategy (by acquisition, licensing or internal development), participate in a web of alliances, or adopt a watch and wait approach? What are the risks and rewards of these and other strategies?

4. *Developing Products for Really New Markets.* How can firms improve their ability to develop breakthrough product concepts that will drive the creation and/or use of emerging technologies, and cope with the ambiguity in market potential, customer requirements and competitive capabilities?

5. *Designing Organizations to Compete in Emerging Technologies.* What organizational structures should firms use to develop and commercialize an emerging technology? What incentives and systems are needed to encourage the innovators and champions within the organization?

6. *Managing Intellectual Property.* How do firms identify opportunities acquire intellectual property and protect it from loss to partners or competitors? What is the value and role of intellectual property in an information economy?

7. *Evolution of Emerging Technology-Based Industries.* How are emerging technology based industries different from traditional industries? How do they emerge and evolve and what are their best practices, strategies and success factors?

PLAN FOR THE BOOK

This book builds off these priorities to examine how managers understand and assess technologies and markets, shape strategy, make investment decisions, and change their organizations to meet the challenges of managing emerging technologies. The opening chapters examine the overall mindset needed to understand emerging technologies, and how this is a "different game," characterized by high uncertainty, rapid change, and shifting competencies. The opening also examines the potential traps that often leave established firms at a disadvantage to smaller rivals, and strategies for avoiding these pitfalls.

With this foundation, we then turn our attention to the issue of assessing fast-moving technologies for development. In Part I, we explore the paths of technology development, frameworks for evaluating technologies and the role of government in the emergence of technologies and industries. After examining the technology assessment process, we turn our attention to the markets for these new technologies. In Part II, we examine how these markets, which are very different from mature technology markets, require new approaches to research and assessment. They also demand an understanding of the intersections between the "lumpiness" of the markets and various technology barriers. Finally, we examine how complementary assets affect the spread of new technologies and the impact these technologies can have on established firms.

The process of crafting strategy is itself turned on its head in environments of rapid change and high uncertainty. In Part III, we explore the demands of strategy making in emerging technology firms, including the need to combine discipline and imagination, the use of scenario planning, and strategies for dividing the joint gains.

Since the goal of developing emerging technologies usually is to produce a return at some point in the future, the next challenge we consider is the issue of evaluating investments in emerging technologies. Part IV offers insights on using real options to help assess the value and potential of emerging technologies for projects in which traditional NPV may be negative. This section also examines a variety of internal and external financing approaches tailored to emerging technology.

Fast-moving emerging technologies exert tremendous G-forces on the organization and call for different models and approaches to management. Part V explores the reshaping of the organization for emerging technologies—both in the development of external "knowledge networks" and alliances as well as through new organizational forms and employee relationships.

The story we tell here—like the industries we study—is a work in progress. It reflects the knowledge of leaders of practice and frontline researchers in the field, but there are no final words or definitive recipes for success. There are hard-won insights, but even these must be constantly challenged. Do they fit your particular situation? Do they hold true for *this* technology? Do the lessons of the past fit with the futures you see? Is there a better way to achieve your objectives and vision?

For those of us involved in this project, it has offered some of the most rewarding and challenging work we have engaged in. There is no greater opportunity to make an impact than to wrestle with the future and its many unknowns. We will continue to learn in this environment and look forward to hearing from readers about their own perspectives in this endeavor. Like the technologies we are studying, the principles of management in this environment are very much emerging.

GEORGE S. DAY
PAUL J.H. SCHOEMAKER

The Wharton School

ACKNOWLEDGMENTS

We first want to recognize the expertise, enthusiasm, and collaborative spirit of our Wharton faculty colleagues who authored the chapters in this book. We are proud to be associated with a world-class faculty who share the commitment of the school in addressing the challenges that face business leaders today. These authors represent a wide spectrum of interests and backgrounds, coming from five different departments. The collective diversity of their contributions gives this book a remarkable depth and breadth of perspective.

All of us owe a large debt to the companies that have supported the Emerging Technologies Management Research Program including Bell Atlantic, Biogen, Cigna, Delphi/General Motors, DuPont, IBM, Knight-Ridder, McKinsey & Co., Minnesota Mining and Manufacturing, Bank of Montreal, National Security Agency, Monsanto, Procter & Gamble, Smith Kline Beecham, Sprint, and Xerox. These industry partners played many roles in this learning network. By helping us set research priorities in light of their needs for practical solutions, they kept us close to the day-to-day realities of emerging technologies. By serving as sounding boards, advisers, and speakers at more than 20 workshops and conferences, they shared their experiences and helped us to test our ideas. We are especially grateful to them for challenging us to understand the issues better and identify best practices in grappling with these issues. They were truly partners in our collective enterprise.

We cannot adequately recognize all the insightful managers who have guided us. We do want to single out six for special contributions: Steve Andriole of Cigna and now Safeguard Scientifics, Terry Fadem of DuPont, Steve Rossi from Knight-Ridder, Harry Andrews from 3M, Larry Huston of Procter & Gamble, and Mark Meyers from Xerox. We have also been privileged to have the wise counsel of our senior fellows who bring a wealth of industry experience: Don Doering, Warren Haug, Adam Fein, Bob Hershock, Roch Parayre, William Pipkin, and Bob Riley.

The Emerging Technologies Management Research Program is an initiative of the Huntsman Center for Global Competition and Innovation. The special mission of the Huntsman Center is to be the focal point for Wharton research on competitive strategy, innovation, and emerging technologies. Our activities are a tribute to Jon Huntsman, a visionary entrepreneur and distinguished alumnus of the school, who had the foresight to establish the Center in 1988. This center has flourished in the intense intellectual climate of the Wharton School. Special thanks go to Tom Gerrity who gave us unstinting support during his tenure as Dean and provided the seed funding to get this initiative started. We are also very appreciative of the support and encouragement that Janice Bellace gave us during her tenure as Deputy Dean.

The intellectual direction for this research program came from a core group of faculty: Bill Hamilton, Graham Mitchell, Harbir Singh, Jitendra Singh, Sid Winter, and the undersigned. This book celebrates the unique contributions of a diverse group of scholars who shared our vision of emerging technologies being a different game. We began meeting regularly in 1994 to hammer out a strategy for the program, resolve conceptual and practical problems, and build a network of scholars that now spans the world. Our core group developed the basic structure of this book and played an integral role in working with authors. A key event in this process was an all-day retreat where authors shared the basic concepts of their chapters and made connections to other chapters.

The underlying ideas in this book were shaped by many people. We especially want to acknowledge our partners in other Wharton research centers who collaborated on workshops and projects: Ned Bowman and Bruce Kogut of the Reginald Jones Center, Ian MacMillan of the Sol C. Snider Entrepreneurial Center, and Jerry Wind of the SEI Center. We are deeply saddened that Ned Bowman passed away recently and pay tribute to his profound influence on the research climate of the school. Many students of management from outside Wharton helped us by unselfishly sharing their latest thinking on the issues we were grappling with. Among those who were particularly influential were Raffi Amit (UBC and now Wharton), Clayton Christensen (Harvard), Kathleen Eisenhart (Stanford), Richard Foster (McKinsey & Co.), Rebecca Henderson (MIT), Gary Pisano (Harvard), Richard Rosenbloom (Harvard), Michael Tushman (Harvard), C.K. Prahalad (Michigan), Andrew van de Ven (Minnesota), and Eric von Hippel (MIT).

The Managing Director of the Emerging Technologies Management Research Program, Michael Tomczyk, provided the enthusiasm, energy, and management support to keep the research program prospering. He saw the possibilities for this program during its genesis and worked closely with our partners to ensure those possibilities were realized. His newsletters and conference reports provided valuable input to our research agenda and this book and are highly valued by our numerous industry partners.

Many people helped bring this enterprise to closure. Special thanks go to Robert Gunther for his invaluable editorial help. He played a central role in developing and presenting the themes in the book; integrating the diverse contributions of the authors, and rewriting many of the chapters so they were persuasive and accessible. We are grateful to Jeanne Glasser our editor at Wiley for guiding this project on its long and winding road to publication. Our sincere appreciation to the staff at Publications Development Company for helping to turn this many-headed project into a book under very tight time constraints. Throughout the project, we had the good fortune to have the assistance of Michele Klekotka, who cheerfully managed the complex flow of communications as outlines and drafts moved around the school. We also benefited greatly from the graphics skills of John Carstens and Meredith Wickman in the Wharton Marketing Department.

The heroines of this book are our wives, Marilyn and Joyce, who provided the essential support, encouragement, and understanding to keep us going. We have both been blessed by our companions for life. We dedicate this book to our children and grandchildren, whose lives may be profoundly transformed by the various technologies discussed in this book. We hope the managers entrusted with these technologies will use them to improve their world and life experience.

Finally, we acknowledge the extended Wharton community with its unique culture of collaboration, excellence, and enterprise that made this venture possible. We hope this book reflects the true depth and excellence of the School and its outstanding faculty.

G.S.D.
P.J.H.S.

Bryn Mawr, Pennsylvania
Villanova, Pennsylvania

Contents

Part V
RETHINKING THE ORGANIZATION   333

CHAPTER 1

A DIFFERENT GAME

GEORGE S. DAY
The Wharton School

PAUL J.H. SCHOEMAKER
The Wharton School

Gene therapy, electronic commerce, intelligent sensors, digital imaging, micromachines, superconductivity, and other emerging technologies have the potential to remake entire industries and obsolete established strategies. This is exhilarating for the attackers who can write—and exploit—the different rules of competition, especially if they are not encumbered by an existing business. For incumbents, however, emerging technologies are often traumatic. Most feel they have no choice but to participate in the markets that emerge. Their first reason is defensive, driven by the fear that the newcomers are plotting to use the new functionalities or modalities to attack their core markets. Their second reason is the converse of the first: If the emerging technology realizes its potential, it will create market opportunities that are too attractive to ignore.

The signs of technological turmoil are widespread. The rise of the Internet has created a growing anxiety among traditional retailers and other businesses as they struggle to master the different rules under which they must play against aggressive young competitors who cut their teeth in this new environment.[1] The emergence of home banking via the Internet, for example, fills most bankers with trepidation as they contemplate the reshaping of the banking industry. The breakthroughs in biotechnology have reshaped the world of traditional pharmaceutical and chemical companies, forcing them to restructure, create new alliances, and develop new strategies.

Some companies are closer to the center of the gale of creative destruction caused by new technologies, but very few will escape the disruptive impact of these new forces completely. Information technologies are transforming many industries, while genetic and materials research promises to

1

What Are Emerging Technologies?

Emerging technologies are science-based innovations that have the potential to create a new industry or transform an existing one. They include discontinuous technologies derived from radical innovations (for example, biotherapeutics, digital photography, high-temperature superconductors, microrobots or portable computers) as well as more evolutionary technologies formed by the convergence of previously separate research streams (for example, MRI imaging, faxing, electronic banking, HDTV, and the Internet). Each of these technologies offers a rich source of market opportunities that provide the incentive for risky investments.

The term *technology* is used broadly in business and science to refer to the process of transforming basic knowledge into useful application. Science might be thought of a *know-what* and technology as *know-how*, while markets or businesses focus on *know-where* and *know-who*. Here, we define technology as a set of discipline-based skills that are applied to a particular product or market. The technology can focus on a component, an entire product or an industry. Emerging technologies are those where (1) the knowledge base is expanding, (2) the application to existing markets is undergoing innovation, or (3) new markets are being tapped or created.

It is also useful to distinguish technologies that are new to the firm or a unit of the firm from those that are new to the world. Many managers struggle with how to scan, experiment and integrate externally available technologies (either new or mature) into their existing products as well as to create new products. Our focus in this book is on how to transform technologies that are still emerging (either within or outside the firm) into value creation within existing or newly emerging markets.

---

have an impact across areas as diverse as pharmaceuticals, food production, and forensics. In industry after industry, managing emerging technologies is becoming essential to success.

WINNERS AND LOSERS

The failures of incumbents are so widely recognized that conventional wisdom holds that attackers from the outside have an advantage when an emerging technology threatens an existing market or technological regime.[2] New entrants have used technology in desktop copiers, electronic calculators, minimill steel making, videotape recorders, and hydraulic earth-moving equipment to take the market away from incumbent firms. The computer industry has evolved from competition among vertically integrated stacks controlled by DEC, IBM, Wang, Amdahl, Nixdorf, NEC, or Matsushita to a horizontal industry model where competition occurs between component providers. Few of the leaders in the horizontal model, such as Dell, Cisco, Microsoft, or Intel came from the ranks of the old vertical industry.[3] The same story has unfolded in the hard-disk drive industry where market leadership changed with each successive product generation.[4] In electronic games, Atari was displaced by Sega, and Nintendo may be leading in the next generation.

The loss of market leadership by incumbents is not inevitable. Some firms manage to overcome the challenges associated with incumbency and avoid being marginalized by a new technology. Notable recent examples of large companies that have successfully embraced emerging technologies include:

- Microsoft has moved aggressively to overcome its late entry into the Internet by redirecting a large portion of its R&D budget to Internet-related projects, and investing heavily in the production of digital content.
- Monsanto has undergone a complex makeover by shedding its cyclical chemicals business to concentrate on pharmaceuticals and agricultural and food-ingredient operations that exploit 20 years of several billion dollars of investments in agricultural biotechnology. It will be a rocky road, with deep concerns about genetically modified foods, but the long-term outlook looks very promising.
- Intel was able to exit the semiconductor memory business and successfully devote itself to microprocessors, despite deep-seated beliefs that memories were the backbone of the firm.[5]
- Charles Schwab, the original discount investing firm, successfully launched an investing web site despite the high initial cost of switching to low online commissions for all trades. The shortfall was soon made up with productivity gains and new revenue generated by giving customers personalized information in real time.
- General Electric is in the midst of a profound transformation, with Jack Welch preaching and pushing for revolutionary change. Every business within GE was asked to create a "DestroyMyBusiness.com" initiative under the motto that either you must change the business model yourself or someone else will do it for you.

Some of these firms had to "stare death in the face" before making the radical shifts in their organizations and strategies needed to succeed. In every case, they had to fundamentally rethink their approaches to the business to meet the challenges of these new technologies. Only by deliberately departing from their tried-and-true strategies were they able to meet the test of this new crucible of competition.

Incumbents bring many resources to emerging technologies: established infrastructure and processes, visible and respected brand names, and deep pockets. They can and do invest heavily in technology development. The 15 largest spenders on R&D are all large corporations. They accounted for 50 percent of the $130 billion in non-governmental R&D in 1990, although most of this is devoted to sustaining innovation.

For all their advantages, incumbents are often impotent in the face of a radical technological disruption. Their size, usually an advantage, now gets in their way. Commitments to facilities, people, and channel partners restrict their flexibility. Equity markets expect continuing earnings from them, while start-ups are valued for their future earnings and are rewarded with large market capitalizations they can use to fund growth.

Incumbents are especially disadvantaged by their structures, capabilities, and mindsets. These firms have mastered the established game in their industry. But their finely honed instincts, slowly acquired heuristics and embedded skills make it tough to deal with an emerging technology that presents a new game. Success in this new arena depends upon a different set of capabilities, tools and perspectives. These new frameworks, viewpoints, and organizational approaches are the focus of this book.

A DIFFERENT GAME

The problems that often befuddle incumbents are rooted in the technological uncertainties, ambiguous market signals, and embryonic competitive structures that distinguish emerging from established technologies. Whereas the technology, infrastructure, customers, and industry are relatively well defined for established technologies, a fog of ambiguity surrounds emerging technologies (see Table 1.1). These differences demand new skills, new ways of thinking and innovative managerial approaches to cope and eventually prevail. Three challenges in particular have to be faced and embraced by an established firm if it is to have any chance of success with an emerging technology: coping with great uncertainty and complexity, keeping up with accelerating change, and developing new competencies.

Table 1.1
Contrasting Emerging and Established Technologies

	Established	Emerging
Technology		
• Science basis and applications	Established	Uncertain
• Architecture or standards	Evolutionary	Emergent
• Functions or benefits	Evolutionary	Unknown
Infrastructure		
• Value network of suppliers, channels	Established	Formative
• Regulation/standards	Established	Emergent
Market/Customers		
• Usage patterns/behavior	Well-defined	Formative
• Market knowledge	Thorough	Speculative
Industry		
• Structure	Established	Embryonic
• Rivals	Well-known	New players
• Rules of the game	Known	Emergent

Coping with Great Uncertainty and Complexity

As recently as 1995, interactive television dominated corporate radar screens as the hottest new electronic marketplace. Dreams of interactive television have faded into the background as the World Wide Web reveals the power of connections through networks, so images and information can flow freely. Does anyone really know what tomorrow will hold in this environment? Which of the flashy new technologies in the labs today will become the hot new products of tomorrow? Which ones will be embraced by the market and which ones ignored?

How can companies move forward in the face of this kind of uncertainty? New technologies often produce a major disruption on the established trajectory of technical advances by drawing on new or different science bases, and thus require the arduous development of new competencies. But in the earliest stages of development, it is not evident they will achieve a decisive relative advantage. Indeed, one of the most confusing aspects of emerging technologies is that consumer usage patterns and behavior are exploratory and formative, while market knowledge is scant and the structure of competition is embryonic.

How is this different from the uncertainty that is present in all industries, even the most stable and predictable? Uncertainty in a stable environment is manageable because there are usually only a few discrete outcomes that define the future and robust strategies can be devised to adapt to these possibilities. The character of uncertainty created by an emerging technology is profoundly different. The risks are not just external but also internal, relating to the biases and limitations of people's thinking frameworks. The German philosopher Jurgen Habermas refers to this as "epistemic risk" (i.e., the risk of not knowing what one does not know).

There are so many unpredictable and volatile conditions interacting in unanticipated ways in the early stages that there is no sound way to predict the future. Yet it would be folly for managers to throw up their hands and fail to analyze the situation. It would be a mistake to give up on creating a coherent strategy for maneuvering through the uncertainty. Choices must be made about initiatives to support the emerging technology, alliances to pursue and human resources to develop. However, the analysis and subsequent strategy must include a humble learning component and acknowledge multiple futures. Even in the most uncertain, embryonic technologies, there are a host of factors such as performance trajectories that may be unknown but can be studied if the right frames of mind are adopted.[6]

Keeping Up with Accelerating Speed

Internet time, like dog years, passes far more rapidly than standard time. This acceleration can be seen in the increasing compression of technology adoption curves in Figure 1.1. While the Internet is at the leading edge of this intensification in time pressure, similar forces are at work with other technologies, reflecting the overall rate of technological progress and the desire to gain a first-mover advantage.

The recent surge of activity in network technologies, biotech, advanced materials and nanotechnology, to name a few, is the culmination of years of research in disparate fields that is reaching critical mass. The first successful gene-splicing occurred in 1973, and the Internet was first conceived (as the ARPANET) in the 1960s (see Chapter 5). Moreover, advances in one domain are feeding and reinforcing progress in other domains, as when powerful computers facilitate combinatorial chemistry and gene manipulation techniques. The overall trends toward free markets, globalization, and deregulation is only adding fuel to the accelerating rate of technological change.

When the pace of technological progress was more measured, it was possible for firms to passively watch for an occasional discontinuity or wait for

Figure 1.1
Adoption Rates of Various Communication Technologies

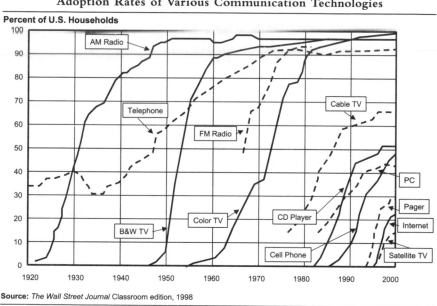

Source: *The Wall Street Journal* Classroom edition, 1998

other firms to take the development risks. This "fast-follower" strategy, which once seemed prudent, is now becoming a risky option in "winner-take-all" markets,[7] where there is a large and widening gap between the leader and the followers. For example, it is axiomatic that in electronic commerce markets, early dominant players such as AOL or Yahoo get most of the rewards. By 1998, one study found that the top 5 percent of all web sites garnered more than 74 percent of all traffic.[8]

Developing New Competencies

A discontinuous innovation can either enhance or destroy the existing competencies of an incumbent player. In the printing industry, the evolution from manual to mechanical typesetting built on the competencies of the existing players and hence incumbents often led the change (as discussed in Chapter 8). But often, emerging technologies do not fit the competencies of incumbents and undermine the slowly acquired skills, knowledge, and assets that were needed to master the established technology that is being replaced.[9] The Friden and Monroe companies found their extensive manufacturing, and strong service networks for mechanical calculators were made obsolete by highly reliable electronic calculators. Likewise, Western Union's telegram transmission competencies were by-passed by fax, e-mail, and overnight delivery services.

These patterns of "creative destruction" are not a recent high-technology phenomenon.[10] Attackers dislodged incumbents when diesel-electric loco-motives prevailed over steam locomotives, ball points supplanted fountain pens, and vacuum tubes gave way to transistors.[11] In the early part of the century, there was a long and bitter standards battle between Edison and Westinghouse over DC versus AC electricity distribution.[12]

While the management of emerging technologies is thus not a "new game," it is a "different game" that managers of established organizations are not well equipped to play. Apart from the three general challenges dis-cussed above (managing high uncertainty, keeping pace, and developing new competencies), the different game of emerging technologies entails more specific challenges within each business. In some cases, a new distri-bution model must be devised or fundamental changes in the production process are called for. In other instances, the technology will require new product features, or new dimensions of competition (for example, shifting from competing on cost to competing on innovation, services or alliances).

It can be the cumulative effect of numerous small changes in many busi-ness dimensions or a few quantum changes in some major dimensions that make the game different. Enough changes of degree can add up to a change in kind. This is the challenge of managing emerging technologies, which can be as different for managers as the transition from regulated business into a free market, or the extension of a firm's products into foreign or emerging markets. If managers fail to recognize that the game is different, they may already have lost it before the first round of play.

THE NEED FOR A NEW MANAGEMENT APPROACH

Given these differences between managing emerging and established tech-nologies, what are the requirements for managerial success? For the most part, the approaches we teach in our MBA programs for strategic planning, financial analysis, marketing strategy, and organizational design have been based on assumptions of continuity, with a corresponding focus on equi-librium, rationality, and optimality.[13] Even in cases where these core tenets are not present explicitly, it is a widely held belief that the manager's role is to control and manage the uncertainty. These assumptions are directly challenged by emerging technologies, which are characterized by disequilibrium, profound ambiguity, and a rate of change that often defies standard analysis. Some of the most successful players in emerging technologies have not managed uncertainty as much as navigated and exploited it.

Traditional frameworks and perspectives continue to serve as a good guide for large organizations in more stable markets. But like Newtonian physics, they begin to break down as we approach the "speed-of-light" intensity and complexity of emerging technologies. In the face of this reality, some managers have walked away from their analytical frameworks altogether. But this is also dangerous. What is needed, instead, is a new set of advanced frameworks and tools that is better matched to the disruptive character of emerging technologies.

While we recognize that there are significant differences among emerging technologies—the challenges of biotechnology, for example, are quite different than those of information technology—we focus on common principles of management. Just as businesses in diverse industries can use common accounting approaches, we have sought tools and perspectives that can apply across a broad set of emerging technologies. In applying these, managers should take into account the specific characteristics of their own technologies, markets, and organizations.

Different Rules

This different game—with its high uncertainty and rapid, competence-destroying change—undermines the old rules used in managing established technologies. The Internet and biotechnology consistently defy NPV analysis, and other traditional approaches. On the other hand, if managers abandon their old tools and rules, it seems to invite chaos. If we are going ahead with projects with a negative NPV, on what basis are we making decisions? If we cannot know when and if a given technology will be developed to the point where it can be commercialized, how can a firm possibly build a business around this dream of an idea whose time might never come? If we cooperate with our rivals, how can we compete?

To managers in incumbent firms, this shift often appears as a move from order to chaos. In reality, there is a different set of rules and underlying structure to this different game of emerging technologies, as summarized in Table 1.2. Understanding this new approach will not necessarily reduce the risks and challenges involved, but it may make it easier to manage them.

Table 1.2
A Different Game

Domain	Established Technologies	Emerging Technologies
Environment/Industry	Manageable risk and uncertainty (a few discrete outcomes define the future)	Volatile and unpredictable (no basis to predict the future), high complexity and ambiguity
(a) Texture	Stable and predictable	Turbulent and uncertain
(b) Feedback	Linear and structured	Causally ambiguous
(c) Players (e.g., suppliers, rivals, customers, channels, regulators)	Familiar	New or unknown
(d) Domain of play	Clearly defined	Formative/evolving
Organizational Context/Climate		
(a) Mind-set/routines	Accepted rules, known comfort zones	No rules, conventional wisdom, irrelevant or misleading
(b) Boundaries	Rigid, well-defined boundaries, with a reliance on existing capabilities	Permeable boundaries with an emphasis on outreach, use of partners to overcome lack of capabilities and a reliance on external resources
(c) Decision making	Well-established procedures and processes, conflict avoidance	Accelerated decision making that puts a premium on constructive conflict and intuition
Strategy Making	Focus on gaining advantage and leveraging resources, present timetable "traditional" strategy tools, convergent thinking	Focus on creating a robust and adaptive set of multiple strategies; real time, issues-oriented process; scenario development; divergent thinking

Table 1.2 *(Continued)*

Domain	Established Technologies	Emerging Technologies
Resource Allocation		
(a) Criteria	Traditional discounted cash flow/pay-back period or shareholder value creation	Real options value; heuristic
(b) Process and responsibility	Well-specified procedures (explicit risk/reward trade-offs)	Informal and iterative (small initial commitments)
(c) Monitoring	Clear yardsticks	Seasoned judgment
Market Assessment	Structured research in a defined context, with known attributes, identifiable trade-offs and known competitors; focus on primary demand	Experimentation and probe-and-learn approaches; latent-need research; lead users analysis; focus on secondary demand
Development Process	Formal stage-gate process that aims for replicability, defined steps, fixed specifications, and time-to-market pressure	Adaptive process for early-stage development through experimentation, carrying multiple alternatives forward and elastic time frame
People Management	Traditional recruitment, selection, supervision, promotion and compensation	Novel/emphasis on diversity, rule breaking, new compensation systems, and so on
Appropriating the Gains	Gains appropriated through sustainable advantages based on durability, causal ambiguity, barriers to imitation, and credible threats	Gains appropriated through mechanisms such as patents, secrecy, lead time and control of complementary assets

The chapters of this book present a variety of new tools and perspectives to help managers play by these new rules. Some of the shifts in thinking and practice mangers need to make include:

- *More fluid organizational context.* Established technologies rely upon accepted rules and well-defined boundaries and capabilities. Decision-making procedures and processes are well established and designed to avoid conflicts. In contrast, emerging technologies challenge existing mindsets and routines, rely upon permeable boundaries and partnerships. Decision making should be accelerated and designed to encourage constructive conflict.

- *More robust and adaptive strategy making.* Strategies for established technologies focus on gaining structural advantage and leveraging resources. They rely upon a more linear approach following an established schedule that leads to convergent thinking. In contrast, emerging technologies are managed by pursuing multiple trajectories. For example, Microsoft in the late 1980s was simultaneously pursuing Windows, OS/2 and Unix platforms. What might have appeared to be an unfocused strategy in existing technologies proved to be a "robust adaptive strategy" in emerging technologies.[14] In contrast to more linear planning processes, managers create portfolios of strategies, using divergent thinking and scenario planning to explore multiple options (see Chapters 9–11). Planning should be viewed as learning rather than a mechanism for control or predictability.

- *Staged resource allocation.* Financing emerging technologies is based on a real options perspective rather than just discounted cash flows or simplistic payback periods. Whereas traditional analysis discounts uncertainty, options analysis places a value on it (see Chapter 12). Resource allocation should be more informal and iterative, with small initial commitments that are reassessed over time rather than more simple go-no go decisions.

- *Market exploration.* Managers in existing technologies can test known attributes and tradeoffs with customers who are very often familiar with the technology. In emerging technologies, managers need to identify market demand for a product that most customers do not understand or recognize. The focus moves away from the technology itself to identifying latent needs and studying lead users. Companies should also use experimentation and probe-and-learn approaches to explore the market rather than perform a detailed assessment of its potential (see Chapter 6).

- *Adaptive technology development.* Research and development of emerging technologies also proceeds differently. In contrast to the formal stage-gate process for extensions and improvements to established technologies, research is more flexible in emerging technologies, with the use of adaptive research strategies, a variety of small experiments and the pursuit of multiple alternatives. Once the exploratory stage is finished, the development projects proceed with the discipline needed for any large project.

PLAYING THE GAME

How do managers successfully play this different game? One of the premises of the Wharton program has been that by studying winners and losers in this environment, we can gain insights into best practices. Consider insights from the following two examples from information technology and biotechnology.

Sleight of Hand

In 1992 Apple Computer CEO John Sculley boldly proclaimed with great fanfare from the cover of major business publications that the Newton "personal digital assistant" would be the herald of a new $3.5 trillion digital information industry.[15] That same year, a tiny startup, Palm Computing, Inc. was founded.

By 1998, the PalmPilot had become the fastest selling consumer electronics product in history, shipping more than 1 million units in its first year and a half. That same year, after spending $500 million on the Newton, Apple pulled the plug on its failing product.[16]

The field of PDAs was crowded with many powerful and well-funded competitors. Besides Apple, Motorola, AT&T, Bell South, IBM, Hewlett-Packard, Novell, Casio, Sony, and Microsoft all announced plans to build or invest in these revolutionary little devices.[17] There were many casualties. In addition to the Newton, Go Corporation, a startup funded by Kleiner Perkins Caufield & Byers, went through $75 million before it also went off line. Palm claimed to have spent just $3 million to develop the prototype for the Pilot.[18]

How was Palm Computing founder Jeff Hawkins able to shape and lead this emerging technology? How was Palm able to turn a technology that many others also recognized into a commercial blockbuster? Among the lessons:

- *Incumbents can be at a disadvantage.* Tiny companies can run circles around much bigger rivals who are pursuing essentially the same market opportunity because current systems and mindset of established companies limit their initiatives and progress. Chapter 2 examines some of the "traps" incumbent firms regularly fall into when approaching new technologies and discusses possible solutions to them. Chapter 8 explores the advantages complementary assets and customer relationships can sometimes provide to incumbents.

- *Knowledge assets outweigh physical assets.* Against the resources of much larger rivals, Palm matched the resourcefulness of its understanding of human cognition and insights into how customers used the machine. The study of human cognition was a lifelong pursuit for Hawkins and this gave him unique insights into human thinking. This knowledge helped him develop a practical handwriting recognition program. It is not just knowledge within the firm that is important in developing emerging technologies, but actually networks of knowledge. Chapter 15 examines the development and management of knowledge networks in emerging technologies.

- *Understand how customers use the technology.* It is not the technology itself that is important but the interface with humans. Hawkins realized that it was easier to teach humans a modified alphabet than to build a machine that could recognize a wide variety of handwriting styles. He also realized that users wanted to coordinate their PDA information with their computers, so he designed an easy system for transferring the information. By understanding the true benefits customers were seeking and the compromises they were willing to make, he could move from geewhiz technology to a product with commercial potential. Hawkins' most important decisions were in determining what technology to leave out (such as personalized handwriting recognition and e-mail) and which ones to add (such as a cradle to allow easy calibration with a desktop computer). Chapter 7 examines how companies can decide what technology to pursue in "lumpy markets" with diverse customer preferences. Chapter 6 shows that it is possible to learn about markets even before they emerge by studying latent needs and lead users, and then continue to learn with selective market probes.

- *Learn from experiments.* Palm's first product, the Zoomer, was slow, heavy, and expensive ($700). Arriving just after the Newton, it fell flat along with all the other offerings in the market. But it provided important lessons to Palm when it took a second run at the market with the Pilot. Chapter 9 presents a framework for planning and learning through "disciplined imagination," balancing the discipline of careful, linear planning and the opportunity for creative leaps of imagination.

- *Don't go for the big market all at once.* The explosion of a new technology depends not only on the technology itself but on the intersection of the technology and its market application. This development often occurs in stages, in which specialized niches that are tolerant of the rough edges and high price of early prototypes provide a testing ground to refine the technology and build the market. Some of the early PDA failures such as Go Corporation and Apple's Newton went to the larger market before the product was ready and suffered humiliation as a result. Palm was able to walk to the market on their carcasses (and its own prior failure). Chapter 3 examines patterns of emerging technology development and the process of "technology speciation," in which a technology moves from one application domain to another environment where it rapidly develops.

Riding the Roller Coaster of Biotech

Founded in 1979, Centocor led the way in the euphoric rise of biotechnology and then held on for the wild ride of developing a new industry. In 1997, after nearly two decades, the company crossed an important threshold: profitability. Chairman and cofounder Hubert J.P. Schoemaker wrote: ". . . I knew that it would be a challenge to establish and manage an early-stage, innovation–oriented company. I could not have imagined how great the challenge would be or how long it would take to realize our goals. Neither could I have imagined the magnitude of the rewards."[19]

The challenges were great. A new drug can take nearly 15 years to develop at a cost of more than $350 million. Estimates are that only one drug in 10,000 makes it to market.[20] Despite the risks involved, the expected payoffs were high enough that biotechnology investments soared in the 1980s. As companies wrestled with the realities of the science and ambiguous regulation, investors rushed into the market for this technology that had the potential to affect a wide range of industries, including pharmaceuticals and medical diagnostics, fish farming, agriculture, chemicals, textiles, household products, manufacturing, environmental cleanup, food processing, and criminal forensics.

By 1991, investors had poured more than $2.5 billion into biotech firms and Ernst & Young estimated that revenue for the industry, then only $4 billion, would hit $75 billion by 2000. Centocor had sold the marketing rights of its first major product based on monoclonal antibody technology to Eli Lilly for about $100 million. The Food and Drug

Administration, however, shot down Centocor's Centoxin application and later refused approval for several other biotech drugs as well, taking the wind out of the industry's sails. Between January 31 and September 1992, the Hambrecht & Quist Biotech 100 index lost 43 percent of its value. Centocor's stock dropped from over $60 a share to below $6 a share during this period. By 1991, the FDA had cleared fewer than 30 biotech-related drugs in biotech's first decade.[21] With shifts in both technology and investor interest, the biotechnology roller coaster seemed headed downhill.

For those companies such as Centocor that survived the ride, the rewards can be great indeed. Centocor developed therapeutic and diagnostic drugs for cancer, cardiovascular problems, and other diseases. By the end of 1994, it had been cleared by regulators in Europe and the United States to market ReoPro, an anti-clotting drug for angioplasty patients. By 1998, Centocor and partner Eli Lilly were selling $365 million of the medicine per year and analysts expected sales to soar to about $700 million in a couple of years.[22] Although Eli Lilly would have lost more than $100 million on the first Centocor drug, it scored big on the second drug thanks to having received an option from Centocor to its commercial rights in case the first drug failed.

To meet the high costs and long road required for participation, many smaller firms such as Centocor joined forces with large pharmaceutical companies. In addition to its marketing pact with Eli Lilly, Centocor teamed up with Warner-Lambert, Glaxo Wellcome, and Schering-Plough.[23] Finally, in 1999, Centocor was purchased by Johnson & Johnson for about $4.9 billion.[24] To sustain the drug discovery course requires very deep pockets.

Among the lessons:

- *Understand the role of government.* The fate of many emerging technologies and companies within them often rests with government decisions. The Food and Drug Administration's decision on Centoxin left the whole industry reeling and dramatically reduced investments in biotech. In drug therapies, FDA approval can be a life or death issue for a firm, but government also shapes the playing field in many emerging technologies. Regulators shape the infrastructure and playing field of competition at the outset, as they did with the development of the Internet, or they may step in later in the process to protect perceived public interests. Chapter 5 explores the impact of the government and examines strategies for managers.

- *Use partners.* Many new technologies, even for established firms, are very difficult to pursue on their own. Alliances and partnerships help large companies such as Johnson & Johnson and Eli Lilly to take advantage of advances in biotechnology. They also help small firms build the resources they need to bring their breakthroughs to market. Chapter 15 discusses the importance of "knowledge networks" in developing and commercializing technologies and Chapter 16 examines strategies for participating with partners.
- *Use flexible planning.* The development of the underlying science, the moves of rivals, the acceptance by the market and the actions of government contain unpredictable elements. Centocor and its peers had to be able to prepare for multiple futures. Chapter 10 explores the uses of scenario planning to meet the challenges of uncertainty, complexity, and paradigm shifts in emerging technologies.
- *Employ new strategies for financial assessment.* Assessing the potential of a company that essentially loses money for more than a decade with highly uncertain potential for returns is a significant challenge. Managers need new approaches to identifying the value of these future businesses built on emerging technologies. Chapter 12 examines how the "real options" framework can be used not only to *evaluate* technology options but also to create and manage them. Chapter 13 explores internal and external financing strategies for emerging technologies, while Chapter 14 explicitly addresses the unique financing challenges in biotech.

A Caution about Lessons

Stories of successes and failures, such as the discussion of Palm Computing and Centocor above, are perhaps misleading because they are told with the benefit of hindsight. In actuality, managers are forced to make decisions about pursuing new technologies with highly imperfect information. The reality is that some technologies succeed and some fail, and managers can never know for sure beforehand whether a technology will be a dud or the next killer application. We tend to celebrate cases of companies that aggressively pursue a new technology that turns out to be wildly successful, and wag our fingers at the companies that took a passive stance when the technology in fact succeeded.

But what happens when the technology fails? This creates a whole different set of cases, as shown in Table 1.3. Then the "foolish" company that

Table 1.3
Decisions under Uncertainty

Decision Posture	Which State of Nature Will Ultimately Prevail? Technology Succeeds	Technology Fails*
Aggressive Stance (Commitment and Options Creating)	Cell #1 Microsoft and PCs 3M and Post-It Notes Sony and Walkman Sun Microsystems and RISC Chips Amgen and Biotech	Cell #2 RCA and videodisk Investors in cold fusion Citicorp and home banking
Passive Stance (Wait and See)	Cell #3 Encyclopedia Britannica and CD roms RCA and FM radio US Steel Co. and Minimills DEC and PCs Sony and VHS format Kodak and copier products	Cell #4 Many utilities and nuclear power Monsanto and prefab plastic homes Other banks and home banking

* Failure could occur because (1) the technology didn't work or couldn't be scaled up to commercially viable production rates, (2) a better technology superseded it, or (3) it was ahead of its time.

didn't invest in the technology appears wise, as with utility companies that decided to wait and see on nuclear power. And the "smart" company that aggressively pursued the new technology feels foolish as it becomes the first one over the cliff, as in the case of investors in cold fusion or supersonic air travel. Any advice based on best practice merits an important caveat about the above "selection bias." This bias runs deep and is hard to avoid, owing to the way history is recorded and the interest of the human mind. The error of commission simply receives more attention in the popular press, as well as in research, than the error of omission, even if they are equally grievous. This selection bias in turn may cause serious distortions and limitations in any advice based on best practice, popular wisdom, or the singular insights from any one entrepreneur who succeeded against the odds.

As managers seek to assess the potential of new technologies, they also should do so with humility. It is easy to make mistakes. Western Union turned down a chance to buy Alexander Graham Bell's patent on the telephone, and the inventor himself saw the patent only as an "improvement on the telegraph." The jet engine in its early inefficient version was dismissed by the National Academy of Sciences as impractical, the transistor was seen as a way to develop better hearing aids for the deaf. IBM initially figured that five orders would satisfy the entire worldwide demand for its clunky vacuum-tube machines. (The number five was deduced by IBM President Thomas Watson by estimating the total number of calculations being performed at that time in the world per year. What he totally underestimated was how the supply of fast computers would create its own demand.)

To make the situation even more uncertain, while these radical new technologies are gaining acceptance, new generations of different technologies may emerge that leapfrog them. Or, alternatively, complementary technologies can lead to the rapid growth of existing ones, as the fax machine and Internet have led to increased use of telecommunications. Stanford economist Nathan Rosenberg commented, "One of the greatest uncertainties controlling new technologies is the invention of yet newer ones."[25]

It is not just moving with the right technology, but also at the right time. DuPont once sank more than $1 billion into acquiring businesses in digital imaging, framing, and publishing. It hoped to capitalize on the emerging imaging technology, but the infrastructure was not yet in place. As Terry Fadem of DuPont commented, "The vision was correct, but the timing was poor."

LIVING WITH PARADOX

While the chapters of this book focus on frameworks, approaches, and perspectives that help us answer some of the management questions raised by emerging technologies, perhaps the greatest lessons come from the unanswered questions. The lessons of emerging technologies are not simple. Managers need to keep a healthy sense of doubt about any simple answers to the complex challenges presented by these technologies. We would be wise to heed newspaper editor H.L. Mencken's observation that, "There is always an easy solution to every human problem—neat, plausible and wrong."

Managers of emerging technology businesses need to become more comfortable with high levels of complexity and paradox. Among the key paradoxes:

- *A strong commitment is necessary but you also have to keep your options open.* On the one hand, there are persuasive arguments that investments in emerging technologies should be viewed as creating a portfolio of options where the commitment of additional resources is subject to attaining defined milestones and resolving key uncertainties. These options are investments that give the investor the right but not the obligation to make further investments.[26] Additional funds are provided only if the project continues to appear promising. If prospects are no longer promising, or the level of uncertainty stays high, then management can either let the option expire, by not making further investments, or delay until the prospects are more attractive. In an options framework, management and financial commitment is conditional and guarded. Presumably this limits the loss exposure to the amount of past investments, while the upside potential is not constrained.

On the other hand, there is compelling evidence that long-run winners are often first movers who committed early and unequivocally to a technology path. Andy Grove[27] of Intel argues that it takes all the energy of an organization to pursue one clear and simple strategic aim—especially in the face of aggressive and focused competitors—and that hedging by exploring a number of alternative directions is expensive and dilutes commitment. For these reasons, Intel decided against working on enhancing the TV set and put all its energy behind developing the personal computer chip so it had the visual and interactive capabilities to offer a universal information appliance. Taking flexible, option positions seems to be prudent, but perhaps undervalues the power of commitment. These cases seem to prove William James' statement, "Often enough our faith beforehand in an uncertified result is the only thing that makes the result come true."

Managers need to be able to balance commitment with flexibility. One way to mitigate this problem is to make the organization more flexible so the investments needed to make a strong commitment or to change that commitment are relatively small. The timing of commitment can also help avoid this paradox. Best practice suggests it is desirable in the early stages of exploration of an emerging technology to keep a number of options open by only committing to investments in stages, following multiple technology paths, and delaying some projects. Once uncertainty has been reduced to a tolerable level and there is a widespread consensus within the organization on an appropriate technology path that can utilize the firm's internal development capabilities—as in the case with Intel's choice of personal computer over television as the preferred information appliance—then full-scale internal development can begin.

- *Winners are often pioneers, but most pioneers fail.* A paradox that follows on the first one is that the only way to arrive first in a new territory is to be a pioneer, and yet pioneers more often than not end up at the bottom of a gulch with arrows in their backs. The big rewards come from being a pioneer, so long as you survive, but patience is needed to increase the odds of success. Pioneering is inherently risky, but there is no need to take foolish risks. As noted in the discussion of PDAs, some companies failed because they raced too quickly into an undeveloped market with an undeveloped product.

  There is a fine line between gambling and calculated risk taking as Alfred Sloan emphasized when he transformed General Motors from a functional form to a multi-divisional organization. By moving forward in stages, you can build settlements and supply posts along the way. This way you won't outpace your lines of supply or get trapped out in the wilderness alone. The outposts of small early niches provide a testing ground for strategies, an opportunity to learn how to survive on the frontier and to develop partnerships that are essential to success. But companies in emerging technologies have to live with the balance between the promise and perils of pioneering.

- *Strategies should build on existing competencies but organizational separation is often required for success.* As will be discussed in Chapter 2, the very characteristics that have made large organizations successful create traps for them in managing emerging technologies. To avoid contamination of the new emerging technology business, companies often set up incubators—separate spinoffs or divisions where these businesses can grow and develop without the burdens of the mature parent. These separate organizations usually have different cultures, compensation structures, organizational designs, and performance measures.

  The problem is that the more separate these operations become, the less they can draw upon the strengths of the parent. Xerox PARC engaged in creative leaps that led to innovations such as the graphical user interface, but it was not Xerox that benefited from it but Apple and Microsoft. Saturn created a very different model for its organization but the hoped-for impact on the General Motors organization overall never materialized and in some ways the new initiative was held back by the parent. IBM, in its quest to develop a truly new PC, set up a separate unit around 1980 that failed to tap into any of IBM's formidable technological competencies. The IBM PC was an assembled product, without any real proprietary technology, and consequently quickly attracted clones. There can be too much separation between the venture and the parent.

There needs to be enough separation so the technology's progress is not stifled but enough involvement to look for synergies with the competencies of the existing organization. Like parenting a teenager, the new venture has to be given enough freedom to experiment but still be kept within the family. Given the distinctive personalities of these new internal ventures, this balance is always a challenge but it is critical to avoid either stifling the emerging technology business or eroding its usefulness to the firm.

In a sense, companies also need to develop new forms of organizations that are ambidextrous.[28] Such organizations are able to run the existing technologies or traditional businesses with the right hand while running the emerging technology business with their left. This requires a particular balancing act, that can have tremendous payoffs if done successfully. As always in such cases, it is important for the left hand to know what the right hand is doing and vice versa. One of the big mistakes companies make in setting up these separate organizations is not paying adequate attention to the connections between them and the parent firm. Companies that do recognize this challenge can create formal coordinating mechanisms and structures to keep both parts of the organization in synch.

- *Competition is intense and brutal, and yet winning requires collaboration.* Competition for emerging technologies can be brutal. With winner-take-all markets and firms that have staked their entire future on success, failure often is not an option. At the same time, no emerging technology company is an island unto itself. The success of a new gene therapy may depend on far-flung networks of researchers in specific fields. The success of a new information technology standard depends on upstream and downstream adoption by suppliers and customers. Managing alliances and other partnerships is one of the central activities in successfully developing and commercializing emerging technologies. And the structure of these relationships determines the payoffs from the process (as IBM found out in its partnership with

Microsoft and Intel). Very often the same companies that are collaborators in one arena are competitors in another. For instance, Sony and Philips who are intense rivals in the consumer market, are working together on setting standards in optical media and supplying components to one another. Companies are entering into technology-sharing pacts, in which competitors agree to share future technology developments. For example, IBM might have an agreement with Hitachi to license future innovations in disk drives for a set fee.[29] A central part of guiding emerging technologies is to manage the complex webs of relationships with the proper mix of cooperation and competition that is so crucial to their development.[30]

Embracing Paradox and Ambiguity

Some of these paradoxes can be mitigated through approaches tailored to emerging technologies, but can never be fully eliminated. An important part of managing emerging technologies is the ability to live with paradox, and its associated ambiguities Simple, absolute answers are few and far between. And if there were simple answers, the rewards of winning in this new game could not be great since many players would master the necessary strategies and tactics. It may be the ability to live with these ambiguities and to continually identify them and think through them that is one of the most important skills of managing emerging technologies. It is the very complexity of the game, and its associated skewed payoff structure, that makes it worthwhile for established organizations to learn how to play it well.

CHAPTER 2

AVOIDING THE PITFALLS OF EMERGING TECHNOLOGIES

GEORGE S. DAY
The Wharton School

PAUL J.H. SCHOEMAKER
The Wharton School

The market position and resources of incumbent firms should give them an advantage over newcomers. Yet many incumbents have a poor track record in developing and managing emerging technologies. What goes wrong? This chapter examines some of the major traps for incumbent firms that have been identified through research and discussions with managers. These include delayed participation, sticking with the familiar, failure to fully commit, and lack of persistence. Can these traps be avoided? The authors present four strategies that can help companies steer clear of these pitfalls: attending to signals from the periphery, building a robust learning capacity, maintaining strategic flexibility, and designing the proper degree of organizational separation.

Despite their superior resources, incumbents often lose out to smaller rivals in developing emerging technologies (Table 2.1). Why do these incumbents have so much difficulty with disruptive technologies? How can they can anticipate and overcome their handicaps? Our studies of emerging technologies find that incumbents suffer from four common traps that put them at a disadvantage. Incumbents may resist participation at the outset, or pursue the wrong technological path. And even if these traps are sidestepped, many established firms are unwilling to make a full-fledged commitment to persist in the face of adversity.

Nonetheless, some established firms such as Intel, Charles Schwab, Monsanto, and General Electric have been able to avoid the "incumbent's curse"

Table 2.1
Consequences of Technological Innovation*

Who Is the Lead Innovator?	*Who Gains Eventual Market Leadership?*		
	Incumbent	*New Entrant*	*Neither/Both*
Incumbent	Electronic data processing Float glass Synthetic fibers Electronic cash registers	Helical scan videotape recorders	Optical data recording Videodisc Wankel auto engine
New entrant		Desktop xerographic copiers Electronic calculators Hydraulic earth-moving equipment Minimill steel making Portable transistor radios Programmable motor controls Radial tires in North America RISC microprocessors Semiconductor electronics Steamships	
Alliance above	Monoclonal drugs		Recombinant insulin

* We are grateful to Arnie Cooper of Purdue University and Clayton Christensen of Harvard Business School for researching some of these examples.

and embrace disruptive technology changes. How did these firms and others avoid the four traps? In the second half of this chapter, we combine lessons from best practice with broader principles of strategic management to propose four possible solutions: (1) attend to signals from the periphery, (2) invest in a learning capability, (3) maintain flexibility by adopting an options perspective, and (4) maintain organizational separation.[1] Each of these solutions enables an established firm to sidestep or escape the traps posed by an emerging technology. These, however, are not easy, off-the-shelf prescriptions. They require careful tailoring to the culture, context, and capabilities of the firm.

TRAPS FACING ESTABLISHED FIRMS

The emergence of a challenging technology such as interactive computing or electronic commerce is seldom a surprise. Most managers attend industry conferences, read the trade press, buy consulting studies, talk with customers, and generally monitor developments in their field. The problem is that each of these sources tends to offer conflicting opinions that are reflected in divergent views within the firm. The inherent ambiguity of an emerging technology and the new markets it may create, coupled with the dominance of traditional decision frameworks, make established firms vulnerable to four sequential traps. Although these four traps are related, they occur at different stages of the decision process, involve different causes, and generally require different remedies.

Trap One: Delayed Participation

When faced with high uncertainty, it is tempting and perhaps rational to just "watch and wait." A watching brief may be assigned to a development group, or a consulting study may be commissioned to examine the implications. Whether there is any organizational energy behind these probes depends critically on whether there is a credible champion for the emerging technology within the firm, who offers an alternative paradigm for interpreting the weak external signals.

Managers operate from mental models that impose order on ambiguous situations to reduce uncertainty to manageable levels. These mindsets usually are sensible adaptations to what the managers have learned from past experience. However, humans tend only to see what fits these mental models and either filter out or distort what does not fit. The mental models that prevail in established firms are helpful for incremental innovations within familiar settings, but may become myopic and dysfunctional when applied to unfamiliar situations, such as emerging technologies.

Because of the limitations of their mental models, managers may not appreciate the opportunities that are latent in an emerging technology. For example, when IBM considered adding the Haloid-Xerox 914 copier in

1958, the primary concern was whether the existing electric typewriter salesforce could handle the product. The focus was on spreading the selling cost of this division over two product lines, rather than viewing it as an entirely new business for IBM. Since copiers did not look attractive within this rather narrow frame, the opportunity was rejected.[2]

Emerging technologies are often framed as suitable only for narrow applications that are not yet required by existing customers. It is quite easy to dismiss such unproven technologies on the grounds that small markets will not solve the growth needs of large firms. Of course, all large markets were once in an embryonic state with their origins in limited application. IBM at first did not see the great opportunity in PCs. They were deemed to be entry systems from which buyers would eventually move to mainframes. IBM's ill-fated decision in the early 1980s not to obtain an exclusive license on Microsoft's operating system is testimony to this major blindspot, as is IBM's failure to exercise its purchase option for 10 percent of Intel (whose stock price increased 40-fold between 1985 and 1997).

The immature, early iterations of most emerging technologies often make them easy to dismiss. We tend to compare these imperfect and costly versions against the refined versions of the established technology. Pictures from electronic cameras initially lacked the resolution of chemical emulsion film; and the first personal digital assistants had disappointing limitations as with the Apple Newton in 1993; the first electronic watches were bulky, heavy, and unattractive; and the Internet continues to frustrate users because of slow response, junk messages, limited interactivity, privacy concerns, and inefficient search engines.

Visionary companies understand that infant technologies—which initially may "have a face only a mother could love"—can grow up to become important mature industries. When managers consider a new technology, they need to focus on the ultimate potential of the technology, and not be blinded by its current look and feel or the current shape of the market. The market will change. The technology will evolve. Managers need to see the technology not as it is today, but as it could become. This requires foresight and imagination. At the same time, however, firms need to be realistic about how long it will take to get there. Then, they need to carefully weigh the potential value of that market, the potential moves of rival firms and the costs of not moving forward. This evaluation may still lead the incumbent to delay participation, but then it will be for the right reasons, rather than the result of a failure of imagination or a desire to protect the status quo.

Key Questions for Trap One.   To engage in the kind of thinking that helps avoid the trap of delayed participation, managers should ask themselves the following questions:

- What could this technology look like in the future? What scenarios can be envisioned?
- How quickly might they develop? What changes would customers need to make for the technology to emerge? What breakthroughs are needed in the technology? What complementary technologies or market changes could accelerate or delay its progress? What competitive technologies might surpass it?
- What are the risks of waiting? Which competitors are poised to move more quickly? How much of the market can we afford to cede to them? Are there first-mover advantages in this market?
- What are some creative ways to develop strategic options to stay in the game without fully committing to it? How can these options be used to avoid delayed participation?
- Assuming that the status quo will not last, how might the above forces create quantum changes in our business? What is the worst case scenario for the established business? Who would get hurt most?
- Are there similar cases from the past where we fell into the trap of delayed participation? In hindsight, what should we have done then? What does that suggest for now?

Trap Two: Sticking with the Familiar

When the Encyclopedia Britannica was approached in the late 1980s to license its content for CD-ROM delivery, it elected to stay with the familiar printed-page technology. The company lost 50 percent of its revenue between 1990 and 1995 due to inroads made by CD-ROM technology.[3] Even if incumbents initially defer participation, they must at some point choose whether and how to participate in the emerging technology. Often they choose to stay with the familiar far too long even when there are compelling arguments for making a change. This preference for the known increases if the potential payoffs of the unknown technology are underestimated due to the limited imagination discussed in trap one.

Most people are averse to risk and dislike ambiguity. We tend to prefer a known probability over an unknown probability of equal expected value.[4]

For emerging technologies, sticking with the known is even more appealing because the leap into a new technology often is fraught with uncertainty. There are doubts about the technology path, whether the technical hurdles can be overcome, and which standard or architecture might prevail as the dominant design. The problem is most acute with emerging technologies derived from radical innovations such as photorefractive keratectomy, high-temperature superconductivity, or recombinant DNA technology.

The challenge is even more complex when multiple technologies are advancing simultaneously. The emergence of revolutionary innovations derived from the convergence and recombination of previously independent streams of existing technologies are very difficult to predict. For example, an extrapolation of computer disk storage costs suggests that optical technologies will overtake magnetic technologies within 10 to 15 years. However, a third technology based on solid-state memory storage is developing at an even faster rate and may overtake optical within 20 years. Meanwhile, there is uncertainty over whether magnetic storage technology has reached the top of the "S curve" (which plots performance improvements against time) and is entering the zone of diminishing marginal returns.[5]

The most vexing technology choices are those where multiple versions are vying to become the dominant design, as occurred historically with the light bulb, and more recently with VCRs, modems, digital wireless telephones, and high-definition televisions (HDTV). A design becomes dominant when it commands broad allegiance of the market, so that competitors are forced to adopt it if they want to participate. This represents a milestone in the emergence of a technology because it enforces standardization that enables product economies to be realized, while removing a major inhibitor to the wide-scale adoption of the technology.

Often there is explicit competition among firms to set the industry standard in hopes of gaining an enduring advantage. Wars for standards involve complex issues of lock-in effects, network externalities, positive feedback loops, increasing returns, switching costs, and installed bases.[6] Above all, each firm must know which kind of standards battles it is facing, which in turn hinges on the compatibility of its own new technology with prior technology as well as that of rivals. The stakes in standards wars can be very large, for if another design or standard prevails, the losers are trapped.[7] The dilemma for managers is that they must commit aggressively to pushing their design or standard, without diluting their efforts with other approaches, to have any chance of prevailing.

The odds of picking a familiar but wrong technology path goes up when:

- Past success reinforces established ways of problem solving and decision making leading the firm to search in areas that are closely related to their current skills and technologies. Often, a firm's capabilities and past history limit what it can perceive and develop effectively.
- The firm lacks in-house capability to appraise the emerging technology fully. Thus, it may be underestimated or overly feared. Running a branch banking network is very different from electronic commerce, for example. Consequently, banks initially shied away from offering electronic services.
- A proprietary mind-set gets in the way. The instinct of a large company with a proprietary position in its core market is to find a similar proprietary position with the new technology that will lock in customers. Such a move makes customers suspicious, however, especially in today's open-system environment. The new technology may not even permit such a proprietary niche. For example, newspapers have long enjoyed a strong local franchise in their communities, based on their advertising savvy and local information coverage. However, in the world of the Internet, a similar local franchise may not be possible.

Our preference for the familiar is deeply rooted in human psychology.[8] First, most people register losses more vividly than equivalent gains. Second, they do not like to stick their necks out. Taking action entails visibility, vulnerability, and accountability so the burden of proof is on those wanting to change. Third, an act of commission (as opposed to omission) carries a greater risk of regret, reproach, or censoring by an oversight committee. Last, change entails transaction costs and effort. For these and other reasons, most humans (and by implication firms) are unduly bound to preserve the status quo.

Key Questions for Trap Two.   To address the trap of sticking with the familiar, managers can ask themselves the following questions:

- How strong is the status quo in your business and what standards prevail in your industry? What are the fundamental assumptions upon which your business is based?
- How can your firm bring in perspectives from experts in unfamiliar technologies, markets and strategies? How can managers draw upon

new sources of ideas (periodicals, conferences, etc.) to break out of their narrow perspectives?

- What are the new ideas that could challenge your assumptions and your status quo? What are the new standards battles that are just heating up?
- If you were to start your business from scratch today, what would you be doing that you are not doing today? Should you be doing it anyway?
- How have managers been treated in your firm when they opted for the unfamiliar? Who are the senior champions for "trying something different?" What is their power base and track record?
- Is this the time to do something different? How stable is the technology base that underpins your industry or market? What lurks on the technology horizon?

Trap Three: Reluctance to Fully Commit

Even when companies overcome the tendency to delay participation (Trap One) and avoid sticking only with the familiar (Trap Two), they may still make just a halfhearted commitment. One study of 27 established firms found that only four entered aggressively while three didn't participate at all in a threatening technology.[9] The majority made a modest initial commitment that gave the entrants from outside the established industry time to secure a strong market position.

Some of these small bets may actually be well-considered real options, as discussed in Chapter 12. The company may want to stage its investments to limit the risks while keeping its hand in the game. But incumbents tend to err on the side of being too cautious. They may rationalize their fear of aggressive commitment as constituting an optimal options strategy. Instead they need to be vigilant and honest about the true risks they take by making small investments and the true options that can be realized in the future. This is not an easy issue to assess, yet it is one of the most important points incumbent managers must confront in their own decisions. To what extent are the decisions shaped by carefully considered prudence versus an unconscious aversion to risk?

Why are leading firms repeatedly unable or unwilling to make aggressive commitments to an emerging technology once they decide to participate? First, managers are rightfully concerned about the possibility of cannibalizing existing profitable products, or about resistance from channel partners, and thus may hold back support. Both IBM and DEC were reluctant to push distributed networks that would undercut their highly profitable mainframe computers. Major insurance companies, such as New York Life or Metropolitan Life, are understandably concerned about selling policies via the phone or Internet in light of their investment in a large, loyal, and productive network of agents. Kodak's first instinct was to enter digital imaging with its Photo CD, which allowed consumers to transfer conventional film images onto compact disks and thus avoid cannibalizing its core business.

Second, there is a paradox in managerial risk-taking, toward *bold* forecasts on the one hand and toward *timid* choices on the other.[10] Bold forecasts can stem from overconfidence in general, or more specifically, a limited ability to see arguments contrary to one's prediction. Timid choices reflect an inclination toward risk-aversion and a tendency to look at choices in isolation (rather than from a portfolio perspective). There are a host of other cognitive biases whose impact on managerial judgment and choice often results in sub-optimal behavior.[11] Much more research will be needed to connect these general individual-level biases to the specific organizational traps encountered in the management of emerging technologies.

Third, when the profit prospects are unclear, and possibly less attractive than the current business, investments are often difficult to justify under strict ROI criteria. The customary decision processes and choice criteria are biased against risky, long-term investments. There is a "certainty effect" in that relatively sure returns from investments in incremental product improvements for today's market are valued higher than riskier investments for tomorrow's markets.[12]

Furthermore, the projected returns from an emerging technology are often worse than those from established technologies that address the specific needs of current customers. It was hard for Encyclopedia Britannica to consider replacing a $1,300 printed encyclopedia set with a CD-ROM that would sell for $80. Likewise, the gross margins in successive generations of hard disk drives narrowed, and required different business models before profitable participation was possible. When gross margins for disk drives in mainframes were 60 percent, the margins for disk drives in PCs and notebooks were only between 15 and 35 percent. These slimmer margins were a real deterrent to enthusiastic participation by the established players, almost all of whom lost their lead in the next generation.[13] Similarly, Kodak had to learn how to contend with much lower margins for its digital imaging products than for chemical emulsion film. Drug companies likewise wrestle with investments in new genetic technologies as they contrast the returns from niche treatments with the generous profits from the large markets for prescription drugs. Managers have a strong incentive to sustain the old, higher margin business for as long as possible.

A fourth reason why half-hearted commitments are often made is that the attention of established firms is primarily focused on meeting the needs of their current customers rather than looking to serve new needs and new markets. Thus, they dismiss or overlook new technologies that seem applicable only to smaller market segments they do not currently serve or don't understand.[14] For example, the large copying centers that were the core of Xerox and Kodak's traditional market failed to appreciate the value of small, slow table-top copiers for homes or small businesses. This oversight opened the way for Canon. Sears underestimated the threat from Wal-Mart's discounting strategy because it was first implemented in small and seemingly uneconomic local markets that could not support conventional stores.

Existing customers requirements often receive disproportionate attention and so the ability of the emerging technology to better meet the needs of noncustomers is overlooked. The manufacturers of 5.25-inch hard disks who were highly attuned to the demands of desktop PC manufacturers for greater memory capacity, grossly underestimated the appeal of 3.5-inch disks that were smaller, lighter, and more rugged and which, in turn, enabled the market for laptops to emerge. Often the potential for improvement of the "disruptive" technology is underestimated until the mainstream market starts to migrate from the established technology. At this point, the incumbents are willing to participate, but the pioneers of the new technology may already be too far ahead.

Finally, and most importantly, successful organizations are not naturally "ambidextrous." They encounter numerous debilitating problems in balancing the familiar demands of competing in markets presently served with the unfamiliar requirements of an emerging and potentially threatening technology. Within the core business, there is usually a close alignment among the strategy, capabilities, structure, and culture, which in turn are supported by well-established processes and routines for keeping these elements in balance. This gives the organization a great deal of stability that must be overcome before the new routines and capabilities needed to compete with the emerging technology can be developed.[15]

The more successful the firm, the more closely the elements of strategy, capabilities, structure, and culture are aligned and the more difficult and time-consuming discontinuous changes become. Inertia also stems from "path dependency" in which decisions and problem-solving approaches used in the past frame and color decisions about the new technology. Chapter 3 provides more detail about path dependency and evolutionary changes.

The above five forces reinforce each other to produce a weak commitment to the new technology. The net effect is to impair decision making, erode the necessary enthusiasm of the advocates, and cause firms to hesitate or hedge before making major commitments. These afflictions do not inhibit the new entrants who may sense the opportunity earlier, better comprehend or believe in the benefits of the new technology, do not have any misleading history to contend with, nor have any core business to fall back on. In a sense, the newcomers have much more at stake in making the new technology successful than the incumbents. The paradox here is that the established firms shy away from the very risks that the upstarts eagerly embrace, even though a large firm can more easily bear these risks given its size and portfolio diversification.

Key Questions for Trap Three.   Managers of established firms need to guard against the bias toward a weakened commitment by asking themselves:

- Is the failure to fully commit deep down a result of fear of cannibalism? If so, how long can you really hold onto your existing product markets? What are the chances competitors will eat your lunch if you don't do it first?
- Are you truly creating and managing strategic options or is this an excuse for making a small commitment?
- Which groups within the company would like to see the new technology fail and how has their direct or indirect influence affected the decision process?
- Are you making bold forecasts but timid choices? Are you too risk-adverse?
- How does your current profit model limit your interest in investing in lower-margin new technologies? Are your present margins sustainable?
- Are you limiting your analysis to your current markets? Are you perhaps too close to your present customers? What markets outside of your traditional ones could you serve with the new technology?

- Is your organization too well fitted to the core business, and incapable of operating out of equilibrium? An optimal fit to the present business may not be an optimal configuration for long term survival.

Trap Four: Lack of Persistence

Suppose, however, that an established firm has managed to avoid the first three traps and made significant investments in a new emerging technology. Will it have the fortitude to stay the course? Large companies, when facing pressure for quarterly results, soon loose patience with adverse results. One study found that 8 of 21 established firms that entered markets in which emerging technologies were succeeding subsequently withdrew, and most did not resume their efforts until the viability of the new product was demonstrated by outsiders.[16] At that point, it was usually too late to achieve leadership.

Yet, missed forecasts and dashed hopes are commonly experienced during the gestation of new technologies that eventually succeed. Market demand may not materialize as soon as expected, too many competitors may crowd into the market, or the technology may veer off in a new and unexpected direction. In time, the initial enthusiasm may be replaced with skepticism about when—if ever—the new technology will become a profitable business reality.

This trap of low persistence is the flip side of another well-known trap—the sunk cost fallacy. The irony is that the very firms that are overly committed to their core business (the sunk cost trap) are often too quick to pull the plug on investments in emerging technologies. Knight Ridder, Inc. (a major U.S. newspaper company) repeatedly has withdrawn from early moves into emerging technologies such as electronic publishing, online news, and interactive services in response to negative financial results (Table 2.2). Time Warner, another media company, abandoned its recent investments in cable and electronic commerce. Such hesitancy is likewise occurring in computer-mediated interactive home shopping, where questions are being raised about how to surmount some critical logistics problems. How can an order be delivered, managers ask, if there is no one at home to receive it? Will the costs be acceptable to the consumer? What kinds of order-fulfillment capabilities will be needed to support home shopping? Can the firms that participate afford these investments?

Those who truly appreciate the possibilities of the emerging technology and feel enthusiasm for any given new project are often deep down in the

Table 2.2
Knight-Ridder's Strategic Moves

Date	Strategic Move
1974	Merger between Knight and Ridder newspapers
1978	Purchase of three television stations in three states
1981	Formation of cable company with TCI
1982	Launched Vu-Text (timesharing computer link)
1988	Purchased Dialog (electronic data base)
1989	Sold television stations★
1993	Purchased Data Star (online service)
1995	Purchased a small stake in Netscape
1996	Sold Financial Information Services (for $275 million)★
1997	Announces plan to divest KR Information★
1997	Purchased four more newspapers from Disney ($1.65 billion)

★ Retrenchment by company from emerging or unfamiliar technologies.

organization and may have little influence on high-level strategic thinking. Thus, if a company's core business begins to struggle and senior managers are looking for ways to cut costs or reduce assets, the new venture is an easy target. After all, the real payoffs from the new venture may not accrue until after senior managers retire. Thus, the political support for the venture evaporates along with the funds to nourish growth and keep abreast of fast-changing developments. For this reason, the CEO should be the champion of long-term investment in new ideas. When Gannett launched its innovative new paper *USA Today,* it experienced over 10 years of losses before it became a winner. Luckily, the concept was championed by CEO Al Neuhardt against the "better judgment" of many.

Patience for continuing losses will be further strained if the firm has already suffered losses with related ventures, and the corporate memory starts to draw unflattering parallels to the latest failed undertaking. CitiCorp and McGraw-Hill were understandably cautious about the promise of electronic commerce after having been burned 10 years earlier in large-scale electronic global petroleum trading. Their joint company, called Gemco, utilized a private network to connect 70 information providers with hundreds of traders and a bank. The network was used for gathering information and negotiating deals, as well as making payment and clearing trades. Although the technology worked as intended, Gemco eventually

failed because the widely dispersed traders did not want to change their interaction patterns to conform to the central network's dictates.

Key Questions for Trap Four.   How can managers avoid falling into the
trap of pulling the plug too soon on promising new technologies? They
should ask themselves the following questions:

- How can the company protect the new technology venture from the
  profit pressures of the parent organization (for example, separate
  organization, different accounting systems, different performance
  criteria)?
- Is there commitment from top-level leadership for the project or is it
  driven only by those who have little control over resource allocations?
- Suppress organizational memory when inappropriate comparisons are
  made with the past. Just as firms must learn, they should also try to
  forget. Some degree of corporate amnesia is desirable since the past can
  be misleading when facing a new world.
- On what basis will the company assess whether the venture is a success or failure? Does it emphasize wealth creation or short-term ROI?
  Will the prevailing criteria give the initiative a fighting chance or
  doom it to premature abandonment?
- Don't agonize over each quarterly loss as it is reported, but aggregate
  the losses into yearly figures, and prepare the organization to expect
  these losses as part of the "rocky road" scenario associated with investing in an emerging technology. (Psychologically, it hurts less to
  lose $1 million in one lump sum than to lose $250,000 each on four
  separate occasions.)

CRAFTING SOLUTIONS

While being aware of these four traps can help avoid them, the best defense
may be a good offense. The established firms that do prevail follow a more
aggressive path that balances flexibility of posture with sustained commitment and follow through. In this section, we offer four proven solutions:
widening peripheral vision, creating a learning culture, staying flexible in
strategic ways, and providing organizational autonomy. These solutions do
not correspond one-to-one with each trap, but rather address several of
them at a time. Think of these four strategies as ingredients from which an
overall solution approach can be fashioned that uniquely fits your firm.

Widening Peripheral Vision

Emerging technologies signal their arrival long before they bloom into full-fledged commercial successes. However, the signal-to-noise ratio is initially low so one has to work hard to appreciate the early indicators. This means looking past the disappointing results, limited functionality, and modest initial applications to anticipate the possibilities. Many signals are available to those who look; other signals can only be seen by the prepared mind. As the philosopher Kant noted, we can only see what we are prepared to see. The winners are those who hear the weak signals and can anticipate and imagine future possibilities faster than the competition. Scenario planning, discussed in Chapter 10, has proved a powerful tool for amplifying weak signals and imagining different futures.

The weak signals to be captured usually come from the periphery, where new (and previously unknown) competitors are making inroads, unfamiliar customers are participating in early applications, and unfamiliar technology paradigms are emerging. But the periphery is very noisy, with numerous tangential technologies that may or may not be relevant. This background noise can be seen in the rapidly converging entertainment, telecommunications, information, cable, and computer sectors, where a myriad of technologies such as interactive TV, Web TV, CD-ROM, desktop video, and satellite transmission combine to create new products and services. What is background noise to one player may be a strong signal to another.

The first step in deciding which signals and trends to scan is to define which emerging technologies are strategically significant (as discussed further in Chapter 4). This requires shifting the corporate focus from the characteristics of its existing products or services to features that provide benefits to customers or help them solve their problems. For example, customers did not want X-rays or CAT scans per se, but desired more accurate images of tissues and bone structures that would identify anomalies. By focusing on potential benefits, managers can more readily identify emerging technologies that might offer new ways to provide or improve those benefits. To consumers, a compact disc is just one way to store and reproduce music that suits a particular mood and occasion so they are open to new technologies such as MP3 players that store music downloaded from the net. Companies also can study lead-users who are ahead of the curve to see the promise of a new technology, or work jointly with lead users (in say, industrial markets) to create the next generation of products.

Once underlying functionalities have been defined, the next question is how well the emerging technology can deliver new features that meet customer needs and budgets, relative to other competing technologies. Such a strategic projection of the performance potential of an emerging technology entails more than a linear extrapolation into the future. First, we should take into account the typical S-shaped relationship between performance and cumulative development.[17] It usually requires a period of significant effort before any results are seen, followed by rapid progress for relatively little effort before eventually reaching a limit. These S-curves apply both within the firm (for proprietary technology development) as well as at the industry or market level. Hence, to arrive at useful projections (which might consist of point estimates as well as ranges), it is necessary to estimate the rate of competitors' spending on development. Also, the projection should take into account possible improvements in the established technology, since what matters most is relative performance. Hence, the projection has to be made against the moving target of the old technology.

Once a technology trajectory has been projected, the challenge then is to estimate the rate of market adoption and the potential size of the market. When it is not yet apparent who the customers will be, and lead-users have neither experienced the level of performance nor the features that are possible, such estimates are hard to make. Conventional market research techniques are seldom applicable to the uncertainties of embryonic markets.[18] Approaches such as sample surveys, concept tests, and conjoint analysis were designed to address better defined problems in existing markets. A different approach is needed when the market concept is ill-formed, the technology is barely ready, and questions of relative cost, availability, and performance are unresolved. Rather than ask customers directly about their response to or interest in new products or services, early market research should concentrate on lead-users and improvements in functionalities. Prospective customers may not know whether they want holographic PC TV as a technology, but they can usually assess how much more they value its functions and benefits relative to the present TV offerings.

Xerox's strategy for estimating the potential market for fax machines in the 1970s illustrates how customer benefits and functionality can be used to develop market estimates for an embryonic technology.[19] Managers based their estimates on the extent and frequency of urgent written messages, their time sensitivity, and the form and size of the message (number of pages, use of graphics, etc.). Then they contrasted the promise of fax capability with existing solutions such as mail, telephone, telegrams, express delivery. With this approach to estimating the latent demand for fax-type features as opposed to studying responses to concept statements, Xerox foresaw a business market of approximately 1 million units in the early 1970s. In hindsight, this number proved to be much too low, but it was much larger than people first thought.

Unfortunately for Xerox, they chose the wrong technology path because they preferred to stick with familiar computer-to-computer transfers (falling into Trap Two). The company developed a system of sending facsimile messages from one computer to another, with the receiving computer printing out a copy using standard imaging technology. This turned out to be a much less attractive approach than having cheaper and more ubiquitous specialty devices linked by telephone lines (even though its thermal wax paper was inferior to plain paper). Eventually, however, Xerox's view of PC-based faxing may prevail over stand-alone fax machines.

The choice of methods to assess a market for an emerging technology should be guided by three key principles that are discussed in more depth in Chapter 6:

1. *Paint the big picture.* The management questions addressed need not be elaborate or ask for precisely calibrated results. At the very earliest stages in the exploration, the issue is simply whether the potential market could be "big enough" to support the development activities. Questions of market potential, timing, and growth rate are premature.

2. *Use multiple methods.* All market estimation methods are flawed or limited in some important respect. Thus, analogies with markets for technologies with similar characteristics are suspect because the situations may not be comparable in critical but unknown regards. Similarly, surveys of experts using Delphi methods to assemble composite forecasts of demand may be no more than a pooling of collective ignorance. While any one method is limited, a combination of methods—each asking the same question in a different way and thus being prone to different biases—may yield conclusions that are directionally sound. The process of triangulation of results looks for common themes and patterns after accounting for specific biases by averaging them out across methods.

3. *Focus on needs not products.* As discussed, prospective customers can't envision radically new products based on discontinuous innovations. However, they can be eloquent about their needs, problems, usage or application situations, and changing requirements that will

dictate their eventual acceptance—but only if the right questions are asked.

Creating a Learning Culture

A second approach to the traps of emerging technologies is to keep learning. And the challenge here is *collective learning,* not just individual learning.[20] In his book *The Living Company,* Arie de Geus recounts a classic study by Allan Wilson of two species of garden bird in England.[21] One species (red robins) was highly territorial while the other species (the titmouse) would flock together. Both species liked to skim cream from the open milk jars that would be placed on the doorstep by the milkman before World War I. Between the two world wars, some new technologies emerged in the form of milk bottles with aluminum seals, which meant that the birds no longer had easy access to the cream. Some innovative birds (entrepreneurs) however, figured out that they could simply pierce through the tin foil with their beaks and get to the cream that way. Among the territorial birds (red robins), this innovation was isolated and infrequent. Among the other species (titmice), this innovation spread rapidly since they communicated and flocked together. The point is that individual learning only translates into collective learning in a culture that shares information and communicates frequently.

As discussed earlier, the diverse sources of information flowing from the periphery create plenty of noise. This flow of information can lead to confusion and immobility rather than insight and action if there is no process for sharing or joint learning. The information must be absorbed, communicated widely, and intensively discussed so that the full implications are understood. This requires an organizational learning capacity that transcends individual learning. This capacity is characterized by:

- An openness to a diversity of viewpoints, within and across organizational units.
- A willingness to challenge deep-seated assumptions of entrenched mental models while facilitating the forgetting of outmoded approaches.[22]
- Continuous experimentation in an organizational climate that encourages and rewards "well intentioned" failure.
- Mastering the skills of deep dialog and strategic conversation.

Encouraging Openness to Diverse Viewpoints.   The uncertainties of disruptive emerging technologies require thorough debate. The initial analysis should encourage *divergent* opinions about technological solutions, market opportunities, and strategies for participating. As learning evolves, one or multiple views emerge as a basis for *convergence* toward a few commercializable solutions that can be tested The tone of this extended debate should be set by senior management through their willingness to bring in outsiders with nontraditional backgrounds, immerse themselves in the stream of data, and by asking challenging questions. Senior managers must wander outside their offices and comfort zones, and start having conversations with informed insiders, outside experts, and customers. They must study competitive moves and analogous situations, float ideas, and seek collaborations. This can be done in diverse forums including team meetings, outside conferences, and electronic bulletin boards. Top-down involvement will only be productive if there is active bottom-up participation. Employees from different levels bring different points-of-view and expertise, and are typically closer to the realities of marketplace and the technology. As these debates proceed and arguments become more refined, the facts and positions will get in much sharper focus.

This was the process orchestrated by Hugh McColl who built Nationsbank into America's twentieth-largest bank through an aggressive series of acquisitions.[23] The traditional model of consolidation leading to large banks that reap huge economies of scale is being challenged by new technologies for delivering financial services. Banking can be done by phone or home computer, via "smart cards," or interactive television with an in-home ATM screen. In this new world, banking might become a low value-added service activity that simply handles transactions for a high-profit electronic "store" controlled by an outsider like Microsoft. In that case, valuable client relationships would erode as the customers search the Internet for loans, mortgages, and low-cost or faster financial services. To deal with these profound unknowns, McColl had to change his previous mode of reliance on an inner circle of advisers, to bring in outsiders, listen to ideas from diverse sources within the company, and invest in a strategic technology group of 95 people (operating as internal consultants) who tracked trends and studied possibilities. The aim is to have a growth strategy to which the whole organization can respond.

As uncertainty increases, so does the potential for confusion and paralysis, which may result in a "wait-and-see" posture of delays and lack of commitment. Organizations need a mechanism for coalescing and focusing the ongoing dialogue while reducing the various uncertainties to manageable chunks. Scenario analysis (see Chapter 10) achieves this through a process of collectively envisioning a limited set of plausible futures that are internally consistent and detailed. Each scenario can be used to generate strategic options, evaluate prospective investments, and assess their robustness.[24]

Challenge the Prevailing Mind-Set.   Diverse viewpoints will not have an impact if the prevailing mind-set of the organization prevents it from absorbing these insights. Expansive thinking about the future is readily subverted by the rigidities and restrictions of the prevailing mental models, industry success formulas, conventional wisdom, and false analogies from the past. The limiting and simplifying operations of deeply embedded mental models raise serious questions about whether even scenario approaches are appropriate for dealing with profoundly disruptive and discontinuous change. The concern is that the scenario building process anchors on the present—as shaped by the prevailing mind-set—and then projects forward to what *might* happen. By contrast, firms that successfully exploit discontinuities may have to separate their thinking from current beliefs and realities to envision what *could be,* and then work back to what must be done to ensure that this aspired future will be realized. Ackoff, as well as Hamel and Prahalad,[25] emphasize that this kind of foresight requires deep and boundless curiosity, a willingness to speculate without conclusive evidence, the liberal use of analogies from comparable situations, and deep insights into customer needs, requirements, and behavior.

Mental models can impede the unbounded thinking needed to envision disruptions, surprises, or improbable developments because they are so grounded in past experience, reinforced by ongoing commitments, and shackled by the inertia of the status quo. Before the constraints of the prevailing mental models can be challenged, the texture of these mind-sets should be described through a cognitive mapping exercise that makes explicit the views and core assumptions of the management team about important forces for change and their consequences.[26] Once these views have been displayed, key uncertainties can be surfaced, and the underlying assumptions challenged. The danger of truly unbounded thinking, however, is that it becomes unbridled fantasy that is disconnected from reality. What senior managers should strive for is disciplined imagination, as discussed in Chapter 9.

Experiment Continually.   The best way to learn is through continuous experimentation. Successful adaptation to the vagaries of emerging technologies requires a willingness to experiment and an openness to learn from the inevitable failures and setbacks. There are several facets to the call for continual experimentation. Sometimes it means a willingness to create a diverse portfolio of technological solutions, by endorsing parallel development activities. For example, Intel undertook research on RISC chips even as CISC microprocessors were being emphasized. It may also mean quick prototyping, by introducing early and immature versions of the product into a plausible market segment, learning from the experience, modifying the product and marketing approach, and trying again in a process of successive approximation. This is how Motorola entered the cellular phone market and GE participated in the CT scanning business.[27] In other settings, more ambitious, formal market tests may be conducted, such as the extended tests of interactive television by the Baby Bells.

Continually experimenting and improvising with a new technology produces insights about the possibilities and limits of the technology, the responses of diverse market segments, and the competitive options that customers consider. Once important uncertainties are resolved, such learning organizations are ready to act. For example, if influencing standards is a key to success, they can better judge which elements to endorse.

Experimentation requires a tolerance for failure. The trial-and-error learning that relies on experimentation is easily subverted if there is a fear-of-failure syndrome. Organizations that reward those who play it safe and blame risk takers even for "well-intentioned" failures will discourage learning. The path of learning is marked by serendipity, and the knowledge gleaned from careful diagnoses of the possible reasons for failure. Unfortunately, there is often little patience for failure, and few incentives to study the failure. In counterproductive environments, without well-developed learning capabilities, a diagnosis that might distinguish causes from effects is seen as a way of assigning blame. It takes concerted leadership to create a more open climate that rewards improvisation and makes learning from failures possible. British Petroleum accomplishes this through post-project performance reviews prepared for the chairman of the board.[28]

Deep Dialog.   The successful application of the above remedies requires that organizations master the art of strategic conversation. Howard Perlmutter has identified various impediments to deep dialog ranging from lack of trust to unresolved problems in the past to cultural misunderstandings.[29] Perlmutter defines deep dialog as both a meeting of the mind and the heart, which in turn fosters bonding among people, bridging of differences, and blending of ideas. Kees van der Heijden, former head of scenario managing at Royal Dutch/Shell, similarly views strategic insight to emanate from thoughtful conversation.[30] He advocates various techniques and approaches, such as scenario planning, to create contexts that are conducive to a deeper meeting of the corporate minds. Organizations may have to (re)learn how to engage in productive strategic dialog in order to pierce the fog of emerging technologies. Individual brilliance is not enough. What really matters is the firm's collective wisdom and emotional intelligence, which can only be enhanced through constructive social interaction and strategic dialog about the unknown market opportunities that lie ahead.

Staying Flexible in Strategic Ways

One of the paradoxes of emerging technologies, as discussed in Chapter 1, is that while it is prudent to make limited investments, sometimes a strong commitment is what leads to success. One way to reduce the severity of this dilemma is to increase organizational flexibility. The greater the organization's flexibility, the lower the cost of making a commitment and the lower the cost to reverse direction. This is similar to using flexible rather than fixed manufacturing systems. A flexible system in an auto plant can be used to make a commitment to a certain vehicle model, but this commitment can easily be changed if the demand for another model unexpectedly increased. In contrast, making the same commitment through a dedicated facility would mean very expensive retooling for every model change. It is important to recognize that commitment is not always measured in absolute dollar value, so there are ways to conserve resources while still making a commitment to a particular technology.

At first glance, commitment seems to be the opposite of flexibility and you may not be able to have it both ways.[31] However, only if the commitment is irreversible does it directly contravene flexibility. For instance, if you make a commitment to a cruise voyage and pay the full amount upfront, it may seem that your flexibility has been diminished. However, if you also purchase cancellation insurance (in case of illness or a death in the family), you do preserve considerable flexibility to change course when needed. This is the art of options management: It involves creative hedging and an ability to imagine diverse scenarios. The only downside of creating flexibility is that it may reduce the strategic signaling value of commitment, which truly requires irreversibility to be credible (e.g., when making a preemptive technology move or when building a new plant to scare off others). All these factors need to be balanced within the options perspective.

Microsoft is a prime example of an established firm maintaining flexibility in the face of uncertain technologies. Its much celebrated "turn on a dime" response when confronted with Netscape's Internet browser is just one instance of its pattern of purposive agility. Microsoft was already playing many bets as early as 1988.[32] At that time, Apple was at the peak with its superior graphical user interface for the Macintosh, making Microsoft's DOS look like a distant second. However, Microsoft was operating on multiple fronts in 1988. On one front, it was developing Windows; on another, it was pushing OS/2 which it codeveloped with IBM. And, at the same time, Microsoft was introducing various application software packages, including Excel and Word, for both Windows and Apple's Macintosh. Last, Microsoft was in partnership with SCO, the largest provider of PC-based Unix systems.

In essence, Microsoft had developed a strong hand of cards to play in a variety of worlds that might emerge. Its portfolio of options was commensurate with the uncertainties surrounding hardware and software development at the time. Questions of standards, features, channels, and delivery modes (PCs versus servers) were still to be settled. In addition to developing a robust hand of cards, Microsoft developed a culture that could quickly change strategy. Bill Gates proved a bold leader when he changed course midway once the threat and promise of the Internet became apparent to him.

Sun Microsystems and its allies had long been pushing for a more universal programming language (Java), which held the promise of quickly developing software code and downloading application programs from servers. When the Internet was added to this thrust as a possible new medium for software delivery (in addition to servers and work stations), Gates recognized that Netscape might run away with the Internet opportunity. Netscape was built in record Internet time. Two months after release of the Navigator in December 1994, Netscape had captured 60 percent of the browser market, representing an installed base of more than 38 million users.[33] Netscape reached annual sales of more than $500 million in slightly over three years, a level that took Microsoft 14 years to reach. When AOL purchased Netscape in March of 1999, it was valued over $10 billion.

Netscape was a threat to Microsoft on two fronts. Its browser was poised to create a universal interface for the networked world by offering cross-platform capabilities (from personal computers to cell phones to intranets). Second, Netscape pursued an open-standard strategy that could sway the momentum away from the Wintel standard. To its credit, Microsoft moved quickly to compete head on with this formidable new force, shifting its strategy in a few months. Hundreds of software development projects were stopped midway, to redeploy people, money, and time toward Microsoft's own browser and new Internet strategy. The Microsoft lessons—keep your options open, study your rivals intently, and change direction boldly and quickly when needed.

Managing Real Options.   Best practice suggests it is desirable in the early stages of exploration of an emerging technology to keep a number of options open by only committing investments in stages, following multiple technology paths, and delaying some projects. Once uncertainty has been reduced to a tolerable level and there is a widespread consensus within the organization on an appropriate technology path that can utilize the firm's internal development capabilities—as in the case with Intel's choice of personal computer over television as the preferred information appliance— then full-scale internal development can begin.

But what if there are many plausible technology paths, the risks of pursuing one to the exclusion of others are unacceptable, and the firm lacks the necessary internal capabilities? This situation was faced by Rhone-Poulenc-Rohrer, which attempted to create a highly flexible alliance structure to pursue more opportunities in gene therapy. To realize therapeutic rewards, many technologies would have to be integrated—ranging from the discovery of the genes and the vectors for delivering the gene to understanding the biological mechanisms involved in cancer, cardiovascular diseases, and nervous system disorders. Not only was it impossible to master all these technologies, but the approaches were changing rapidly. Their solution was to form a separate subsidiary, RPR-Gencell, to serve as the hub of a network of 19 technology partners including universities, biotech firms, and others. Each partner provided distinct research capabilities in therapeutic paradigms according to an integrating design. The network remained fluid and open as partners came and departed, depending on whether they could achieve performance milestones based on comparisons with what had been achieved by the best available competitors. Once a therapy becomes commercially successful, the partners would share royalty payments.

It is doubtful whether the particular model developed by RPR-Gencell will work. Their particular alliance approach worked well in the knowledge creation phase, but experienced stress in the commercialization phase. Nonetheless, the strategy of developing a more flexible organization is vital to creating the proper balance between commitment and options needed to successfully develop emerging technologies. The strategy of collaboration makes good sense when exploring numerous possibilities, but may collapse if issues of ownership of intellectual property are unclear. Also, RPR-Gen-cell played perhaps too central a role in the network they created, an issue address in more detail in Chapter 16.

In addition to creating organizational flexibility, the pacing of commitments can also provide strategic flexibility. Many large pharmaceutical firms facing the same uncertainties as Rhone-Poulenc-Rohrer have adopted a "wait-and-see" posture or engaged in cautious "learning by doing." Some firms such as Merck, Schering-Plough, Eli Lilly, or Novartis have made investments that are significant in the industry, but small as a percentage of their total development spending. The assumption is that the pioneers will either fail or, if they succeed, will become available as acquisition candidates. For example, Johnson & Johnson purchased the biotech firm Centocor for $4.9 billion, 20 years after its founding. Pfizer also pursues a late-entry strategy. Such a strategy is most likely to be feasible if the growth in the market is very slow, there are many competing technologies with little likelihood that a dominant design will ever emerge, and whatever technologies do prevail will be difficult to protect or easy to imitate. Ultimately, the optimal approach (ranging from betting-the-farm to wait-and-see) depends on the choice set a company and its competitors face. The more flexibility in the organization, the lower the costs of making a commitment and the lower the costs of changing that commitment.

Providing Organizational Autonomy

The fourth strategy to avoid the traps of large incumbents is to take the emerging technology business outside of the main corporation into a separate unit. The more the emerging technology initiative can operate from a smaller, entrepreneurial mindset, the less it will be held back by the inertia, controls, risk-avoidance, and big-firm mindset that leads to the four traps discussed earlier. By creating an isolated nursery, the company protects the emerging technology business from infection by microbes that, while not dangerous to the large firm, can be deadly to the new venture.

Many large companies set up separate organizations dedicated to pursuing new endeavors (such as GM's Saturn division, IBM's PC unit, or Roche's Genentech investment). The objective of "cocooning" the new business is to create a boundary that enables the new group to do things differently while still permitting the transfer of resources and ideas with the parent. This also permits separate objectives, recognition of long development cycles and continuing cash drains, as well as different measurement criteria so the performance of managers in the rest of the organization is not jeopardized. Above all, it creates flexibility.[34]

There are many degrees of separation. Some companies take the separation approach as far as to create new spin-offs: newly-formed companies with their own stock, board, and management teams, in which the parent retains some ownership position. Such equity carve-outs have been pursued by Thermo Electron (over 14 spin-offs), Safeguard Scientifics (over 10 spin-offs), as well as Enron, Genzyme, and The Limited (2 spin-offs each). A McKinsey study reported an annual compound return of 20 percent for such carve-outs after three years, which is 9.6 percent better than the Russell 2000 index for that period (which measures the performance of a broad array of small public companies).[35] An equity carve-out offers access to capital (via a public stock offering), strategic value from the corporate center, operating independence, development of talent, and higher motivation for key personnel (through stock options and greater freedom). As John Wood, CEO of Thermedics, put it, "subsidiary carve-outs allow us to develop new business we would not otherwise have developed." Figure 2.1 summarizes Thermo Electron's structure circa 1999. Since then, it has undergone significant simplification and reorganization because it became too

Figure 2.1
Thermo Electron's Corporate Structure circa 1999

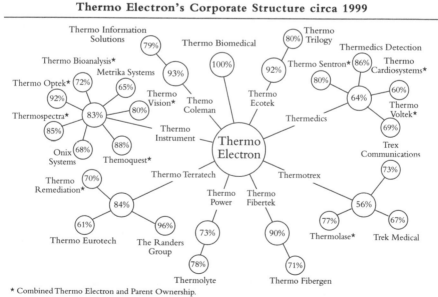

* Combined Thermo Electron and Parent Ownership.

unwieldy. The stock had come under pressure, in part because of weakness in the Asian market, but especially because there was too much organizational separation and complexity. It is a difficult balancing act between control and autonomy, especially if the parent is also to benefit from the equity carve-outs.

Kodak's experience with electronic imaging highlights the strategic importance of organizational separation (in whatever form). Originally, electronic imaging activities were dispersed among Kodak's chemical imaging facilities in Rochester. This had a number of negative consequences. Managers of the film business continually interfered with electronic imaging projects which were perceived as threatening the existing customer base. The company policy that all engineers be paid the same meant Kodak could not compete for higher priced electrical engineers, even if they could be attracted to upstate New York. Because the digital imaging projects were scattered throughout the company, there was no cohesive vision and limited accountability for market performance.

When George Fisher moved from Motorola to become CEO of Kodak, he assembled all the digital imaging projects within one autonomous division and demanded that the division launch new products. In a departure from a traditional "go it alone" strategy, he also initiated a number of alliances to jointly develop digital imaging projects.

How much independence is optimal?   This depends on the magnitude of the technological discontinuity and whether it threatens to undercut or make obsolete the competencies of the core business, the extent to which the activities and customers of the two businesses are different, and differences in profitability. The greater these differences, the more important it is that the new business not be evaluated using the lens of the old.

For completely new and disruptive technologies both *physical* and *structural* separation—by setting up a separate division that reports to senior management—may be necessary. Even when such a full degree of separation is not warranted, it is still desirable to have separate *funding* and *accounting,* so that losses from the new projects are not carried by an established business unit.

The new venture also needs *distinct policies* that match the realities of building a new business. The new venture must be able to attract the best personnel, have the latitude to do fast prototyping and probe ill-defined markets, while keeping restrictive controls and burdensome overhead to a minimum. They must be exempted from much of the routine planning and budgeting required of their more mature siblings. And above all, the new unit should be allowed and indeed encouraged to cannibalize the established business.

Cannibalize yourself.   This was the choice of Bank One when they launched Wingspan.com as a totally separate and unrestricted virtual bank. The commitment of the CEO, John B. McCoy, to the venture was wholehearted. Over $150 million was spent on marketing in the first year to launch the new bank (compared to $1.2 million spent by Internet pioneer Net.Bank). Starting fresh meant no time was wasted hooking up with Bank One's massive systems, detailed controls, and personnel policies, or its pricing and branding policies. This also meant they couldn't directly access the existing customer base. The need for innovation at the pace of the Internet world simply dominated considerations of efficiency and compatibility with legacy systems.

What about "synergy?"   When should the two structures cooperate? One view is that internal competition and some redundancy should always be encouraged, with different business units championing different business models.[36] A more nuanced view is that the separate venture should be able to leverage the parent's strengths while avoiding absorption or subservience. It is ironic and instructive that IBM, in its quest to develop a truly new personal computer, set up a separate and geographically removed unit in 1980 that failed to tap into any of IBMs formidable technological capabilities. As a consequence, the IBM PC was essentially an assembled product without any proprietary systems or semiconductors, and it quickly attracted clones.

Consider the potential for resource sharing between traditional retailers and their online offshoots. Established firms such as L.L.Bean, Wal-Mart, or Barnes & Noble bring visible and established brand names (that translates into much lower marketing costs), an established distribution and fulfillment system, as well as substantial cash flows, and consumers like the convenience of returning goods bought online to a physical store. Yet bringing together the disparate cultures of online and physical retailing will be difficult. The online business needs to be at arm's length, to avoid being mired in the slow-moving policies of the physical stores, as well as to attract and reward skilled employees who are being wooed by start-ups with stock options.

At the extreme the new venture may absorb the parent if the emerging technology spawns a business design that obsoletes the old model. This happened when Charles E. Schwab entered online brokerage with e.Schwab.

Their new web site enabled investors to electronically trade online any securities available in a regular account for a flat fee of $29.95 for any stock trade up to 1000 shares. By the end of 1997, they were the dominant online broker with assets that were 10 times those of the challenger E*Trade.

Meanwhile, customers of Schwab's discount brokerage were still paying an average of $65 per trade, but getting personal service. As online trading boomed the tensions created by a two-tier pricing system that forced customers to choose between service and price were mounting. The decision was made that all trades would cost $29.95 despite the net damage to short-run profits and revenue, after allowing for the additional assets garnered and heavier trading volume. Not only did this help them compete against online rivals, but it made Schwab a bigger threat to the full-service brokers, because they could now deliver personalized investment information to the customers at a very low cost.[37]

CONCLUSIONS

How can established firms compete, survive, and succeed in industries that are being created or transformed by emerging technologies? Success requires continuing support from senior management, separation of the new venture from continuing activities, organizational and strategic flexibility, and a willingness to take risks and learn from experiments. There should be a diversity of viewpoints that can challenge dominant mind-sets, misleading precedents, and potentially myopic views of new ventures. The best innovators seem to be able to think broadly and to entertain a wide range of possibilities before they converge on any one solution.

These prescriptions are directionally correct, but need considerable tailoring to the distinctive character of each emerging technology and the particular characteristics of the organization. While this is not an exhaustive list of traps or solutions, these pitfalls and remedies have been prominent themes in academic research and discussions with managers. We are a long way from understanding the full intricacies of either the traps or ways to avoid them, but the broad outlines of common mistakes and practical solutions are becoming more apparent. By being aware of the four traps and tailoring the four solutions to their particular organization, managers have the opportunity to learn from the mistakes of others and not be doomed to repeat them.

PART I

ASSESSING
TECHNOLOGIES

How does an established chemical company make the transition to biotechnology? As John Hume, Monsanto's director of corporate planning, described the shift during a conference at Wharton, it took "faith, hope, and $2 billion." It also took something else: It demanded a change in mindset, a new way of looking at technology and the business. "We've made a huge bet and we're creating a new company around it," he said. "We're building on faith and hope. Now the question is how to measure how productive you are and how valuable the innovations will be in science and the market?"[1]

The march of new technologies is relentless and each one raises new challenges for established firms. For example, research into the human genome offers a genetic map that could point the way to new products for Monsanto. With diverse technologies emerging, the company needs to decide if these new footpaths through the wilderness will run off cliffs or become the next superhighway. "We will have to make choices, to decide whether to investigate the genome of an organism," Hume said. "We could also be experts in dealing with proteins. The field is so large, we have to make difficult choices concerning our involvement in science, our focus and the business potential."

This new perspective of emerging technologies had to be gained at the same time the company sustained its core business. As Hume describes the challenge, "The way we think about it, we're working on the tail of the airplane as we're flying, and we hope to have it done before we have to go anywhere too far." Especially critical at the time of this writing is the acceptance of genetically modified foods in societies around the world.

THE NEXT BIG THING

Emerging technologies are a foreign territory for many managers of established firms. The risks and payoffs are huge, the uncertainty is extremely high. Assessing and managing these technologies—developing those that will define the future and avoiding the technological sinkholes that could swallow the company—are a significant challenge. Part I offers insights into patterns of development, frameworks for identifying and assessing technologies, and perspectives on the role of government in their development.

Technology development emerges from a complex interaction among research, consumers, and businesses. The Internet was created to serve a military and academic community but exploded with a relatively minor technological development of the web browser. Like biological developments where a species undergoes rapid growth and develops new characteristics when transplanted to a new environment, shifts in application can lead to a process of "technology speciation." This process is discussed in more detail by Ron Adner and Dan Levinthal in Chapter 3. Their biological analogy offers a powerful metaphor through which to understand today's technologies and holds up well when tested against the past.

In Chapter 4, Don Doering and Roch Parayre examine strategies for selecting technologies to develop and evaluating those selections. Companies have limited resources to invest in technologies and their choices can have significant implications for the firm. They present a framework for scoping, searching, and evaluating potential new technologies. They examine the technology decisions of AquaPharm, a biotech startup in the aquaculture industry, to explore lessons of the firm's good and bad technology choices.

Finally, no participation strategy can ignore the substantial role of government in the development of emerging technologies. As examined by Gerald Faulhaber in Chapter 5, the public sector plays a role from funding basic research to regulating commercial products and services. This chapter explores the role of the government in shaping the Internet throughout its development. He offers lessons from that experience for managers of other emerging technologies who have to work with, or around, this imposing force.

These chapters provide frameworks and perspectives on assessing and managing technologies, but technologies are not managed in a vacuum. They have to become products and services that are sold to markets that do not yet exist. The technology management challenge is intimately intertwined with the task of assessing and managing the emerging markets. The technologies and customers emerge together and managers have to keep their eyes on both. Assessing, building, and managing these emerging markets will be discussed in Part II.

CHAPTER 3

TECHNOLOGY SPECIATION AND THE PATH OF EMERGING TECHNOLOGIES

RON ADNER
INSEAD

DANIEL A. LEVINTHAL
The Wharton School

Sometimes the overnight success in emerging technologies has been in development for decades. The revolution of emerging technologies is often not a result of a major scientific breakthrough as much as a shift in the domain of application of the technology. For example, the radical development of the Internet was not fundamentally the result of a technological revolution. Rather, the relatively small development of the web browser shifted the domain of application of the technology from government and academic researchers to the mass consumer market. Drawing parallels to evolutionary biology, the authors discuss this process of "technology speciation" and its impact on the commercialization of emerging technologies.

From a technological perspective, the key elements of wireless communications technology were largely in place by the 1940s. It was more than four decades later, however, that the "cellular revolution" took place. This revolution was not primarily the result of laboratory triumphs but rather changes in regulation and market applications. At each stage in the development of wireless communications technology, the initial prototype of the technology was readily derived from the existing state of knowledge. Although the science of the radio signal followed a fairly continuous pattern of development, there was a radical discontinuity in the domain of application. And there was a radically different commercial value for a radio signal that travels across a laboratory versus one that travels across cities.

Wireless communication technology has, at many junctures, been heralded as revolutionary, including the introduction of wireless telegraphy, radio broadcasting, and wireless telephony. However, beneath these seemingly radical changes was a gradual technological evolution. The dramatic changes were the result of the application of existing technology to new domains:

- *Laboratory device.* The technology started as a laboratory device used by German physicist Heinrich Rudolph Hertz to test Maxwell's theories on electromagnetic waves. The critical functionality in this domain was the reliable measurement of electromagnetic waves.
- *Wireless telegraphy.* The development of wireless telegraphy was driven by the ability of Italian engineer Guglielmo Marconi to generate financial backing for a corporation to pursue the commercial application of electromagnetic waves as an alternative to wired telegraphy.[1] For the second application domain of wireless telegraphy, a new functionality of distance was required. Researchers focused on enhancing the power of transmitters and increasing the sensitivity of receivers. The effort to develop superior receivers for wireless telegraphy (and an effective repeater for wired telephony) ultimately led to the development of the vacuum tube.[2]
- *Wireless telephony and broadcast radio.* The vacuum tube provided the basis for a continuous wave transmitter, a technology that was readily applied in the new application domains of radiotelephony and broadcasting. Wireless telephony was initially used for public safety purposes, such as police and emergency services. Only in recent years, with the shift to mobile communications, has wireless technology penetrated more mainstream, mass consumer markets. The application of the vacuum tube to broadcast radio was facilitated by the commitment of resources by the already established corporate entities of Westinghouse, RCA, and General Electric to its refinement.

These shifts in domain of application are significant breakpoints in the technology's development. The shifts affected technology development because they signaled a shift in selection criteria, or the critical functionality of the technology. In Hertz's lab, the ability to transmit over large distances was relatively unimportant whereas it became a central focus of technology development for Marconi. Entering a new application domain not only changed the selection criteria, it also radically changed the resources available to support the development of the technology.

These shifts in applications were made possible by wonderfully creative laboratory developments that commanded tremendous amounts of time and financial resources. However, these efforts were supported within existing application domains. The miracle of broadcast radio was initially demonstrated by ham radio operators, before the large corporate entities took the wireless technology and created the infrastructure needed for commercial broadcast radio. The continuous wave transmitter that facilitated broadcast radio and wireless telephony was originally developed to enhance the distance and clarity of wireless telegraphy and to create an effective repeater for long-distance, wired phone service.

The Process of Speciation

Evolutionary biology offers important insights into this process of taking a relatively small development in one domain and seeing it take off in a new direction when it is moved to another. Evolutionary biologists Stephen Gould and Niles Eldridge describe a process of "punctuated equilibrium," that is similar to the gradual change in underlying science of emerging technologies and apparent discontinuities in their commercial applications.[3] Gould and Eldridge confronted a fossil record that seemed inconsistent with the gradualist interpretation of Darwin's ideas of an evolutionary process of descent with modification. They identified periods in which there seemed to be bursts of evolutionary activity.

Their resolution of empirical evidence and Darwin's Theory was to note the importance of "speciation events." A new species is formed when one population becomes separated from the larger population by some event such as a climate change, formation of a physical barrier such as mountain range or an invasion of the habitat. This change creates a new environment in which the species develops very different characteristics than the antecedent population.

There are two critical features of speciation. One is that it is "genetically conservative," that is, speciation is not triggered by a transformation of the population from within. Second, the speciation event allows the two populations of homogenous entities to grow quite distinct as a result of their now different selection environments.

What can this framework tell us about the evolution of new technologies? How can it help resolve the discordant images of both gradual and radical technical change? In particular, how can it help us identify those critical transition points when emerging technologies realize commercial importance?

The analogue of speciation in technological development is the application of existing technologies to a new domain of application, as when radio technology moved out of the lab into wireless telegraphy. Technological discontinuities are generally not the product of singular events in the development of the technology itself. As in the biological context, the critical factor is often a speciation event, transplanting the existing technological know-how to a new domain of application where it evolves in new directions. The technological advance associated with the shift in domain is typically quite minor; indeed, in some instances, there is no change in technology.

Evolution of the Internet

This speciation process can be observed in the development of the Internet. As illustrated in Figure 3.1, Internet technology evolved over a long development process starting with the U.S. Defense Department sponsoring the development of protocols to facilitate distributed computing. It was initially used by DARPA scientists and continuing with IBM scientists in Switzerland who wished to facilitate the sharing of a physics database among a larger population of scientists. It was not until Netscape engaged in relatively minor technical developments to create a user-friendly HTML interface that the technology leaped from its small group of technical users to the mass market. Moving the technology from the narrow domain of research scientists to the broad population of commercial and individual users had an enormous impact that is still reverberating through the economy.

In this new environment, the technology took on a completely different character. The new commercial technology was virtually unrecognizable as related to the old species of technology used by technologically savvy researchers. Yet the underlying technology was not tremendously different, although it had evolved in new ways. Once the technology was ported over to this new domain, tremendous amounts of resources were devoted to its subsequent development. Further, these developments, such as video streaming, were focused on aims that were largely irrelevant to the initial application domain of research scientists.

In general, a technology undergoes a process of evolutionary development within a given domain of application. At some juncture, that technology, or possibly set of technologies, are applied to a new domain of

Figure 3.1
Speciation in the Development of Technology

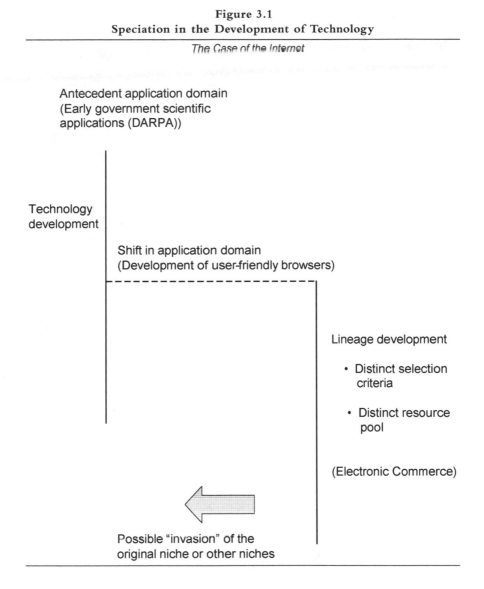

The Case of the Internet

Antecedent application domain
(Early government scientific
applications (DARPA))

Technology
development

Shift in application domain
(Development of user-friendly browsers)

Lineage development

- Distinct selection
  criteria

- Distinct resource
  pool

(Electronic Commerce)

Possible "invasion" of the
original niche or other niches

application. The technological shift necessitated by this event is modest. Just as biological speciation is not a genetic revolution—the DNA of the organism doesn't suddenly mutate—technological speciation is not usually the result of a sudden technological revolution. The revolution is in the application. The distinct selection criteria and new resources available in the new application domain, however, may result in a technology quite distinct from its technological lineage.

Framing technology evolution in terms of speciation leads us to differentiate between a technology's *technical development* and *market application*. This distinction is useful in understanding broad patterns of technological change, and leads to specific strategic implications for technology management.

The first uses of a technology are often quite different from its ultimate primary applications. "The earliest steam engines pumped water from mines, the first commercial use of radio was for the sending of coded wireless messages between ships at sea and from ships to shore, and the first electronic digital computer was designed to calculate firing tables for the guns of the U.S. Army."[4] The revolution is not so much in the technology as the shift in application.

Not all emerging technologies exhibit this pattern. Some development efforts, such as genetic engineering, take place in the context of research laboratories and have no commercial precursor prior to their initial, dramatic commercial applications. But this pattern of speciation may be far more common than is widely realized. Many emerging technologies that are viewed as having appeared dramatically and rapidly on the business landscape, such as xerography,[5] actually have a long prehistory of technical development occurring in relatively small and peripheral market segments.

Lineage Development: Selection Criteria and Resource Abundance. As shown in the cases of the development of wireless communications and the Internet, this speciation process involves more than simply transferring the technology from one domain to another. In the process, the technology itself is changed. First, there is a process of adaptation to the new environment. Spanning large distances became important when wireless telegraphy emerged, so researchers began focusing on this area. The technology is adapted to the particular needs of the new niche it is exploiting. Second, the new environment may have an abundance of resources that supports the rapid development of the technology.

Technology development adapts to the niche in which it is applied, focusing on the particular elements of functionality that are valued by that environment.[6] Some of these attributes may have been largely irrelevant to the prior domain, just as video streaming was not important to the early scientific users of the Internet. In the computer disk drive industry, for example, it was not until the application domain of portable computers became more important that the attributes of size, weight, and power requirements began to be a central focus of development.[7] These attributes had little relevance in desktop computers

Rapid technological development also results from a resource-rich environment. Just as natural resources such as food and shelter promote biological development, abundant resources that are found in new application domains also lead to the rapid development of new technological forms. It is the combination of distinct selection criteria and the availability of substantial resources to support innovative efforts associated with the new application domain that results in a speciation event with dramatic consequences for subsequent technological development.[8]

Gradual Evolution and Creative Destruction. This perspective of speciation can help reconcile contrasting perspectives of technology evolution. On the one hand, we have arguments that technological change is gradual and incremental.[9] In contrast, others have offered the image of technological change as being rapid, even discontinuous,[10] advancing through "waves of creative destruction."[11] From the perspective of speciation, the scientific developments may be incremental but the shift in application domain is where the discontinuous "creative destruction" takes place.

As the emerging technology grows and adapts rapidly in its new environment, it often will reach a point at which it begins to displace an existing technology. The technology that emerges from the speciation event is ultimately able to successfully invade other niches, possibly including the original domain of application. For example, the 3.5-inch computer disk drives that were initially developed for the niche of portable computers ultimately became viable for the mainstream desktop market.[12] Radial tires were initially developed in the distinct niche of high-performance sports cars.[13] The resources made available from the success of radials in this niche lead to increased efficiency in the production process. That reduction in cost, in conjunction with a different attribute of radials—their greater longevity relative to bias-ply tires—allowed radial tires to penetrate the mainstream niches of replacement tires and ultimately the original equipment market of automobile manufacturers.

This successful "invasion" of the mainstream niche is the dramatic event on which commentators tend to focus. However, that dramatic invasion is the outcome of a substantial period of development in a relatively isolated domain. Prior to any lineage development, there is little possibility that the new technological form can out-compete the refined version of an established technology in its primary domain of application. Even those technologies that ultimately became widely diffused general purpose technologies initially focused on the needs of a particular application domain.[14]

Such "invasion" of the original, or predecessor, application domain need not occur. It depends on the resources available to support technological development and the distinct selection criteria in the new application domain. For instance, the teletype endured even with the full development of telephone technology, because it had the attribute of providing a written record that was valuable in business transactions, as well as allowing for asynchronous communication. Wireless telephones have not initially been seen as a substitute for wired communications in developed countries. Switching costs also tend to inhibit the spread of the new technology.[15]

Sometimes the invasion of the existing markets is only a matter of time. The teletype eventually met its demise with the development of an alternative form of written networked communication—improved facsimile technology and the development of large scale computer networks. Substitution of wireless for wired communications systems is being seen in emerging economies that lack a developed wired system. Also, as price drops and performance rises in cellular phones, it is conceivable wireless communications may make further inroads into wired networks. Already, companies are promoting their wireless phones and low-priced calling plans as an alternative to installing a second wired phone line in the home.

Melding of Technological Lineages—Convergence and Fusion.   The emergence of a new technology is actually more complex than the migration of a single technology from one application domain to another. The "family tree" may actually draw upon several lines of technology that come together in a new application. New lineages may emerge as the result of melding or hybridizing two formerly distinct technologies in a common application domain.[16] This may be a process of *convergence,* in which the common domain is an application domain in which one of the two antecedent technologies is already applied. For example, the development of the CAT scanner for medical imaging drew upon X-ray technology, which was already applied in that domain, and computer technology, which had been applied to data processing (Figure 3.2). The combination of these two technologies created a new technology that was applied to the old domain. In this case, only the computing technology jumped to a new domain but the result was a radical shift in the technology application.[17]

Figure 3.2
Technological Convergence in CAT Scanning

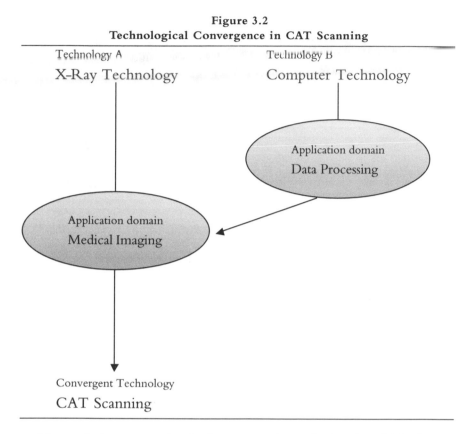

Technology A
X-Ray Technology

Technology B
Computer Technology

Application domain
Data Processing

Application domain
Medical Imaging

Convergent Technology
CAT Scanning

Alternatively, two technologies may undergo *fusion,* in which the resulting technology is applied to a new domain. For example, magnetic recording technology and the technology of optical signal processing, which had been applied extensively to audio tape recorders and television broadcasting, respectively, for some time, were brought together in the new domain of fiber-optic communications technology (Figure 3.3). Technology fusion "blends incremental technical improvements from several previously separate fields of technology to create products that revolutionize markets."[18] Unlike biological processes, technological evolution is not restricted to natural reproduction processes. Agents of technological change are continually generating the "creative recombinations" of which Schumpeter spoke. Many of these creative recombinations produce new forms that are not viable in the marketplace. Witness the many variants of pen-based computing and personal digital assistants (PDAs) that have been commercial failures.[19]

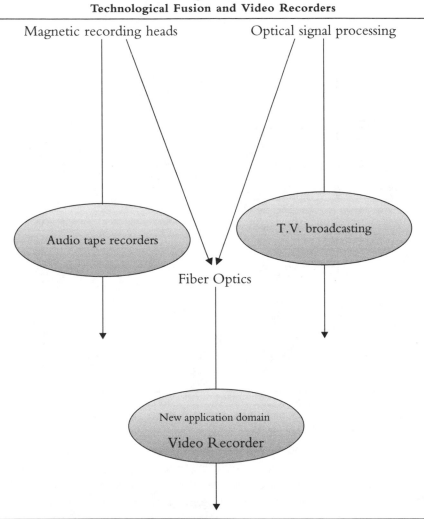

Figure 3.3
Technological Fusion and Video Recorders

Patterns of Technology Evolution

At least initially, commercial discontinuities in wireless communications and the Internet took the form of providing services where none existed rather than replacing existing services. The existing technologies were too highly refined in their primary applications. A new technology will be viable if it out-competes existing technologies on some performance criteria and thereby achieves a relative advantage for that specific application.[20]

For example, early wireless communications technology, despite high cost and poorer sound quality, outperformed the more refined wired systems for mobility and flexibility. This allowed it to build an initial niche in military and police communications where mobility was valued enough to outweigh the other weaknesses.

In most cases, the early domains of application are ones that not only value the distinctive functionality, but also ones that can tolerate relatively crude forms of the new technology. For example, minimally invasive surgery can be used in gallbladder surgery but not heart surgery. Early pen-based computing could be used for structured forms but not for tasks that required handwriting recognition.

At a certain point, the technology crosses a threshold as it moves from satisfying the needs of its initial niche to a broader market. As it develops functionality or cost is reduced, the technology is able to penetrate larger, more mainstream niches. The development of these particular attributes not only involves trade-offs (between price and performance, for example) but must cross a minimal threshold of functionality to be viable in a given application domain.[21] A horseless carriage that is likely to breakdown after a quarter of a mile is a novelty, not a substitute for a horse. But once the Model T crossed the threshold of price and performance, this unreliable toy for the rich became the ubiquitous automobile for everyone.

The above argument suggests a possible pattern of industry development from a small niche to an ever broader set of niches through a series of speciation events. For example, the history of the video recorder shows a movement from a peripheral niche to the mass consumer market (Figure 3.4). The technology was initially introduced for the distinct niche of broadcasters. As the manufacturing process was refined and the product design simplified, it was possible to penetrate a new niche of industrial and commercial users.[22] Finally, this development continued to the point that the product was able to penetrate the mass consumer electronic market. At each point in its development, video recording technology was commercially viable and profitable within the niche in which it was operating.

In ecological terms, we might think of this as the artifact shifting from a specialist to a generalist. No longer must the video recorder look for resources for its survival and development from the narrow niche of television broadcasters. The entire set of households has now become a basis for resources.

Figure 3.4
Technology Evolution and Penetration of Application
Domains by Video Recorders

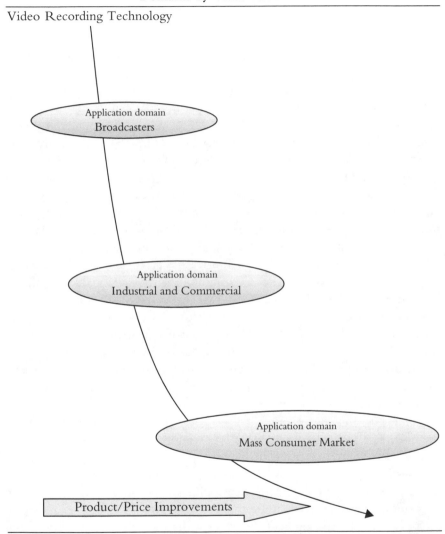

In some cases, the technology never goes beyond the initial, peripheral market, but remains an isolated "island of application." For instance, gallium arsenide was heralded as a replacement for silicon in semiconductors in the early 1980s based on the superior speed that it provided. However, the technology has proved commercially viable only in the context of supercomputer applications and communications devices.[23] Recently, the demand for gallium arsenide has increased, but this demand in still in the context of communications devices, not mainstream computing applications.[24]

Implications for Firm Strategy

How can managers use this process of technological speciation to their advantage? Biological speciation often occurs through external events that separate one population from another. But once the process is understood, it can be, to some degree, actively managed. Similarly, once managers understand the process of technological speciation, they do not merely have to wait for it to happen, as so often happens. They can make it happen by actively looking for ways to move the speciation process forward. Among the specific strategic implications for managers are:

- *Focus on the intersection of markets and applications.* Technology speciation distinguishes between a technology's *technical development* and *market application*. Because of the belief that technological revolutions usually occur in the lab, managers sometimes undervalue the importance of applications. Some writers have focused on the impact of technological milestones on the rate of progress of development.[25] While these "supply side" considerations are important, it is critical to consider constraints and thresholds on the demand side as well. The leap to new application domains affects the attributes of the technology that are developed as well as the resources available for its development. The implication is that there are probably many technological developments that could take off, *in the proper application domain.*

    Managers need to focus attention on the issue of potential application domains. There may be a variety of technologies that are sitting in the laboratory that would begin to emerge if transplanted into the right application domains. Discussions of the management of emerging technologies emphasize long-term vision and patient investment as the key to developing nascent technologies to the point that they can have a real impact on the existing technological order. These discussions pay little attention to the market contexts in which innovations are exploited and the impact these market interactions may have on exploration activity. The wireless communications revolution and the emergence of electronic commerce came as a result of recognizing the potential applications of technology. Which technologies in your labs might have a similar potential if they were moved to the right domain?

- *Focus on selecting market contexts for a product, rather than selecting products for a fixed market context.* The question should be: Where can I place this fledgling technology where it will thrive? There are many examples of elegant products based on relatively crude technologies which serve intermediate markets while biding their time to enter the mainstream market: solar powered calculators were proving grounds for solar cells; digital watches and calculators were platforms for early liquid crystal displays; inventory management units and simple signature capture devices were predecessors of pen based computing. Even in these technologies, which have to date only partially fulfilled initial expectations, firms such as Sanyo, Sharp, and Casio were able to introduce profitable products into the market that allowed them to learn about and refine the technology. Contrast the experience of those firms with companies that kept development in-house until such time as they felt they could address their mainstream customers rather than focusing on a small target segment to refine the technology. For example, ARCO's investments in solar energy power stations and Apple's investments in the Newton pen-based computer were prematurely transplanted to the mainstream market in which they were unprepared to thrive.

- *Understand market heterogeneity.* While increasing attention has been paid to the influence of market feedback on technology management,[26] the implications of the simultaneous presence of a *diversity* of such feedback to development strategies remain relatively unexplored. Exploiting market opportunities at early stages of technology development requires closer consideration of market heterogeneity. Different consumers have different requirements for purchasing products and use different criteria when evaluating their options. Different facets of the commercial market have different thresholds of viability. These differences may be differences in magnitude, such as the level of script recognition required of pen-based computers for inventory management versus word processing applications; or they may be differences in kind, such as the relative importance of price and performance preferences of space science application of computers and that of home users. An initiative that fails in one market subset may still be highly successful in another. For example, the first users of the xerographic process were specialty printers who used it to make offset masters.

Early machines required a 14-step process to make a single copy, which had prevented them from penetrating many markets. The requirements of specialty printers were sufficiently low, and their complementary skills sufficiently high, that they were able to derive benefit from the product despite the cumbersome technology. Further, because of their understanding of the printing process, these specialty printers aided Haloid (later renamed Xerox) in expanding the market for xerography to other printing subfields which ultimately led to the corporate mainstream.[27]

- *Expand your selection criteria.* While introducing greater experimentation, the company also needs to diversify the selection criteria it uses to evaluate initiatives. Because firms do not have the same internal diversity as the broad market, they cannot match the richness of the market selection environment in their own internal selection processes. Indeed, it is impossible that the internal selection environment of an organization, which is governed by a hierarchical structure, could reflect the diversity of selection criteria of the independent consumers that compose the market. This is why companies often tend to overlook potential opportunities for applications of technology. Viewing the market as an amalgamation of independent selection environments, the challenge is not to accurately determine the needs of the market, but rather to recognize the variety of evaluation criteria being applied by the market's components. Many firms generalize the criteria of their current market segment to the global market, which drives firms to reject proposed initiatives that do not satisfy their focal markets.

- *Study lead users.* Rich feedback may also be provided by users with very high requirements, such as the Hollywood special effects studios that helped Silicon Graphics to understand and satisfy the sophisticated graphic needs of high-end users. This helped the company

expand and extend the market for its workstations.[28] The extreme of
market feedback is illustrated by lead users who actually create or design the sort of products they and other consumers like them would
like to purchase, and then educate and recruit firms to produce them.[29]
Such users tend to be entirely enmeshed in their task environments
and go beyond supplying market feedback, to actually providing market feed-forward which not only directs firms' development activities, but actively supports them in these efforts.

- *Be careful where you look for market insights.* Because of the diversity of markets, the lessons managers take away about the potential
applications of new technology may, in large part, depend on where
they look. As companies "probe and learn" about markets, what they
learn may be directly related to where they probe.[30] For example,
when disk drive manufacturers evaluated the merit of pursuing 3.5-
inch hard disk drives, manufacturers who looked to assemblers of
desktop computer systems for feedback were told that the new drives
were of insufficient capacity and should not be pursued. Other manufacturers who looked to assemblers of portable notebook computers
received a positive signal from this segment, which valued their small
size and reduced power consumption. Thus, different drive manufacturers developed a different set of beliefs regarding technological
opportunities depending on the set of assemblers to which they were
attuned. When 3.5-inch drives became the standard choice for the industry (with improvements in capacity), those firms that had focused
on the desktop segment, and therefore did not engage in 3.5-inch development, were generally excluded from the restructured market.[31]
Because learning and adaptation are feedback-driven processes,[32] decisions regarding the sources of feedback have significant implications
for learning and directing change.

- *Learn by doing.* Engage in exploration through exploitation. Production activity not only leads to the manufacture of a sellable product, but also moves the organization down a learning curve that
reduces costs and increases reliability.[33] By engaging the market, firms
not only sell product and create revenues, they also gain information
on market size, preferences, and requirements. Flexibility in market
focus allows for a broad set of alternative bases of market support, customer feedback, production experience, and accompanying these, the
increased capacity to learn and improve in subsequent development
attempts. Learning requires action, and different types of actions will

foster different aspects of learning. R&D activity advances a firm's own technical knowledge, but also increases its ability to learn from the research disseminated in the broader environment—or its "absorptive capacity."[34] Arguably, in mature markets, firms can learn about market preferences by observing consumer responses to other firms' products. However, for emerging technologies which offer new functionalities and functionality bundles, understanding consumer preferences strictly on the basis of this kind of vicarious learning is highly unlikely, especially given the difficulties experienced by firms who are directly engaged in the market.[35]

- *Look for opportunities for convergence or fusion.* Some significant opportunities for technology speciation are created from the combination of several distinct technologies, as we saw with the combination of X-ray and computing technologies to create CAT scanning. Exploring technologies that are far afield and creatively combining diverse technology may offer insights into new application domains.

- *Accelerate the evolution.* Managers can look for opportunities to accelerate the evolution shown in Figure 3.4 by identifying technologies in small niches and finding ways to move them to broader markets. Are there relatively small technological changes or complementary technologies such as web browsers that will open the technology to a whole new level of development? Firms face a choice regarding when to move a technology initiative from the lab to the market, making the shift from exploring the technology's possibilities to exploiting them. The challenge of identifying applications in the early stage is driven not only by the limitations in technology performance, but also by the fact that attention focused on a search for market application is attention diverted from immediate development. Therefore, firms have incentives to develop technologies in-house rather than attempt to exploit their possibilities in an elusive market. This is, in a sense, the reverse of the more common criticism leveled at the management of mature technologies that a lack of research focus is attributed to concern with the short-term over the long-term.[36] Early on, the market search process is likely to lead to blind alleys and, because negative knowledge is not highly regarded, the outcomes of such a search are not seen as being of high value. As such, in the case of emerging technologies, short-term results are easier to show for research and development activity than for market activity.

ORIGINS OF NEW TECHNOLOGIES

New technologies, like new genetic species, undergo periods of evolution and revolution. They involve technological development and the transfer of the technology to new domains of application. They can be created through the convergence or fusion of existing technologies. Beneath the revolutionary emergence of new technologies is often a process of shifting application domains and rapid subsequent growth in the new domain. By understanding this process, managers can better use it to their advantage.

An investment process based on feedback is implicitly making assumptions about the speed of feedback relative to the pace at which financial commitments must be made. To the extent that large fixed investments must be made prior to the realization of any market feedback, the process suggested may not be appropriate. (Or it may need to be combined with the real-options approach described by William Hamilton in Chapter 12.) But this pattern has been used successfully in a number of innovations, including xerography, video recording, wireless communication, machine tools, and electric power.[37]

Another concern is protecting secrets from rivals. By participating early in markets, the company may give away information to competitors about the technology, the markets, and the relevant production techniques. While concerns over appropriability are relevant to any discussion of technology development, they are less pressing than they might first appear.[38] Early markets for technology tend to be small and, as such, do not appear on the radar of many firms—consider IBM's rejection of xerography as late as 1960,[39] or Xerox's rejection of the graphical user interface.[40] Consider also, the innumerable cases of technologists, frustrated with what they perceived as organizational hesitancy regarding their projects, who left the organization to start their own firms. Appropriability is a valid concern, but not a reason to abandon a search for new applications.

As discussed in this chapter, the transfer of technology to new application domains can be a powerful force in developing markets for the technology. It also can lead to subsequent changes in the technology that make it even more valuable to the market. Managers need to focus on this combination of technological development and market applications to truly understand and manage the process of development of emerging technologies.

CHAPTER 4

IDENTIFICATION AND ASSESSMENT OF EMERGING TECHNOLOGIES

DON S. DOERING
The Wharton School

ROCH PARAYRE
The Wharton School

Significant emerging technologies are easily seen after the fact, and companies are then congratulated or castigated for their decisions to pursue them or ignore them. But rarely are the winners clear at the outset. Yet, this is the challenge managers face. From a swirling sea of technological possibilities, they must identify commercial potential and choose whether, how much, and when to invest. This chapter focuses on the early stage of technology development: the identification and assessment of an emerging technology, and postures for commitment. Using biotech firm AquaPharm as an example, the authors describe an iterative process to help managers sift through the chaos of early stage technologies to identify the most promising new technologies to pursue.

AquaPharm Technologies Corporation was an early-stage agricultural biotechnology company that produced health and nutrition products for the $30 billion global aquaculture (fish farming) market.[1] The rapidly growing industry, which accounts for almost 25 percent of the seafood consumed by humans, was ripe for applying new technologies. AquaPharm was founded to commercialize health and nutrition products, including a highly effective, controlled-release formulation of a peptide that successfully induces captive fish to reproduce.

The invention, ReproBoost™, was well into the international patent process and field-tests of the prototype demonstrated both its very low cost

The authors acknowledge Roy Lubit for his valuable assistance and insight.

to manufacture and its very high value to commercial fish hatcheries. However, at $15 million, the annual market for ReproBoost was too small. Seated in front of a venture capitalist, managers were asked: "So, what's the big banana here?"

The company desperately needed a major product in the short-term as well as a broader portfolio of technologies in the long-term to attract significant venture capital investment. How could they find other technologies they could use to serve this market? Which technologies had the greatest chance of success? How could the company evaluate and develop these technologies successfully? The firm's long-term survival depended on evaluating these technologies effectively. As a start-up, backing a dog could pull down the whole company. This chapter explores a process for technology assessment—projecting the future commercial value of scientific and engineering discoveries—and shows how AquaPharm applied this process (or failed to apply it) to its technology decisions.

THE TECHNOLOGY ASSESSMENT PROCESS

Early detection of emerging technologies is crucial. Studies of the semiconductor industry in the United States, Japan, and Korea suggest that the competitive advantage "now often goes to the companies that are most adept at choosing among the vast number of technological options and not necessarily to the companies that create them."[2] It is this ability to choose among technologies, based on an integrated understanding of market potential, that has been credited with driving the resurgence of the U.S. semiconductor industry in the 1980s and 1990s.

A 1992 study by the R&D Decision Quality Association identified 45 best practices for R&D decision making (the top practices are summarized in Table 4.1). These were further tested against 79 R&D organizations commonly recognized as excellent.[3] Technology identification and assessment, as the first step in an R&D process, lays the foundation for many of these best practices and long-term competitive advantage may derive from the framework and culture developed by the assessment process: assess the portfolio, hedge against technical uncertainty, create frameworks for learning, insist on alternatives, and evaluate projects quantitatively.

The development of revolutionary emerging technologies, the increasing costs of R&D, and rapid product life cycles have raised the risk and value of choosing technologies wisely and developing them quickly. As competitive

Table 4.1
Practices for Excellent R&D Decision Making

Ten Essential Practices	Ten Practices for Competitive Advantage
Agree on clear, measurable goals	Learn from postproject audits
Use a formal development process	Evaluate the portfolio
Coordinate long-range business and R&D plans	Create frameworks for learning
Coordinate development with commercialization	Hedge against technical uncertainty
Understand drivers of industry change	Insist on alternatives
Hire the best and maintain expertise	Measure effectiveness of R&D
Use cross-functional teams	Manage the pipeline
Focus on end-customer needs	Design progression of your technology
Determine and measure end-customer needs	Evaluate projects quantitatively
Refine projects with regular customer feedback	Anticipate competition

advantages and "tolerances" have narrowed, precise and dynamic technology assessment has become essential to success. The high uncertainty of emerging technologies renders traditional static analysis tools for technology assessment ineffective. In this chapter, we propose a more dynamic, iterative approach to technology assessment. To address the greater uncertainty of these technologies, this iterative approach helps reduce uncertainty and preserve options. To address the rapid pace of change in emerging technologies, this approach helps to anticipate the feasibility and prospects of the new technology, and to get to market faster.

As illustrated in Figure 4.1, there are four interrelated steps to the technology assessment process:

1. *Scoping.* Managers establish the scope and domain of the technology search, based on the capabilities of the firm and the potential threat or opportunity from the technology. This scope will be continually changing, as more is learned about the firm and the technology.
2. *Searching.* The firm must determine the information and technology sources to monitor, the procedures to follow, and the organizational

Figure 4.1
The Technology Assessment Process

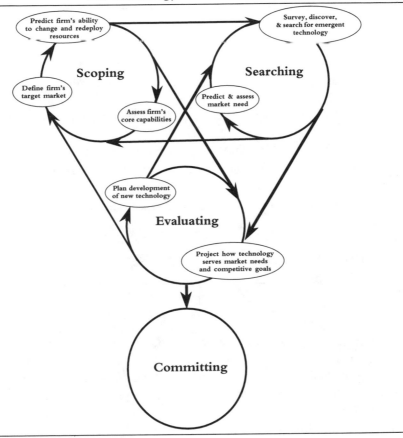

arrangements that will allow it to screen technologies and search for signals of both emergent technology and its commercial viability.

3. *Evaluating.* Candidate technologies must be identified and prioritized, and evaluated against the firm's technical capabilities, the target market's needs, and the firm's competitive opportunities. A technology development and market entry plan must be drafted, and the financial, competitive, and organizational impacts of the new technologies analyzed.

4. *Committing.* The first three steps above are used to determine *whether* to pursue a particular technology. This fourth step addresses *how* to pursue it, by making a strategic commitment to the new technology, in the form of a particular strategic posture.

As indicated in Figure 4.1, this technology assessment process is iterative and combines the *foresight* of probing the uncertainties of the market and technology with *insight* into the firm's own capabilities and resources. With each survey of technologies, their application to market needs and their suitability within the current and future capabilities of the firm is understood in greater detail. The technology assessment process itself educates the firm and this learning informs the process. The sensing of the emergence of new technologies is accomplished by an individual or collective intellect that is prepared for insight by past experience and by study of the firm itself, the market, and technologies applicable to the market needs. Technology strategy ultimately demands a commitment to a specified pathway to the exclusion of other technical approaches but commitment too early in the process or the elimination of a particular approach can have long-term consequences for the firm and the firm's ability to correct its course.

Scoping

It is said that "if you don't have a target, you'll miss it every time." The scope of the technology assessment must be clearly limited. Where the firm looks for technologies depends upon what it is looking for. The scope will be shaped by strategic factors including the firm's strategic intent and capabilities.

Strategic Intent. Technology assessment is only meaningful when performed in the context of the firm's strategic intent. Strategic intent is an "animating dream" or "sense of direction, discovery, and destiny," such as British Airways' goal of becoming "The World's Favourite Airline."[4] This intent may be to meet this year's goal of launching a new $200 MM business unit, to increase product shelf life, or to lower production costs. Technology cannot be considered in the abstract but should be used to answer a question or to address a challenge. The strategic intent of the company is this challenge. If the company is going after a new market or seeking a new level of performance, what technology will take it there? For example, what new information systems could help British Airways become the world's favorite airline? If the technology does not respond to the challenge of strategic intent, then there is no clear strategic purpose for pursuing it. This may sound self-evident, but fascination with technical possibilities, fantasies of new markets, and being "blinded by science" can sometimes obscure the strategic business questions. The firm's strategic intent must be first and foremost in the minds of the technology assessment team.

Firm Capabilities.   In addition to looking at its intent, the assessment team also needs to take a hard look at the firm. According to the resource-based view of the firm, success depends on a firm's resources, particularly its ability to develop and leverage core competencies, rather than upon finding the best position within the market.[5] A core competence could be special technical capabilities leading to development of unique products, the ability to provide a superior service, capabilities for rapid product innovation and development, or particular process capabilities (such as supply chain management or production efficiency) which yield enduring cost advantages.[6] First-rate technology assessment can itself be a crucial core competence in today's economy.

Codifying the firm's technical capabilities is a starting point for a technology assessment process from both an organizational and procedural point of view. One of the key questions that will arise from any candidate technology is whether the firm will possess capabilities or will have to acquire, develop, or create partnerships to attain the specific capabilities needed for technology development and commercialization. This may require divesting some existing capabilities. What is the firm's capacity for such change? Managing a creative and visionary technology assessment process requires balancing the capabilities and constraints of the firm (its financial and technical resources and its ability to adapt to emerging technologies) with its intellectual openness to new technologies and ambitions for growth.

The Scope of the Technology Assessment.   The scope of technology assessment will generally be broader than the capabilities and technical assets of the firm, but it must still be bounded, to establish some limits to the potential new markets and new technologies. The scope includes the target market and target customer, and the existing or latent need that will be served by the new technology. The scope of the technical field must be given some boundaries to eliminate certain technical approaches and products, and to understand the relative risk of different approaches. Boundaries may include the market and customer definition, its technological standards, intellectual property conditions, the technical expertise of the firm, the cost of R&D and commercialization, assumptions of feasibility, organizational structure, and the firm's place in the value chain.

This process of defining the scope is often best accomplished by a multifunctional team, with both technology and market knowledge. Effective teams should include scientists and engineers as well as members with production, marketing, and customer service experience in the target market.

These teams can be a great asset in integrating technological assessment with market needs. In determining the scope, the teams might address questions such as:

- Does our team carry technical biases or limitations that may prejudice the process?
- Is there a new scientific or engineering discovery that can be the basis for a viable commercial opportunity?
- What is the best new technology to meet my target market's current and future needs?
- What are the next generation technologies that will transform my firm's markets?
- Are different technology streams converging to create new opportunities?
- What transformative technologies are under development outside of my firm or my industry?
- To what existing and future markets can we apply the firm's technology?

Sometimes the firm's strategic intent requires that it expand its scope, as with Barnes & Noble's decision to compete on the Internet, which required directions in technology development that were entirely new to the industry. Sometimes the new technology itself requires that the company expand its scope, as many pharmaceutical firms did in appropriating biotechnology or as chemical photography firms did in moving to digital technologies. Sometimes the identification of market needs also requires an expanded scope to include these needs. In other cases, the scope of the firm may be narrowed, because technologies become obsolete or no longer provide a competitive advantage. In Chapter 8, for example, Mary Tripsas describes how the shift from hot metal typesetting to phototypesetting ultimately made about 90 percent of the firm's technological capabilities obsolete and required the development of new capabilities.

As shown in Figure 4.1, this is a dynamic, iterative process. As new information is gained from the search and evaluation process, the scope may be redefined. If a promising new technology is discovered in this process, the scope may be narrowed to focus more intensely on this area. Alternatively, if a promising technology is identified on the outer edges of the original scope, it might be expanded to look more broadly at related technologies.

Searching

Once the firm knows what it is looking for, it needs to determine how and where to look for new technologies. Managers need to systematically survey a variety of sources. These sources vary from field to field and market to market, but generally include:

- *Inside the firm.* The first place to search for new technologies is within the firm. These technologies are likely to be proprietary, and internal expertise for both assessment and commercialization is easily accessible. Within large, innovative corporations such as 3M, DuPont, and IBM, internal discovery generates a constant flow of new technologies in search of applications and transfer to new markets. The search process may include searching of company databases, patent applications, invention disclosures, and libraries of technical reports. The formation of cross-divisional technology assessment teams may promote the cross-fertilization of technical fields and the sharing of corporate memory. The assessment process and technology search may also be assisted through systematic brainstorming sessions throughout the technical community of the corporation. Participation in knowledge networks and collaboration with partners can greatly improve this effort.
- *Public licensors of technology.* A fertile source of new discoveries includes universities, government, technology transfer organizations, and independent research institutes. These organizations publicize their technologies and often make available searchable databases of available technologies. Even with consolidated databases, however, there is still significant fragmentation of technology sources. The assessment process should not only define the sources and how they are accessed but also how sources discovered during the search will be added to its scope. Examples of technology sources in this class include the searchable database of the Association of University Technology Managers (AUTM), the Federal Laboratory Consortium for Technology Transfer, the technology transfer offices of U.S. Federal Government agencies, and databases of federal grantees of research and development awards.
- *Technical and trade literature.* There is a vast collection of technical and trade literature that can be searched through relatively few portals such as MedLine, Dialogue, or Lexis Nexis. Literature searches reveal only public domain information yet can be used to define the

boundaries and subjects of the technology assessment and can be powerful leads toward the individuals and organizations that are expert in each technical area. Similarly, patent databases not only directly describe commercially applicable discoveries but guide the assessment team to the innovators through the patent's inventors and assignees. Finally, the writings of futurist organizations can provide interesting insights into sources of potential new technologies.

Sensing Technological Emergence. Every day, thousands of scientific discoveries are made public through conferences, patents, and publications. Perhaps an even greater number are retained within the confidential boundaries of private firms and government laboratories. From the vast noise of "science-as-usual," how can the technology manager detect the signal of a potentially transformative discovery?

The challenge is to recognize some momentum beginning to form around a given technology. Just as human leaders are defined for having followers, leading technologies can be recognized as *emergent* for their technical "following." Value in science and engineering can be measured by signals of a following such as the degree to which a discovery is cited, duplicated, imitated, and applied. Ultimately, signals of technological emergence—whether the distributed community of scientists and engineers has endorsed a particular technology—must be detected.

This leads to a process of technology forecasting, not by looking at a fixed point and projecting a future, but by looking carefully into the recent past for signals of a momentum building behind a technology and a convergence of technologies toward filling a market need. Signals of technological emergence do not answer the questions of the revenue and profit potential of a technology, but instead point to those technologies for which the firm should ask such questions.

Strong Signals. Strong signals of technological emergence are those that clearly reveal commercial investment in the candidate technology and signal its technical feasibility to serve target market needs. As such, the presence of such signals may indicate lower risk of that technology but also indicates public knowledge and perhaps barriers to their appropriation. Two examples of strong signals of technological emergence and its following are patents and competitors' actions:

1. *Patent and literature citation.* The global intellectual property protection system may be a rich source of valuable technologies as well as signals of technical emergence. The following of a particular technology may be detected as citations to a patent of interest that allow both backward and forward searching from the patent. Such searching not only reveals inventors and assignees that may be valuable partners in technology assessment and acquisition, but also reveals cross-disciplinary citation and the linkage of fields through co-citation of a patent. Cross-fertilization of ideas from one technical field to another is frequently a rich source of new innovations in both basic research and the commercial sector. The *common citation of foundation patents* as well as *the citations of the prior art* searches of the patent examiner can link disparate fields and foster the transfer of proven technologies to new markets. Patents indicate that others have already made the leap from scientific and engineering discovery to commercial application and that the knowledge is publicly available to competing firms. *Co-citation analysis* of scientific literature uses citation databases to identify communities of researchers by their common citation of prior literature. Members of these communities may not be aware of each others' discoveries since they share an intellectual or technical base but may not cite each others' work. Co-citation of a common body of literature may indicate the powerful convergence of disparate fields toward solving a common problem or exploiting a common technology.

2. *Competitors' actions.* The technology strategy and investments of competitive firms are a strong signal for technological emergence. The investments of a competitor may not only validate a firm's analysis but also change the market context of the analysis. In fields where standards and regulations have a powerful influence on the market, joint development of a particular technology may lower the costs of product development and marketing through establishment of a common market standard. Competitors' actions can precede or follow the investment of the firm in a particular technology. Imitative actions may be a signal that an internally developed technology has commercial potential and offers an external confirmation of technical feasibility.

Weak Signals.   Weak signals of technological emergence are those more subtle indicators that a scientific discovery has commercial potential and that independent analysis has recognized this potential—and a following has formed. The endorsers of a particular technology may not be aware of each other's existence. Multiple independent research efforts on the suitability of a technology for the desired applications may indicate that others have not yet had the insight of the true commercial potential of the technology. Such subtle signals may likewise come from internal and external validation or citation and co-citation analysis, but also from confirmation within knowledge networks, corporate intelligence, and parallel discoveries:

- *Confirmation within knowledge networks.* Very important technical information can be culled from formal sources such as journals, industry reports, and online databases. However, it is informal networks cultivated at trade meetings, scientific meetings, and through continued education that may be most valuable to collecting signals of the latest technical advances. Conversely, a record of past failures of different technological approaches may only exist in the memory of the knowledge network. As discussed by Lori Rosenkopf in Chapter 15, these knowledge networks play a significant role in the creation and development of new technology, although most companies do not manage them and are often not aware of their role in spreading information. Informal networks also exist across technical units or product teams within the firm, each acting as "antennas" for certain kinds of technical developments. Whereas external sources of information are costly, time-consuming, and attract much managerial attention, the intra-firm networks for learning and consolidating technical information merit equal cultivation and development. Steve Jobs, the founder and current CEO of Apple, was asked in a recent interview if he still takes time to "troll for new technologies that [he] could turn into new kinds of products"? Jobs replied, "There's a certain amount of homework involved, true; but mostly it's just picking up on things you can see on the periphery. Sometimes at night when you're almost asleep, you realize something you wouldn't otherwise have noted . . . I've always paid close attention to the whispers around me."[7]
- *Competitive intelligence.* This is an invaluable tool for adding technologies for consideration in the technology assessment process and for validating a technical approach. Information on competitors' actions may be gleaned from a careful analysis of public information, both from what is disclosed and what is not disclosed. However, the most valuable information may be derived from the many ways in which competing companies leak otherwise confidential information. Former employees and consultants often reveal technical information with a direct intention of helping another firm, or as an unintentional

means of signaling their own knowledge and expertise. Service providers such as lawyers, vendors, technical contractors, and academic collaborators all may release small items of information that signal technological emergence. Though there are ethical precepts that discourage the overt solicitation of such information, the well networked technology assessment team should be highly receptive to the detection and analysis of such information.

- *Parallel discovery or convergence.* A subtle signal of technological emergence is the recognition of parallel discovery or the convergence of independent researchers upon the same technology. This may be viewed as recognizing patterns of weak signals. Discovery organizations may be separated from each other by boundaries such as technical fields, geographic and political barriers, and organizational type such as industry and academe. A technology assessment process may be able to cross these boundaries and access information that is not easily available to any individual organization involved in technology development.

Knowledge and Information Capture. As the company develops a more complex picture of potential emerging technologies, it needs a system for keeping track of all the information and progress along the various research streams. Implications of information gathered at the start of the technology assessment may only be understood later on in the process. Technologies continue to progress and develop. Different technologies may complement one another, creating opportunities for new hybrids. Codification of this information allows its ready analysis, and its re-analysis in each iteration of the assessment process. As the technical fields change and new expertise is required, compiled information can rapidly bring new team members to date. Not only should the raw information be captured, but also the knowledge of the team and its rationale for selecting certain technologies and rejecting others should be explicitly described to monitor the process and to earn support from the firm's leadership for new initiatives.

The goal of this information and knowledge management is to create the memory of an organizational intellect, a "group mind" that is prepared to make leaps of reason and intuition ahead of the competition. The form of such knowledge capture and transfer systems may be relational databases, compilations of reports, or even oral reports and seminars. The Internet and corporate intranets create extraordinary opportunities to distribute learning throughout an organization, receive critical input, and bring multiple viewpoints to bear upon an issue. A noted example in one of the world's most innovative companies is the Technical Forum at 3M, which since the 1950s has networked top corporate scientists to consider and communicate research issues and opportunities.

AquaPharm's Scoping and Searching. How did AquaPharm address the challenge of establishing its scope and searching for new technology? Like many start-up technology companies, it relied more on serendipity than a systematic process of technology assessment. AquaPharm started with the meeting of its entrepreneurial CEO and the two research scientists developing ReproBoost. As the idea for a new commercial venture developed, the founders sought to create a company with a broader reach than just ReproBoost.

Their initial technology search was close to home. The researchers' interest in seeing their own discoveries commercialized and the accessibility of intellectual property rights made the technology portfolio of the founding scientists the likely place to search for additional technologies. The first potential product identified was based on a proprietary application of ultrasound technology for administering therapeutic compounds to aquaculture species. It was unproven for either humans or fish in anything other than a laboratory model. AquaPharm obtained an exclusive option for the aquaculture application of ultrasound.

This "internal" technology search also identified the concept of creating long-acting aquaculture vaccines using encapsulation (controlled-release) technology. Controlled-release vaccines are a hot topic in veterinary and human medicine and, at the time, the aquaculture vaccine market was in explosive growth as the salmon and trout industries launched the current practice of individually vaccinating every farmed fish. Through the founders' networks, AquaPharm obtained an option to a portfolio of biomedical patents for controlled-release vaccines.

Armed with the option agreements for ultrasound and controlled-release vaccines and the license to ReproBoost, the company obtained seed venture capital financing and began operations in 1991. AquaPharm built laboratories and a testing facility, an atypical capability among technology providers to aquaculture and one that was designed to reduce dependency on contracted field trials. The immediate tasks were to test market ReproBoost, enter the regulatory process for ReproBoost, and to begin experimental technology evaluation of the vaccines and the ultrasound concept. Still, after a year of operation and fruitless fund-raising, the management needed something more to create a compelling investment opportunity that was easily communicated to investors, who were unfamiliar with the largely non-U.S. aquaculture market.

The company stumbled onto a new technology right in its own backyard. A neighboring company had developed a revolutionary food safety test for beef, swine, and poultry. Based on personal friendship and mutual trust, they were happy to give AquaPharm an exclusive option for rights to the technology for seafood testing. Though the product enhanced the forecast bottom line, which helped make the case to investors, the move was not based on a careful technology assessment. The company dangerously moved out of its defined market scope of aquaculture and into the seafood processing market.

As the company neared its two-year anniversary, it finally began to establish a more systematic process of technology assessment. The pressure to build value and to attract capital was rising. In assessing possibilities for technologies that would create large, long-term revenues in aquaculture, managers determined that the only answer was to enter the field of animal nutrition. Feed costs comprise approximately 50 percent of the $30 billion farm gate value of the aquaculture harvest. This realization defined the scope of its search. Here, the young company was at a critical crossroads. Entering aquaculture nutrition meant embarking into a fifth product development area, in a precariously over-committed start-up company. The move also required moving outside the founders' expertise. It was crucial to get this right.

The company hired a nutrition scientist and began a systematic process of searching and scoping the opportunities in aquaculture nutrition and examining analogous opportunities in more mature animal husbandry industries. The nutrition scientist intensively scoured publications, patents, personal networks, trade journals, and technology transfer organizations for technology leads and tested them against a matrix of market needs, AquaPharm's capabilities and its strategy. Through a database search, AquaPharm discovered an expired European patent application that was very close to the product concept that was guiding the technology search. The patent's inventor led, in turn, to an academic lab that tested the core technology as well as to a company with proprietary manufacturing skills for the product. This company was able to confirm the feasibility of the product's manufacture through experience with its related products sold in the cosmetics industry.

Networking to evaluate the technology also led AquaPharm scientists to researchers in Europe and Israel who had each independently arrived at different elements of the product concept and some of its components but not a complete or viable product design. These independent and parallel discoveries—unearthed through testing a well-defined market concept to database searches, co-citation of key work, and networking—gave the company a unique and market-leading forecast of technological emergence. By bringing them all together, AquaPharm was able to successfully sense the emergence of a promising new technology in hatchery feeds.

In short, the first three technologies (ReproBoost, ultrasound, and vaccines) came from the founders and were all within the scope of the market (high-end health products sold to aquaculture hatcheries). The fourth technology, seafood testing, met the need to create revenues and also came from networks, but fell outside of the market scope. The fifth technology of hatchery feeds, though driven to generate forecast revenue, was created after the management had gotten "smart" about scoping and searching.

Evaluating

Now that a scope has been established and a set of promising technologies has been identified (as shown in Figure 4.1), the next challenge is to begin sifting through this set of possibilities. In the evaluation process, managers need to rank candidate technologies according to a set of common financial and organizational criteria. The position in a ranking may be established by weighted scores of these different criteria, financial analysis, or measures of risk. While it is important to limit the technology choice, it is also important to preserve a record of those technologies that did not make the grade. Today's nascent discoveries and inappropriate technologies may be tomorrow's breakthrough products.

To assess the candidate technologies in the context of corporate strategy, managers need to develop technology plans that approximate the path to technology commercialization, required investments, organizational implications, and potential financial rewards. Within these plans will be considerable uncertainties. The discipline of creating a draft commercialization plan will reveal the outstanding questions that must be answered to reduce the uncertainties. Not all uncertainties will get resolved, however, and different strategic postures can be adopted relative to the uncertainties that remain (see "Committing" below). An effective plan describes the technical and market uncertainties and the timing of investments and steps toward resolving each area of uncertainty.

The development of the fax machine illustrates the importance of identifying and continuously evaluating these uncertainties. Xerox had developed the underlying technologies in the late 1960s but could not create a profitable business. Within the following decade, technological advances in component technologies enabled higher speed transmission, improved image quality, automation, cost reduction, and ease of use. During the same time period, increased mail costs and unreliability, and lower phone rates changed the market for facsimile acceptance.[8] Though hindsight is 20/20, if the company had better identified and tracked the technical and market hurdles for the fax machine, it may have recognized more quickly its opportunities for product development and commercialization.

Firm Environments of Technology Search.   The technology must be evaluated in the particular strategic environment of the firm. The technology assessment team must carefully evaluate how each candidate technology serves the need of the market and is within the boundaries of the technical capabilities of the firm and its potential partners. Though such considerations took place throughout earlier steps of the assessment, an in-depth analysis is best performed on only the highest-ranked candidate technologies.

The emerging technology may be new to the company or may draw on existing capabilities. It may be used to serve the company's existing market or open new markets. The probability of success is much greater for products that use current technology to serve current markets. The newer the technologies and the markets, the riskier these growth paths become. Estimates of the probabilities of success range from 0.75 for penetrating an existing market with an existing enabling technology, to 0.25–0.45 when either the market or the technology are new to the company, to 0.05–0.15 when the company diversifies into a new market with a new enabling technology.[9] The risks of new market development, either through new or existing technology, can be reduced through acquisition or joint venture, which helps to mitigate the company's lack of experience in that market (see Chapter 16).

Risk Profiling and Impacts of Technology.   Different types of risk must be considered in the evaluation of the technology. A process of risk profiling offers managers a framework for considering three specific types of risks associated with the technology, as shown in Figure 4.2. The new technology may have market risk, for example, as a result of uncertainty about the size and scope of the market and definition of customer needs. The firm also faces technology risk, as a result of factors such as uncertainty about technical feasibility, emerging standards, or product liability. There may be organizational risks as well, including a lack of fit with capabilities and dependence on new organizational structures or outside partners. These

Figure 4.2
Risk Profiling

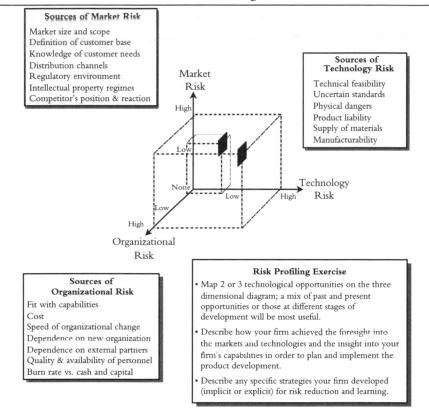

Sources of Market Risk
Market size and scope
Definition of customer base
Knowledge of customer needs
Distribution channels
Regulatory environment
Intellectual property regimes
Competitor's position & reaction

Sources of Technology Risk
Technical feasibility
Uncertain standards
Physical dangers
Product liability
Supply of materials
Manufacturability

Sources of Organizational Risk
Fit with capabilities
Cost
Speed of organizational change
Dependence on new organization
Dependence on external partners
Quality & availability of personnel
Burn rate vs. cash and capital

Risk Profiling Exercise
• Map 2 or 3 technological opportunities on the three dimensional diagram; a mix of past and present opportunities or those at different stages of development will be most useful.
• Describe how your firm achieved the foresight into the markets and technologies and the insight into your firm's capabilities in order to plan and implement the product development.
• Describe any specific strategies your firm developed (implicit or explicit) for risk reduction and learning.

risks are not absolute values, but rather relative risks determined by mapping several different technological opportunities at the same time.

Each of the three types of risks needs to be assessed. Managers will sometimes have a hard time assessing the risk of an emerging technology along a given dimension. This, in itself, also provides important feedback because it shows where more knowledge is needed before the potential and pitfalls of the technology can be accurately determined.

In addition to assessing risks, managers also need to analyze the competitive, financial, and organizational impact of investing in new technologies and commercializing them. Financial analysis at the early stage includes estimates of development costs, market size, product pricing and sales, and market penetration. Competitors' actions may include business actions such as a hostile or friendly takeover, cross-licensing of the technology, legal challenge to intellectual property positions, and price competition. From a technical perspective, competitors may invest in other proprietary technologies, take imitative actions, or upgrade their existing technologies to become more competitive.

The organizational impacts of new technologies not only include the organizations required for production and market channel access, but the organizations needed for product innovation. The development of pharmaceutical and agricultural biotechnology had so many organizational implications for its innovation, commercialization, and marketing that Monsanto chose to divide the company in two and spin off all of its traditional chemical businesses into a new company, Solutia. Other organizational impacts of new technologies such as the formation of new organizational forms and alliances are discussed in Part IV of this book.

Evaluation at AquaPharm.   As AquaPharm grew, most technical efforts were focused on evaluating the technology portfolio and the development of product designs. In hindsight, the company paid a very high price at the evaluation stage when the searching and scoping stages were not rigorous. The Hatchery Feeds program, which was the result of the most exacting technology assessment process, was the least governed by founders' biases, and was soundly based in prior knowledge of the field, rated successfully at the evaluation stage. The program was launched by a collaborative R&D program and a corporate alliance that received a very large U.S. government grant for international joint ventures in new technologies.

The vaccines program, by contrast, started down a path that proved to be a dead end. At the time the company obtained the intellectual property rights, there were no technology assessment guidelines in place to reflect the economic realities of the aquaculture market. The company invested a great deal of scarce time and resources in developing and testing vaccine formulations that never had a chance of reaching price targets for the aquaculture market or of obtaining fast regulatory approval. The licensed formulations also appeared to have a negative effect on the vaccinated animals, revealing the technology assessment's failure to realistically appraise the state of knowledge in the emerging field of fish vaccines. Ultimately, the program obtained very promising results after changing the product design to conform to regulatory guidelines for food safety, to the price and volume targets for the aquaculture market, and to the state of knowledge of fish immunology.

The evaluation stage in the ultrasound program revealed the inappropriateness of the technology, but only after considerable managerial time and significant confusion for potential investors. The reasons were many.

- The requisite engineering technology was outside of the company's core platform in formulation and encapsulation.
- There was no clear market application for the technology.
- There were no clear estimates of cost and how the customer would finance the equipment purchase.
- The preliminary data were not rigorously reviewed for variance and accuracy.
- The technology did not fit with the declared strategy of only commercializing proven technologies.

In its last year of operations, the company ceased all development efforts on ultrasound.

The seafood safety product evaluation was conducted in a virtual way through collaboration with AquaPharm's neighbor. The partnership did require managerial efforts, chiefly the application for government grants to finance experimental work. Though the technology was extremely promising, seafood testing did not fall within the scope of the company's target market and though it attractively improved the projected bottom line, it further diluted corporate resources and investor attention.

The ReproBoost program was very successful and sales of the prototype product brought considerable revenues as well as market exposure for the company. The successful launch of the product reflects, in part, that the prototype was exhaustively tested in research and commercial environments by the licensing academic institutions. The manufacture, field trials, and testing of ReproBoost also brought about the management learning that contributed to the success of the vaccine and feed programs, which shared the same platform technology (controlled release formulation). However, the investments in evaluating the other technologies drew time and funds away from ReproBoost commercialization and made the company (fatally) dependent on additional capital investment.

Committing

Rarely will the potential of an emerging technology be so certain that a firm will fully commit its resources without distinct milestones at which defined results trigger the next level of future commitment or the exit from that technical field or marketplace. Though there is an infinite variety to the plans and actions to commercialize emerging technologies, four general forms of strategic commitment can be described that reflect four increasingly aggressive strategic postures or intention. These depend upon the risk-reward relationship of the technology and its market, the competitive imperative to take action, and signals of technological emergence of the new technology. The postures described below are not in themselves strategies but describe intentions to act:

1. *Watch and wait.* This posture is applied when the uncertainty associated with a new technology is too great to begin its research and development. However, the candidate technology has enough potential that activities for monitoring the emergence of the technology and the development of its market merits an active process. This may mean that part of the technology assessment team remains active and creates systematic ways in which to detect signals of technical emergence such as the weak signals mentioned earlier. However, the company remains on the sidelines and creates competitive barriers. Watch and wait may also be an advisable strategy when a firm has the technical and financial capabilities and market position to be a fast follower and lets the market leader bear the greatest costs of technical development, standards development, and market testing. Although it is a common myth that the chemistry-based pharmaceutical industry "missed" the biotechnology revolution, it is impossible that they could have missed the front-page news developments. Instead, they were pursuing a watch-and-wait strategy, although they may have waited too long. Today, pharmaceutical companies are buying up innovation and capturing much of biotechnology's value. They learned from their late start in the cloning and manufacture of therapeutic proteins (e.g., Genetech's insulin and Amgen's EPO), and have taken a more aggressive approach to moving quickly into genomics and gene therapy.

2. *Position and learn.* When there is less uncertainty associated with a technology, or if the risk of inaction is greater, a firm may choose to take a strategic posture that positions the company to develop the technology and to deter or exclude a competitive threat. This more aggressive posture actively engages the company in the new technology and thus creates a more active learning process. An example of this approach is the execution of an option agreement that gives a company rights of first refusal to a technology during a period of technical assessment or market testing. Smith-Kline Beecham uses its evergreen corporate venture capital fund, S.R. One, as a positioning-and-learning approach. The venture fund independently identifies cutting edge technologies and through equity investment puts SmithKline in an advantageous position to learn of emerging technologies, to discourage competitive relationships, and to develop product development alliances.[10]

3. *Sense and follow.* When a company completes a technology identification and assessment process and chooses to invest in an emerging technology, the strategic posture may be described as one of "sense and follow." The firm is satisfied that there are sufficient signals of technological emergence—strong enough for confidence yet subtle enough to lead the market—to proceed with an active commercialization strategy. The signals may be any combination of those mentioned earlier.

4. *Believe and lead.* When the technology opportunity is very promising, the company may fully commit its resources to commercialization of an emerging technology. Sometimes this is the result of a compelling technology assessment or internal information that convinces a firm of a technology choice in the absence of broader external validation. It is also often the case, particularly for entrepreneurial firms, that even in the absence of signals of technological emergence or with only the weakest signal, a firm truly believes in the technology and leads the technical field and the market application. At the cutting edge of emerging technologies, decision making is often based on the intuition and experience of technology leaders. These are the decisions based on beliefs, cues, and educated guesses that register in the gut instinct and insight of scientists and business leaders—a "second sense" that this is the winning technology.

AquaPharm's Strategic Postures and Commitment. AquaPharm used different levels of commitment to its emerging technologies depending on the technology, market, and operational risks. In the realm of greatest risk, the company adopted a watch-and-wait posture. For example, the company was frequently approached by partners and scientists seeking partnership for developing genetically engineered fish. The management actively monitored the field's development through these discussions, but considered the technology to be very immature, unneeded by the market, and to be very high risk. Through this process of active learning, AquaPharm remained well networked and positioned to move into the field if any of the risk factors were dramatically reduced.

For technologies with more immediate promise, AquaPharm used a position-and-learn strategy. It used this strategy in the case of the seafood safety product in which AquaPharm protected its rights to the product and learned through its partner's R&D efforts. In obtaining exclusive rights to the ultrasound technology and contracting R&D with government sponsorship, the company protected its position and evaluated the technical feasibility of the invention. Initially, AquaPharm was in a posture of positioning and learning with the controlled-release vaccines. After the first phase of vaccine testing, and additional confirmation of the concept from other industries, the technical risk was reduced and AquaPharm adopted a sense-and-follow posture. Industry networks revealed that AquaPharm's vaccine competitors were starting programs in encapsulated vaccines. Coupled with the internal R&D data, the vaccine program moved more toward the center of the company's strategy and efforts. The best example of sensing and following was the previously described sensitive technology assessment that lead to the rapid launch of the development program in hatchery feeds.

AquaPharm used a believe-and-lead strategy for its ReproBoost. It knew the efficacy and market value and its proprietary field trials sent clear signals that the technology merited full commitment to commercialization. The trials also allowed accurate projections of market size, market penetration, and best market entry strategies.

CONCLUSION: HARD LESSONS

There is a fine line between success and failure in managing emerging technologies, and lessons often come with a high cost. By the time AquaPharm had developed a more rigorous and effective process of technology assessment, it had already gone down too many costly dead-ends. Once AquaPharm's managers learned the risks of the aquaculture market through participation in the market with ReproBoost, the company made correct technical and strategic decisions. By that time, the company had missed the opportunity to finance growth with ReproBoost revenues through mistaken investment in technological dead-ends. A series of national and international events quite outside of managerial control contributed to the delay and ultimate cancellation of a significant round of institutional venture capital financing for AquaPharm. Under the cloud of promised financial commitments to the firm that were withdrawn and a patent dispute (eventually resolved in AquaPharm's favor), the managers chose to close the company to recover and preserve the remaining sharehnolder value. However, the sensitivity of the company to those unforeseen events was created by managerial miscalculations. Ironically, the forced closure of the company came at a time of unprecedented technical successes in the product development programs and profitable operations from growing ReproBoost revenues.

In retrospect, the managerial decisions made to create an immediate perception of value in the eyes of potential investors harmed the long-term value creation in the company. The acquisition of intellectual property and the launch of multiple R&D programs to increase revenue forecasts compromised the technological assessment. A more rigorous technology assessment process in the formative days of the company may have prevented "learning the hard way" and also conserved the scarce resources that ran out—one moment too soon. As shown by AquaPharm's experience, the efficiency and elegance of the assessment process counts. Developing successful technology will be a Pyrrhic victory if the cost of getting there is too great.

A rigorous and iterative process for assessing technologies is crucial to managing emerging technologies. Sometimes the race is decided based on the horses companies choose to back right at the gate. The mistakes that are made in the early stages can have tremendously expensive strategic and financial consequences in later stages. These mistakes tend to be compounded later, and if there is no iterative process for self-correction, they will not be recognized until much time and energy has already been invested in them.

In companies and industries transformed by technological change, technology assessment is not a discrete event at the start of product development but an ongoing and critical process. AquaPharm's experience shows that technology assessment must pass through cycles of scoping, searching, and evaluating as a continual learning exercise through the lifetime of the company and its technology. AquaPharm evaluated and committed well, but often on incorrect paths because of errors in scoping and searching. The pursuit of ultrasound technology moved the company away from the traditional scope of its capabilities in product formulation and encapsulation. When it entered the market with ReproBoost, the initial vaccine technologies moved outside of the price and regulatory scope of the marketplace. Had the company adhered more rigorously to the scope of its market and capabilities, it might have avoided the costly tangential foray into seafood safety. Once management understood that the technology assessment process is most successful when it is inviolate, as with the hatchery feeds, it conducted a more thorough assessment and the results added significant value to the company.

The complexities, rapid change, and unexpected combinations in the pool of emerging technologies make any assessment process somewhat imprecise. There will always be a high level of uncertainty in the scope established, technologies discovered, the accuracy of the evaluation, and the decisions to commit. A clear and consistently applied technology assessment process can ensure more thorough analysis, which ultimately leads to better decisions.

CHAPTER 5

EMERGING TECHNOLOGIES AND PUBLIC POLICY: LESSONS FROM THE INTERNET

GERALD R. FAULHABER

The Wharton School

Government is a powerful player in the game of emerging technologies. It can help develop truly revolutionary innovations, such as the Internet, before the private sector. It can also inhibit the deployment of emerging technologies, such as the response to genetically modified foods. From supporting basic research to building infrastructure to establishing regulations, policymakers can define industries and affect the fortunes of individual firms within them. Using the development of the Internet as an example, this chapter explores some of the many ways public policy can influence emerging technologies and some of the issues that lead to increasing regulations. It offers lessons for managers about how to effectively work with this unpredictable and unavoidable partner.

In the rapidly changing world of emerging technologies dominated by both entrepreneurial activity and powerhouse firms with substantial R&D capacity, is there a role for government? Many perceive government to be a problem, not a solution, to the enhancement of a nation's technological capabilities, with its power to tax,

I wish to thank the Annenberg School's Public Policy Center for their financial support for this project. I have also benefited from comments by Christiaan Hogendorn, Wharton School, on an earlier draft. I am also indebted to my colleagues at INSEAD's Business Economics Seminar, Fontainebleau, France, for their comments. Some of the material in this chapter was previously published in the *Journal of Law and Public Policy*. Internet: faulhaber@wharton.upenn.edu; www: rider.wharton.upenn.edu/~faulhab regulate, and otherwise burden innovation at every turn. On the other hand, many new technologies that are quickly snatched up by the private sector for commercialization have been created in research laboratories funded by governments. However managers feel about government, policy plays an essential role in determining the rate and direction of growth of innovation in virtually every country in the world. How governments manage their involvement with science and technology can make their countries technology leaders or laggards.

Every country has a technology policy, even if only by default, and its impact is substantial. Policymakers help to build infrastructure, sponsor basic research, and develop military technology (see box). Government also shapes the emerging technology through directives and subsidies. The government role is complex. The full range of government involvement is far more than could be covered in a single book chapter, but we will explore some of the ways policymakers weave in and out of the process by examining the development of the Internet.

POLICY LESSONS FROM THE RISE OF THE INTERNET

The development of the Internet offers an illustration of the complex dance between policymakers and businesses in bringing a new technology to market. U.S. Internet development can be divided into three phases: (1) the early years, 1969–1993; (2) the high-growth years, 1994–present; and (3) the future, in which the deployment of national, perhaps global, broadband networks is likely to play a key role. During each of these phases, the role of the government is critical but quite different. In examining the public policy role in each phase, we can gain an appreciation for scope of government actions.

This account of Internet history is U.S.-centric, because until quite recently, the Internet has been almost exclusively a U.S. phenomenon. This is rapidly changing. As the Internet becomes more globally dispersed, the cross-national regulatory challenges are becoming some of the most significant and tangled issues for companies. For example, views of privacy reflected in emerging regulations are quite different in Europe than in the United States.[1] The U.S. experience illustrates some of the ways government can shape the development of new technologies and key insights for managers.

The Pervasive Role of Government in
Developing Emerging Technologies

The following annotated list (in order of increasing public intervention) gives some idea of the pervasiveness of public policy in setting the stage for innovation:

- *Institutional infrastructure.* Governments can provide legal and public institutions that encourage or discourage innovation. An enforceable intellectual property regime that carefully balances the need to reward innovators with the need to encourage follow-on inventions is possibly the most important infrastructure for inventors of patentable products and processes. Innovation is also fostered in the presence of an educational system that produces skilled workers capable of rapid adoption of new technology and a financial system that provides capital over a broad range of firms.
- *Research infrastructure.* Basic research in physics, electronics, microbiology, software, and other fundamental disciplines has the economically awkward property that its benefits are *non-appropriable.* Once a theorem or physical principle is known, it can be used by anyone who knows it. While basic research is essential to progress in developing new technologies for market, few firms are interested in investing in unprotected research that can be appropriated by competitors. Typically, the solution has been for governments to invest in basic research and encourage the results to be widely disseminated by scholarly publication.[*][†]

  Governments have several methods of supporting a research infrastructure. A government laboratory structure that supports scholars and encourages publication is one; in the United States, the laboratories at the National Institutes of Health and the Brookhaven particle accelerator are examples. In Japan, the so-called fifth generation computer project of the early 1980s falls into this category; in Europe, the CERN accelerator also fits. In the United States, funding of academic research by the National Science Foundation and National Institutes of Health is the more important avenue of research support.

  A similar model has been used for government funding of "testbeds," operating systems designed to test the feasibility of particular technologies. The initial funding of ARPANet in the late 1960s, eventually to become the Internet, was justified as a research testbed for packet switched message networks within a university/research community.
- *Military technology.* Probably the most successful and costly technology policy of the postwar era was direct government funding, both in

(continued)

101

the United States and the former Soviet Union, of technologies relating to defense, particularly aviation/space and electronics/communications. In both countries, defense officials took a highly active role in encouraging the development of new technologies to strengthen national military capabilities.

- *Government directives.* This model is more interventionist, in that governments take a direct role in encouraging or protecting the commercial exploitation of well-understood technologies, but do not directly fund it. Examples here would be the assurance of low-cost capital to Korean firms in the late 1980s to build up their microchip manufacturing capabilities, the U.S.-Japan Chip Agreement (1986) which attempted to ensure a market for U.S. DRAM manufacturers, and similar protectionist policies in some European countries targeted to U.S. or Japanese high-technology products.‡
- *Standard setting.* Governments may have a role in standard setting; this has occurred most recently in the United States with the establishment by the FCC of a standard for High Definition TV. However, in most emerging industries (such as the personal computer industry in the 1970s and 1980s), standards typically are set by the market (possibly a dominant firm or a patent holder) rather than through government intervention.
- *Government regulation.* Innovation may also be affected by government regulation. The Food and Drug Administration is an example; all new medications offered for sale in the United States must be approved by the FDA. As a consequence, much drug research is structured around achieving FDA approval for the resulting medications. In other technologies, the influence of government regulation appears to be less significant.
- *Government subsidies.* This model is perhaps the most interventionist of all, in which governments explicitly attempt to "pick winners." Examples include the French Minitel of the early 1980s, the European Airbus Consortium, and the U.S. support of Sematech. In each case, there was no pretense that the government was supporting research; rather the government was supporting a technology commercial rollout via specific firms.

★ This is the very opposite of commercially exploitable development activities, which are kept proprietary by the developers, at least until patent protection is achieved.

† Having a government fund basic research does not fully resolve the issue of non-appropriability, since foreign as well as domestic firms may use government-funded basic research.

‡ Many countries have adopted industrial policies designed to create advantages for domestic industries at the expense of foreign firms. These industrial policies often have a technology component to them, but are not explicitly a technology policy, and so are not discussed here. These policies have fallen from favor as a result of their complete failures in countries such as Brazil and India; more sophisticated industrial policies, often given credit for the "East Asian Miracle," have had somewhat less appeal in the wake of the recent financial crises in a number of East Asian countries.

Internet: The Early Years

Military research was the driving force behind the creation of the Internet in the 1960s. The Defense Department's (DoD) Advanced Research Projects Agency (ARPA) funded the first quite primitive packet-switched network, connecting a small number of universities and laboratories involved in DoD research. At the time, the dominant, indeed only, architecture for communications networks was *circuit switching,* in which each telephone call was assigned a specific set of network facilities (switch points, transmission lines, etc.) for the duration of the call that were completely committed to that call. For military purposes, fixed network routing algorithms had the disadvantage that disruptions by enemy action of a small number of network nodes could separate the network, so that large sections of the network could not communicate with others; such networks were not "survivable."

Some engineers recommended what was, at the time, a radical solution of *packet switching,* in which a message was broken up into packets of digitally encoded information, with each part of the message taking its own path through the network. All the packets were then put back together at the destination. The DoD funded the initial ARPANet in 1969 linking four university sites. These government/academic researchers used this quite primitive net for further research in network protocols, and eventually e-mail and file transfers.

Within the scientific and engineering community, ARPANet was an instant success, creating a surprisingly large demand for this service at other universities and research institutes, eventually spreading to other countries. The government played a testbed role in setting up ARPANet and permitting it to expand. By 1975, the ARPANet experiment was stable, and ARPA turned operational control over its network to the U.S. Defense Communications Agency. The National Science Foundation began the organization of the NSFNet, a network for the conduct of research among universities around the country, and in 1990 ARPANet was shut down and all traffic moved to the NSFNet.[2] Together, these interconnected networks became known as the Internet.

Innovations were added to the basic packet-switching technology, creating services such as File Transfer Protocol (FTP), Gopher and WAIS (methods for accessing remote files), and Archie and Veronica (early search engines). At this point, however, government and academic users dominated, and the Internet had a culture that spurned commercialization. In fact, a network culture developed around the Internet; sharing was the rule, everything was free, and charging for anything was not only forbidden by the administrators of this academic experience, it was universally vilified by all "netizens " For example, the X-Windows system for Unix operating systems, developed and maintained at MIT, was distributed free on the Internet, as were periodic updates. While the network and the network community grew very rapidly in the late 1980s and early 1990s, this phenomenon was still largely invisible to the general public and to major corporations. For many netizens, protective of their "gift economy," this invisibility was both desired and nurtured. However, it was not to last:

> Lesson 1: Government can play a powerful role in shaping the development of a new technology in its earliest research stage.

> *Implications for Managers:* Monitoring research in your field at universities and government labs can be a good early-warning indicator of what opportunities lie ahead. The next Internet technology is probably already incubating, perhaps even being tested, quietly in some laboratory. Careful monitoring and early involvement may give you a headstart in developing and commercializing the new technology before competitors are aware of its significance. Don't let the limited early applications of the technology blind you to its future potential.

The Internet Comes of Age

Three major developments in the early 1990s changed forever the course of the Internet:

1. In the early 1990s, NSF decided that the Internet should be privatized. It notified the midlevel networks of its intent to exit the business, and suggested that the midlevels should migrate to a for-profit model. It also began plans to phase out NSFNet in favor of a private solution to the backbone network. This announcement was met with strong resistance from the Internet community, which perceived the publicly funded "gift economy" disappearing,[3] a perception that was largely correct. This privatization was completed when the NSFNet ceased to exist in 1995.

2. In 1989 a computer scientist working at the CERN Laboratory in Switzerland invented the standards and protocols that constitute the

World Wide Web (www), a method for accessing data of all forms worldwide with a unique addressing structure. Previous methods for achieving this, such as gopher and WAIS, were much less elegant and easy to use. However, it was not until a team of programmers at the National Center for Supercomputing Applications (NCSA) designed a graphic "browser" that worked with PCs that a rich and easily accessible means of using the web was available for the nonscientist PC user. Both CERN and NCSA are publicly funded research centers.

3. By 1993, over 30 percent of U.S. households owned personal computers, generally with a Windows or Macintosh interface.[4]

While we cannot say that these developments "caused" the recent popularity and growth of the Internet, it is reasonably clear that these developments were *necessary* conditions to its growth. Of these developments, two of the three were initiatives from public agencies.

The Internet was no longer the testbed engineering experiment it had been in the 1970s. It was now an operating entity with a wide user base. NSF was quite correct in moving the ownership and management of the Internet out of the public sector and into the private sector. However, in doing so it incurred the wrath of many in its natural constituency of scientific researchers. We can thus draw our second lesson from the Internet experience:

Lesson 2: Withdrawal of government support for a research effort as it gets closer to commercialization is strongly resisted by the beneficiaries of that support (be they universities, scientists, not-for-profits, or private-sector firms).

Implications for Managers: If you are a beneficiary of government support, lobby hard for continued (or new) subsidies. If you are a firm taking advantage of opportunities for commercializing this emerging technology, position yourself as a safe haven for potentially valuable players about to lose government support. You may gain influential allies among lead customers if you can anticipate and carefully manage the transition of the community of lead users into the more mature market.

Despite the predictions of many academic users, the switch to a privatized network proceeded smoothly and seamlessly (not without enormous effort on the part of many service providers). The phenomenal growth of both Internet volume and Internet hosts continued apace; Figures 5.1 and 5.2 illustrate the historical growth through 1995 and its projection into the future.[5]

It might be thought that this astounding growth would have caught the eye of corporate America, particularly in light of the recent privatization of this extraordinary resource. But corporate America's eye was elsewhere. During 1993, the buzzword was multimedia, a catch phrase that included video-on-demand and other entertainment options. Several very large mergers were proposed, the most publicized being the Bell Atlantic-TCI deal, only some of which were consummated.[6] These mergers were predicated, in part, on the expected future market potential of broadband network entertainment delivery systems.[7] At the time, the Internet and web were still perceived by most communications, entertainment, and software firms as at best a predecessor, and at worst a distraction, to the true Information Superhighway.

Nor was their skepticism unwarranted; having seen several false dawns, the unruly hackers' paradise of the Internet hardly looked like the engine of commerce and entertainment that large corporations envisioned as the

Figure 5.1
Traffic on the NSF Backbone
(in billions of bytes)

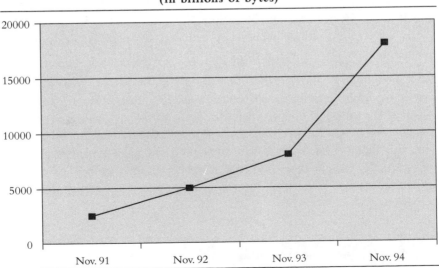

Figure 5.2
Number of Internet Hosts

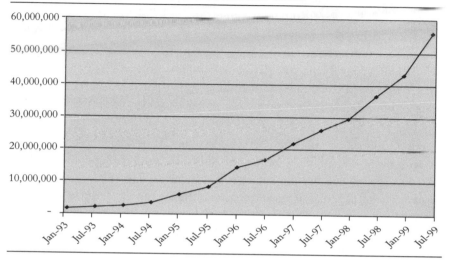

Information Superhighway. However, this skepticism may also be seen as a missed opportunity; for example, if Barnes & Noble had been alert to the challenge of the Internet early on, it may have been able to preempt Amazon.com.

Corporations and users quickly woke up to the potential of the Internet. By 1994, the number of .com commercial sites exceeded the number of .edu educational sites for the first time in Internet history.[8] Both total traffic and total number of hosts on the Internet exploded. Nevertheless, the Internet continued to be viewed by most large corporations throughout 1994 and 1995 as something of a fad, the "oat bran muffin of the 1990s."

Microsoft, for example, thought it could brush aside this "cowboy" Internet simply by placing an icon for its Microsoft Network (MSN) service on the Windows 95 desktop. By December, after the strategy was clearly a failure, the company announced a major shift in strategy that would focus its considerable resources on the Internet.[9] This acknowledgment by the most influential software firm in the world marked a turning point in both public and corporate perceptions of the future of the Internet. No one could ignore it now.

As discussed in Chapter 3, this speciation event of moving to larger commercial markets changed the nature of the Net and shaped its future development. The years since have seen an explosion of new businesses, Internet startups, new ways of doing business, e-commerce, and many other Internet-related activities undreamed of prior to 1995.

With this transformation has come several challenges to the government's role in this technology, including governance, concerns about offensive content, intellectual property rights, and other legal issues.

Internet Governance and the Shift to Private Control.   In its early years, the Internet was "run" by a loose confederation of Internet professionals in the Internet Society and the Internet Engineering Task Force. In the academic/scientific research model, this volunteerism of dedicated professionals worked extremely well; in the highly commercialized, international infrastructure model, this governance structure was no longer appropriate. The U.S. government, acting in consultation with other national governments and the Internet community, attempted to aid in the transition to a more private-sector, international governance structure. This new not-for-profit organization, the Internet Corporation for Assigned Names and Numbers (ICANN), will oversee, among other things, the re-design of the domain name system. This system, for years was run by the legendary (and recently deceased) Jon Postel, an academic from Southern California, who was the very soul of the old Internet culture. Network Solutions, Inc., the domain name contractor for the last few years, has been in continual conflict with ICANN as it attempts to establish rules and procedures for a new domain name system. The stakes involved are quite large, and many players besides Network Solutions, Inc. are lobbying ICANN for their own preferred solution. However, the U.S. government has removed itself from play. This saga is not without its lesson:

> Lesson 3: Government as coordinator can help manage the transition from public (research and education) to private (commercial). Everyone will complain.
>
> *Implications for Managers:* Periods of transition are wonderful opportunities to gain a lasting competitive advantage; lobby hard and lobby smart to have your standard adopted (or at least avoid having others' standards adopted).

New Social Challenges.   The global ubiquity of the Internet and the ease with which anyone, including young children, can access its power and breadth of content have raised issues that will shape the future development of the Net. For example, the availability of sexually explicit material on the Internet to children is a cause for concern to parents, many of whom are less competent at navigating the web than their children. The Communications Decency Act, which became an amendment to the Telecommunications Act of 1996 (and subsequently struck down by the courts) was a political response to that widespread concern. While First Amendment advocates in the United States attacked the amendment, many see this as a legitimate concern that needs to be addressed in some fashion, and that this has not yet happened. Similarly, the presence of information from "hate groups" such as neo-Nazis, white supremacists, and extremist militias has raised concern. While it has always been possible for such groups in the United States to exercise their First Amendment rights, the ease with which the Internet makes this expression available to all is unsettling to many. There is political pressure to "do something" about this problem.

While these issues cause some concern in the United States, the impact on other cultures is far greater.[10] In Islamic countries, the level of sexually explicit material available on the web far exceeds socially acceptable levels, and yet they have no way to control web sites originating in other countries. In Germany, there are strict laws concerning the distribution of neo-Nazi materials, and yet such materials are readily available from U.S.-based web sites. In some cases, the German authorities have taken action against local ISPs through whose facilities this material was accessed. Most would agree that this is not the best solution to this problem, and yet the political demand for solutions is strong:

Lesson 4: Public concern about the affect of new technologies on social mores may lead to demands for political solutions to limit these impacts. These solutions often have unintended consequences.

Implications for Managers: Be prepared both offensively and defensively; political demands can be a profit opportunity if you are prepared to capitalize early. For example, ISPs which provide software that permit parents to screen their children's web activity were well-positioned for profitable sales when children's access to salacious material online became a public issue. Companies that ignore these issues may ultimately face tougher regulations when the public wrath is aroused.

Intellectual Property, Encryption, and Other Legal Issues. The law often has a hard time catching up to the issues raised by emerging technologies. For example, the ability of the web to make information available anywhere instantaneously has been a copyright holders' nightmare. Those who place original content on the web, or whose content is electronically copied and placed on the web, view the Internet as one huge copy machine. This phenomenon is not new; the advent of cheap duplicators in the 1960s caused a similar concern. There is no question that the Internet will certainly change how intellectual property, especially copyrighted material, is distributed and protected. The digital distribution of music, particularly the development of MP3 technology, is sending shock waves through the music industry.[11] Current institutions will have to change,[12] and there will be winners as well as losers.

The Internet also can provide very secure means of transmitting sensitive information via encryption. While this is very useful for doing online banking, it makes police work, combating terrorists, and other law enforcement and defense efforts much more difficult. Attempts by the U.S. government to limit the export of encryption technologies have been met with derision by both the Internet community and the commercial interests for whom security is essential.

In fact, electronic commerce on the Internet has spawned a host of other legal problems concerning how commercial law applies to e-transactions. How do fraud laws apply? Are digital signatures valid? What privacy rights do consumers have when they engage voluntarily in e-commerce? Does the owner of a web site have the ability to restrict unwelcome links from other web sites? This emerging technology has forced legal scholars and practitioners alike to confront new issues in commercial law:[13]

Lesson 5: Both commercial and governmental interests will seek a legal/political response to disruptions created by the technology to their way of doing business. Such responses may evoke disruptions to other parties, thus propagating further demands for legal/political response.

Implications for Managers: Foreseeing all the implications of changing technology is a difficult but necessary part of doing business today and in the next century. Anticipate both corporate and market disruptions and plan how to take advantage of the profitable opportunities that result. Legal/political responses to disruptions will offer great opportunities to those who can influence the political demands and responses, and provide profitable support for those firms who suffer disruption.

The Internet of the Future

The future of the Internet, as with many other emerging technologies, may depend on the availability of a supporting infrastructure. The growth of cellular (and PCS) telephony depended on a network of radio towers, allocation of electromagnetic spectrum, and the existing terrestrial telephone network. The development of genetically engineered drugs depends on the distribution and knowledge infrastructure of existing physicians and the manufacturing capacity of traditional pharmaceutical firms. Earlier developments in electricity generation and distribution, and automobiles and railroads also depended on building an infrastructure to support the new technology. In some cases (such as cellular telephony), government plays a very active role in either developing the infrastructure directly or regulating its operation. The supporting infrastructure of the emerging technology, if required, may thus be a public policy issue.

In the case of the Internet, the supporting infrastructure of the future is the national and eventually global diffusion of ubiquitous broadband networks.[14] The demand for increasing "bandwidth," which refers to the amount of information per unit time that can be processed or transmitted through an electronic medium, promises to be a significant force in driving the future development of the Internet. Narrow bandwidth has been one of the limiting factors in the use of more sophisticated Internet technologies (such as video and resource-intensive graphics). Access to such networks would provide households and businesses with a far richer interactive experience, including two-way video, online demos, and three-dimensional visualizations. Whether this will occur, and how it will play out, are great uncertainties.

Engineers and communications specialists have been predicting the coming of broadband systems with both confidence and regularity over the past 30 years. For those familiar with this technology, the surprise is that it took so long. There have been numerous "false dawns," such as teletext and videotext, and more successfully, Minitel in France. However, despite the enthusiasm of engineers and telephone companies, consumers did not yet have a need to which broadband data networks were the solution. The growth of Internet applications and frustration with the "World Wide Wait" have given new impetus to the push for broadband.

While it now appears that the long-anticipated mass deployment of broadband data networks to support the Internet is at hand, there are still many questions. How fast this will occur? What fraction of households, businesses, schools, and governments will eventually become active users? What technologies will be used? What they will be used for? There is a very wide range of possibilities, from "small impact on a few enthusiasts" to "a fundamental change in the way we all live and work." Which route is taken, and how fast it develops, will almost surely be deeply affected by public policy decisions and through direct encouragement and investment or regulation.

What role will government play in creating this infrastructure? We can gain some insights into looking at the role of government in the development of other types of networks in the past. "Hard" networks, such as road and rail systems, power grids, water and gas distribution networks have been with us for a century. "Soft" networks, such as computer hardware and software, and automobile service and parts systems, depend on shared standards and protocols to link products and their uses and are a barely noticed part of our lives. Telecommunications networks have also been with us for a century, from early telephone networks, local in scope, to the emergence of the current globally connected telephone system. In the 1920s, radio networks emerged, followed by television networks in the 1940s and 1950s. Somewhat later, cable television networks grew, slowly at first, but now available to 90 percent of U.S. homes. In other countries, satellite TV distribution networks perform much the same role. More recently, cellular telephone networks have also grown, illustrating the point that telecommunications networks, though "hard" in the sense used above, can be wireless links, without a continuous physical connection.

Drawing on these past experiences and current debate on the Internet, there are several key public policy concerns that promise to shape the government's role in developing broadband networks, including:

- *Universal service:* Is the service available and affordable to all citizens?
- *Quality of service:* Is the service efficiently provided at a reasonable quality?
- *Monopoly power:* Is the provider earning excess profits from abuse of a monopoly market position?

- *Access to distribution:* Is the distribution system available to all content providers?

Providing Universal Service.   Implicit in the concept of infrastructure is that it must serve most of the population. For example, over 94 percent of households have telephones, over 98 percent have television, over 90 percent can access cable, about two-thirds of which subscribe to the service,[15] and well over 50 percent have personal computers.[16] Most every desk and workstation in U.S. industry has a computer on it, and almost all the growth is now coming from sales to homes, where growth rates are still high. Other related industries have also achieved relative ubiquity, such as VCRs.[17]

The routes each industry followed to achieve "universal service" are quite different. In cable and telephone, universal service was an explicit public policy objective, but different policy instruments were used to achieve it. In telephone, active regulation was the chosen instrument; in cable, the contract terms of a geographic franchise were the chosen instrument. In the case of television, VCRs, and personal computers, policymakers stepped back a bit. Competitive markets drove prices down and market penetration up.

The universal service mandate of regulators has traditionally gone beyond ensuring that service is available to all; it is rather that service should be *affordable* by all. To achieve this objective, regulators have traditionally insisted on pricing practices that involve cross-subsidies from premium services to basic services. This pricing ensures relatively inexpensive access to basic service for all users. Such pricing practices are not unique to the United States, but occur in publicly regulated or publicly owned networks throughout the world.

In both cable and telephone, however, the price of publicly mandated universal service was monopoly. To make it feasible (so it was claimed) for a firm to serve everyone, profitable and unprofitable, the government had to forbid entry by competitors into the firm's market area. Why should this be? Cross-subsidies and the investments needed to support low-cost basic service cannot be sustained in the presence of competitive entry. New firms could enter only those markets in which prices are held above cost in order to subsidize other customers, forcing incumbents to respond competitively with price decreases or lose the business altogether. In either case, the source of internal subsidy would eventually disappear, and the incumbent could no longer afford to serve unprofitable (though allegedly deserving) customers. Therefore, to maintain the subsidies that most regulators use to achieve universal service, regulators restricted competitive entry, either by regulatory fiat or by the granting of franchise monopoly.

The costs of regulated monopoly have been well-documented elsewhere,[18] including reduced incentives for efficient operation, reduced incentives for innovation, excessive resources devoted to "rent-seeking" through the regulatory process, and so forth. With an emerging technology, the threat that the industry could be regulated, could well be enough to stifle the whole enterprise. Yet some very procompetive commentators from the computer industry have suggested that computers should be made available to all, perhaps without realizing the economic and regulatory consequences of this advice. There is no question that paying the price of monopoly is quite high; is it really necessary in order to achieve universal service?

A public policy of universal service need not necessarily lead to franchise monopoly. For example, government could provide direct subsidies to low-income (or high-cost) consumers, perhaps in the form of computer stamps or Internet stamps. Alternatively, nonexclusive franchises could be granted that require each franchisee provide universal service, but let multiple franchisees compete for consumers' favor. For example, two or three broadband network providers could compete in a metropolitan area, all under the requirement to provide universal service. In the case of cable television, potential cable operators claimed that they would only undertake cable investments if they were granted exclusive franchises. Very few municipalities chose to test this claim. Several researchers have subsequently suggested that many municipalities could indeed support more than one cable system.[19]

The universal service issue for broadband two-way networks is currently relatively quiescent. There is little evidence that broadband access from the home (as opposed to broadband access from school or narrowband access from the home) constitutes an essential tool for all Americans to achieve equal opportunities, either in the political or economic marketplace. It could become a valued entertainment distribution channel, but this is hardly a public policy reason to subsidize universal service. There are proposals to provide "basic" broadband service to schools and perhaps libraries,[20] but generally few calls for a broadband link into every home in the United States, an enormously capital-intensive venture.

This may change if there is a substantial increase in the demand for broadband in rural areas or from disadvantaged groups. This demand would translate into political action that could redefine universal service to include broadband, possibly fiber to the home (or curb). Should this occur before the industry has had a chance to form, there could well be public intervention to ensure that all suppliers were required to provide fiber service to all households and businesses.

In fact, if municipalities are permitted to limit broadband fiber providers by monopoly franchising, as has been done in cable TV, this outcome is highly likely. Even more likely is that those firms who believe they have a good chance of winning such monopoly franchises may press legislators toward universal service as a means of justifying monopoly. It could be argued that only a monopoly can ensure that everyone will be served:

> Lesson 6: A new technology which is highly valued by all may lead to political demands for "universal service" from low-income and/or high-cost constituents, which is likely to result in some form of government intervention.

> *Implications for Managers:* Demands for "universal service" can be translated into protection from market competition by government fiat. Cable television companies argued in the 1970s that they would only invest in cabling entire cities if they were given exclusive franchises, to their enormous profit. Similar opportunities are likely to accompany the emerging broadband technology deployment.

Ensuring Quality of Service. In a market with some form of competition, the expectation is that quality of service will take care of itself. Firms will provide the level of quality that customers demand and are willing to pay for, and competition will ensure their responsiveness to customers. In the case of monopoly, however, the incentives for the firm to provide appropriate quality levels may be diminished, so that quality of service may suffer.

The decline of service quality observed by many customers of the cable TV industry during the late 1980s led to re-regulation. Customers complained of receiving shoddy treatment in handling requests and failure to provide reliable, outage-free services. They turned to Congress for a solution, which came in the form of the Cable Re-regulation Act of 1992. As a result of letting their initial quality decline, cable companies now had to work under stricter regulations. In contrast, the cellular industry has been more successful at avoiding price regulation, even when each local market was a duopoly (prior to the introduction of PCS).

While Internet regulation is not far enough along in its development to allow for such action, there is already concern about slow downloads and response times and failures of overloaded popular services such as America Online. If government plays a more active role in shaping the infrastructure, calls for government intervention to assure quality service may be expected.

This type of regulation is problematic. In principle, regulators generally have the legal power to coerce firms to provide the "right" service level. In practice, this is more difficult, as is borne out in Williamson's well-known analysis of cable TV franchise bidding.[21] Additionally, it is not clear that regulators are good at assessing the quality level that customers would demand in a more competitive market. For example, in the precompetitive airline market, most scholars agree that airlines over-provided schedule quality, at the cost of higher fares, as a result of the CAB's regulatory practices. After deregulation, schedule quality "deteriorated" to that for which customers were willing to pay. Another example occurred in telephone; prior to the deregulation of terminal equipment, the Bell System (with regulatory approval) provided rather simple telephones that were virtually indestructible. After deregulation, it became clear that most customers preferred telephones with many more features and a shorter life; the telephone soon became another consumer electronics product. In both cases, regulation led to an inappropriate quality level (as measured against the competitive standard).

The days in which the U.S. government directly provided network capacity or even managed network capacity are long gone. At best, the government can assist in the transition to new institutions, which can then address the capacity problem. The new privatized governance structures discussed in the previous section are precisely the correct public intervention. As the Internet continues its transformation from enthusiast's toy to mission-critical infrastructure, pricing, revenue-sharing, and investment incentives will be put into place to ensure to smooth and rapid response of service providers' capacity to demand changes:

Lesson 7: Dominant firms can often make the mistake of treating customers poorly, which can lead to a political demand for intervention by regulators.

Implications for Managers: Carefully monitor customer satisfaction and emerging concerns. A service level that was acceptable with relatively primitive technology may suddenly be unacceptable as technology develops and expectations rise. Whatever your market power, in today's environment, companies ignore customers at their peril.

Concerns about Monopoly.   The power of individual firms, either through government franchises or "natural monopolies," is another concern of government regulators. The primary focus is on cases in which there is a dampening of competition with a negative impact on consumers, but antitrust authorities may pursue firms with dominant positions even without a groundswell of popular negative opinion, such as the Microsoft case. This case also illustrates the highly effective influence of government antitrust actions by firms such as Netscape and Sun Microsystems.

For some emerging technologies, bottlenecks develop in which a single firm controls the network infrastructure through which consumers access to this technology. In such cases, it may well be that antitrust actions to break up a monopoly would be ineffective in the long run, as market forces would eventually lead to the remonopolization of the industry. Some form of regulation may be justified as a means to control the abuse of monopoly power in such industries, and this is the rationale given by many for the creation of regulated monopolies in network industries. Others argue that these monopolies may not be so natural, but are products of the regulation that seeks to control them. This latter view is somewhat more compelling, in that virtually all regulators protect regulated monopolies with entry prohibitions. As Alfred Kahn said, "If the monopoly is so natural, why does it have to be protected?"[22]

The protection is necessary to maintain subsidizing price structures, which are indeed a product of regulation. Regulators find that control of monopoly power is added to their list of responsibilities, be that monopoly natural or created. Generally, much regulatory attention is devoted to ensuring that the firm's earnings are not "excessive," that is, exceed the cost of capital. In regulated monopolies operating in some markets subject to competition, this concern takes the form of ensuring that power in monopoly markets is not being used to subsidize operations in competitive markets. Both tasks are extremely difficult; concern for cross-subsidy is virtually impossible. For example, as telecommunications competition slowly increased during the 1970s and early 1980s, the Federal Communication Commission devoted very substantial efforts to develop a standard by which to judge whether or not Bell System rates involved cross-subsidy, without success:

Lesson 8: If a new technology threatens to lead to a single firm gaining a dominant market position, government may intervene to control this "natural monopoly," either through regulation or antitrust.

Implications for Managers: Avoid a situation in which customers and competitors can credibly claim that you have used market power to stifle competition and/or innovation. Behavior that is acceptable in a robustly competitive market is not acceptable if you are a dominant firm. Like it or not, you can be held to a higher standard; if you fail that standard, costly regulation or litigation may follow.

The Problem of Vertical Integration (Content versus Conduit).   The "network" of a network industry is a distribution system, a conduit over which something else, *content,* is sent. In telecommunications, this something is telephone calls; in cable, it is video programming; in electric utilities, it is power. In computing, it is possible to think of hardware as conduit and software (which actually delivers what customers want) as content. In both regulated and competitive markets, an important economic issue is the vertical integration of content and conduit.

In some markets such as telephone, content and conduit are separated as a matter of law, generally on First Amendment grounds. In other related markets such as cable and broadcast television, content and conduit can and generally are integrated within each firm.[23] For example, subscribers to a particular cable firm can buy only material that the cable firm chooses to make available. In contrast, anyone can use the telephone network to distribute any information (such as 800 or 900 services); the telephone company has nothing to say about it.

The computer industry provides a prime example of how competitive markets evolve. Prior to the early 1980s, virtually all computer companies bundled hardware and software. An IBM customer had to buy IBM proprietary software, because no other commercially available software ran on IBM machines. This was the era of "closed" computer architecture. In contrast, the PC ushered in the era of "open" architecture, in which hardware vendors encouraged provision of software by as many as possible. The result was a proliferating of both hardware and software, with thousands of companies, many no more than a single person, pumping out tens of thousands of software titles. Many have credited this open architecture with the extraordinary growth and richness of the computer industry of the 1980s and 1990s, compared to the relatively stately pace of innovation in the closed architecture era.[24] However, in the early 1990s, many software firms complained that Microsoft, the firm that controls the dominant PC operating system (the conduit), has used its OS control to unfair competitive advantage in the applications (content) market, such as word processors, spreadsheets, and presentation graphics.[25] After considering such complaints, the Department of Justice did not prosecute, reaching a relatively mild agreement with Microsoft in 1995 that it cease certain practices.[26] No one seriously suggests that Microsoft should not be permitted to compete in the applications software market. However, the example brings home the fact that vertical integration of content and conduit is certain to give rise to contention of market abuses, if not actual abuses, and constitutes a public policy problem, either regulatory or antitrust.

> Lesson 9: If the technology leads to firm dominance in a bottleneck market, there will be a political demand for government to limit the dominant firm's ability to vertically integrate.
>
> *Implications for Managers:* If your firm must use the bottleneck market, you should lobby hard to keep the bottleneck firm out of your market through government-imposed restrictions on vertical integration. If you are the bottleneck firm, aggressive ultra-competitive behavior on your part toward competitors is sure to invite public scrutiny with possibly unwanted restrictions resulting. Realize also that how aggressive the government is in pursuing allegations of anticompetitive behavior changes with political administrations.

The Future of Broadband Policy. Regulation and franchise control have traditionally been the chosen instruments in virtually all electronic network infrastructure industries. Will the same approach be taken in developing broadband networks? As we have discussed, competitive markets can address issues of universal service, service quality, the control of monopoly pricing, and open architectures.

While more limited regulation in broadband seems likely, concerns about issues such as universal service or monopoly could lead to calls for government intervention. What would be the impact of regulation? Competitive simulations of the market for broadband service—under cases in which regulation require universal access or conditions of more open competition—produce interesting results.[27] For competition without requirements for universal access, the market is projected to develop as follows:

- For "reasonable" estimates of cost and demand for broadband distribution, it appears likely that major metropolitan areas may support more than one broadband distributor, but not until demand levels are approximately double present levels.
- However, competitive deployment of fiber may occur in rings, in which the areas of densest population are served by n fiber distributors, the less dense areas are served by $n - 1$ distributors, until the final ring which is served by only one provider. Prices within each ring would reflect competitive conditions. In high-density cities, fiber is likely to serve the entire metropolitan area. In low-density cities, fiber will not extend to outlying areas.
- If competitors anticipate gains from being the largest provider in this "ring" model, then there are gains from preemptive investment in the early years to ensure that the first mover locks in his advantage into the future. This would lead to more extensive investments, and therefore greater geographic coverage, than a static oligopoly market would suggest. Paradoxically, the dynamic competitive chase for long-term profits may lead the market to achieve the universal service regulators previously believed had to be mandated.
- On the other hand, if regulators mandate universal service, this would impose a fixed cost on entrants that would constrain the number of fiber providers who would be willing to enter with a universal service constraint. Simulation analysis suggests that the imposition of a universal service constraint to make service available to (say) 95 percent of households in a metropolitan area will increase the cost of providing fiber sufficiently that initial entry as well as competitive entry are only feasible at greater demand levels than would otherwise be the case. The reason is that the cost of supplying fiber infrastructure to unprofitable customers may be greater than even duopoly profits from the profitable markets.

Thus, we find the surprising result that imposing the universal service obligation may lead, at certain demand levels, to monopoly, even if unconstrained competition could support multiple fiber vendors. The price charged under this scenario would be a monopoly price, substantially higher than most customers would pay under unconstrained competition.

On balance, then, it would appear that competitive provision of broadband access is superior to any form of regulation or franchising. Further, unless there is significant pressure of rural or disadvantaged groups for below-cost provision of broadband access to the home, it should be relatively easy for legislators, regulators, and municipalities to resist vendor demands for monopoly franchises. No matter what broadband networks develop, there is a competing existing infrastructure for delivering Internet-type services to everyone. Most schools and libraries have some form of access, and most households have telephones, which permit slower access, which is satisfactory if not perfect for Internet access, at least at present. The policy direction established by the Telecommunications Act of 1996 should provide a rationale for policymakers to take the competitive option.

It is unlikely that two or more fiber providers will enter the market simultaneously. It is more likely that a single firm will enter, possibly expanding its service area over time, most likely competing with satellite providers. The second fiber firm may not enter for several years, when demand levels are sufficiently high to support two providers. During this interim period, the temptation to regulate the monopoly may be quite strong. It is critical that this temptation be resisted, as efficient and innovative competitors are unlikely to emerge in a regulated environment.[28]

Not only may there be more than one fiber provider in broadband competition, there is likely to be wireless coverage as well. Further, existing infrastructure providers are currently developing technologies that increase the effective bandwidth of their infrastructure. Telephone companies are experimenting with Digital Subscriber Line (DSL), a technology that will permit over 1 Mbps or even higher over existing telephone lines. All these technologies would certainly compete with each other, *provided* they are actually deployed.

And therein lies the concern. It is possible (some would say likely) that after all the grand announcements, alliances, IPOs, and other fanfare, only the telephone or cable companies will actually lay fiber to the curb, and thereby control the only broadband two-way distribution channel into the home. In that case, two problems confront public policymakers. The first is the classic problem of monopoly: a firm takes advantage of its market position to charge prices higher than costs. The second is the problem of access control. The monopoly firm chooses the content its users can access, which limits both its customers as well as potential suppliers. Monopolies tend to be closed architecture systems, with a limited choice controlled by the bottleneck supplier.

On balance, it is likely that the second problem would be more serious than the first. If it is the case that the market can only support a single supplier, then it is likely that monopoly prices are not very much higher than total costs. It could be that the monopoly might be a temporary one, until other firms can deploy resources to compete. In this case, it is particularly important that antitrust authorities be alert to attempts by the incumbent to raise potential rivals' entry costs, or other anticompetitive behavior. This would appear to be a problem for the antitrust authorities.

The second problem is somewhat more difficult. Should a single firm be the monopoly supplier of broadband distribution, it is likely to control content, increasing its profits through price discrimination among content providers. By analogy with the IBM-dominated computer market of the 1970s, we would expect proprietary content provision in a closed architecture, without the profusion of content and access that a more competitive market would provide. If such a monopoly emerges, or emerges even temporarily, how should policy makers respond?[29]

Fortunately, the FCC already has adopted the Open Video Systems approach, in which telephone companies (indeed, any OVS supplier) providing video distribution to the home are required to make its facilities available to other content providers under the same terms and conditions it offers its own content provider.[30] While this approach is not without problems, it does represent a regulatory approach to convert an otherwise bottleneck facility into an open architecture system.

This is a good example of a regulatory intervention that opens up markets to a far richer supply structure than would otherwise obtain, and certainly far richer than would obtain under traditional rate-base rate-of-return monopoly regulation. Should temporary monopoly of two-way broadband facilities become a problem, then this relatively light touch of regulation designed to open access to any content provider is an effective solution to that problem:

> Lesson 10: Regulations that appear to promote competition or policy objectives for emerging technologies can have unintended side effects or even the opposite effect.

> *Implications for Managers:* It is often to companies' advantage to lobby for "minimalist" regulation, that makes maximum use of market forces and gives minimal scope for political/judicial "gaming." If some form of regulation is truly needed, propose solutions to regulators/legislators that are market-friendly. Otherwise, an initially small amount of regulation can become pervasive and destructive in response to political pressure.

CONCLUSIONS

As shown in the development of Internet technology, government plays a role in every stage of the development of an emerging technology—from building its early infrastructure to sorting out its complex social repercussions. At each stage, companies can take advantage of opportunities for government support and anticipate or influence government intervention. The regulation of new technology, once it begins to emerge and become commercialized, is one of the most challenging areas for managers because it is often difficult to predict public and political reaction to an evolving technology.

The United States has established a procompetitive context for the development of the Internet. However, control of telecommunications is fragmented among state jurisdictions plus the federal level, suggesting that progress within this framework and the policies adopted may be quite varied, even contradictory. The process by which the individual states and the nation as a whole comes to understand what needs to be done is likely to be drawn out over the better part of a decade, after which there will no doubt continue to be some variation among jurisdictions. The role of government in the evolution of the Internet and other technologies is not monolithic, but can be expected to involve many jurisdictions, some of whom may be acting against one another at any point in time.

The problem of overlapping jurisdictions becomes even greater when we move to a global perspective. Issues of whose rules apply and where they apply create tremendous uncertainty and complexity for businesses. While the Internet, with its borderless nature, makes this problem particularly challenging, any emerging technology must address the complex texture of global regulation. This can work in the company's advantage if it can use more favorable environments to develop and commercialize the technology and avoid less favorable environments. It can also work against companies that run into regulatory roadblocks as they seek to rapidly build their global positions. How well can we deal with political and legal issues across a wide range of countries, many with divergent interests and cultures? As in the past, the abilities of governments and companies to adapt to dynamic emerging technologies will determine whether or not their potential can be realized.

While the exact flashpoints for government intervention may be difficult to predict, astute companies can learn from the lessons of the past and pay close attention to the potential for regulation. While government may be playing a less direct role in an age of deregulation and privatization, emerging technologies continue to be an area of intense public scrutiny and demands for government controls. This chapter has identified some of the key policy issues for the Internet, but the exact challenges will differ for any given technology. Companies need to carefully assess potential areas of government action and develop strategies to respond to and shape the policies that will shape their emerging industries.

PART II

MANAGING MARKETS

$\mathbf{A}$s voice phone calls are replaced by e-mail, Internet, and other digital traffic, Sprint has had to rapidly anticipate and reconfigure its services to meet the new demands of an uncertain and rapidly shifting market. "These changes are profound, and require new approaches," said Ken Henriksen, chief technology integration architect at Sprint. "The telecommunications industry is becoming a complex mosaic of emerging technologies, standards and market developments." What do these new customers want? What kind of technology will best meet these emerging needs? How big will these markets be? How should services be priced?

"We've got to adapt to changing market needs quickly," Henriksen told a conference at the Wharton School.[1] "We have a tremendous amount of technology choice and there is an ongoing debate in the industry as everyone tries to determine what will best meet these market needs."

Internet telephony is typical of the challenges they faced. In the 1990s, only a few brave souls used the poor quality Internet transmissions to stay in touch with distant countries. What will happen, however, when increased bandwidth leads to much better quality?

When customers are brought together with new technologies the results are often unpredictable. This is amply demonstrated by the Internet—which crawled out of the backwaters of military and technology applications to stomp through the main streets of business. Favorite technologies often get the thumbs down from consumers and fail in the market. And some unlikely prospects end up as hits. Understanding and managing these potential and emerging markets demands a very different approach than managing markets for existing technologies. Part II explores strategies for

125

assessing future markets and pushing technology barriers in ways that have the most market impact. The section also explores factors that affect the commercialization of new technologies, particularly the importance of complementary assets.

MARKETS THAT DON'T EXIST

The challenge facing emerging technology firms is to develop markets that don't exist for products that are not yet fully developed. In Chapter 6, George Day explores a variety of approaches that can be used to learn about nascent markets for emerging technologies. These strategies include watching lead users, gaining insights into latent needs, and anticipating strategic inflection points. He also emphasizes the importance of using multiple methods (triangulation) and of adopting a broad system's view.

At any given time, managers need to decide where they are going to invest their research dollars to have the most impact on the market. For example, in making laptop computers, should the company invest in research to lower its weight or increase its durability? As discussed by Ian MacMillan and Rita Gunther McGrath in Chapter 7, markets are lumpy—unevenly distributed. This means a small shift in one technology barrier can lead to a significant jump in the size of the market that breakthrough opens. The authors offer several strategies for using market lumpiness to seize or merge attractive niches.

Superior new technology alone doesn't assure success in commercialization. Typesetter manufacturer Merganthaler Linotype remained an industry leader for over a century, despite three waves of discontinuous technological innovations. The firm's proprietary typefaces were "complementary assets" that help ensure its leadership, even though competitors arrived with superior new technology. In Chapter 8, Mary Tripsas discusses Merganthaler's success. She points out that it is important to recognize that the primary technology alone is not the only factor in successful commercialization. She offers a framework for examining the interaction between technology, complementary assets, customers, and rivals that shape commercialization success.

CHAPTER 6

ASSESSING FUTURE MARKETS
FOR NEW TECHNOLOGIES

GEORGE S. DAY
The Wharton School

The challenge of assessing future markets for new technologies is to determine the demand for products that don't exist from customers who don't yet know about them. At the same time, the trajectory of technology development and speed of market acceptance are also uncertain. In this market vacuum, there is not enough oxygen to sustain traditional methods of marketing assessment. But there are a variety of approaches that can be used to better understand market potential in this environment. This chapter examines the adoption patterns for new technologies, strategies for continuous exploration and learning, and the "triangulation" of insights about lead users, latent needs and inflection points. These strategies can give character and dimension to the embryonic and evolving markets for emerging technologies, providing clues to their ultimate potential.

What will the market be for *automated highway systems* and when will it emerge? These "smart highways" will enable vehicle control (with collision warning and avoidance and navigation help), automated toll systems, and even automated driving and steering lanes. They will require the integration of technology for automated vehicular control, satellite-based global positioning systems, and roadway sensor systems. If and when these technologies come together, will potential customers be interested? The reactions of potential users to rental cars with navigation aids or automated toll systems while commuting may provide some clues about potential benefits, barriers to use, price sensitivity, and eventual acceptance. As the systems technology continues to advance, however, the big questions about the market remain: How quickly should trials be launched? Who should take the lead role? Will regional governments

127

mandate these systems to solve problems of congestion? Will drivers be willing to pay for the technology once the benefits are demonstrated?

Within a decade, *biochips* (formally known as DNA arrays)[1] that have
the ability to analyze thousands of genes at one time should make it possible to analyze a person's genetic risks for scores of diseases. In the future,
patients with maladies such as a sore throat could have a culture tested with
a disposable biochip with the ability to check for a myriad of microbial
genes and determine exactly the right drugs to prescribe. How big is the
market for these disposable chips and when will it emerge? Technologists
must be able to economically produce biochips that can accurately detect
gene glitches that cause disease. Patients and doctors must be convinced the
tests help by guiding preventive therapy for example, and insurance companies must be willing to pay for the tests. In the meantime, numerous
startup companies are exploring the breadth of applications and big pharmaceutical companies are placing their bets by investing in these companies.

The next generation of *rapid prototyping technologies* use laser cutting
or ink-jet depositions of material to quickly transform complex three-
dimensional CAD images into solid objects of powdered ceramics or metal.
At present, these shapes are models used to guide the product design process, but in the future they can be saleable end products. The technology
could be used to make big objects such as tank turrets or airplane parts or
customize a tennis racket with a grip uniquely configured to fit an individual player's hand. Which applications will prove a feasible basis for a
market?

Each of these technologies has exciting prospects. But the history of
emerging technologies is that early champions held compelling visions about
the future market prospects. Whether the envisioned markets materialized
depended on resolving a series of uncertainties.

THE CHALLENGE OF EMERGING MARKETS

The turbulence and uncertainties of future markets for new technologies
confound the research approaches that have been honed for assessing established markets. Seldom are there precedents or sales histories to study. Because the applications are evolving, it is not clear who will be the most
attractive customers, when and how they will use the product, or what
they will be prepared to pay. Since the industry structure is embryonic,
there are many conflicting views and much speculation about potential rivals or competing technologies.

Assessments of markets for new technologies are further complicated by the interaction between technological development and the rate of market acceptance. Price and performance improvements come more quickly when acceptance is accelerating. But this can only happen when the quality and performance standards are in place and the product can be made, distributed, and serviced.[2] Lack of any one of these elements will slow acceptance of the technology.

Before the technology is proven and cost-effective, and the market is still in a nascent stage, the question is whether the market is big enough to warrant a development project.[3] This spawns many related questions: Does the product satisfy a need or solve a persistent problem of a significant group of customers better than the alternatives? Which segments and applications will be the most attractive? In what order will they emerge?

As the project progresses, a new set of questions emerges that demands greater precision. How large is the prospective market, and how quickly will this potential be realized? Here numerous assumptions have to be made about the technology improvement trajectory, the availability of standards and supporting infrastructure, benefits and costs to target customers relative to competing alternatives, and the collective investment of competitors in market development.

This chapter is about how firms have learned how to answer these questions. Yet those who have lived through the emergence of a market for an emerging technology know that definitive answers are elusive; there are too many qualifications and contingencies and the answer depends in part on the actions of the firm and its rivals who are also trying to answer the same questions. A more realistic goal is to reduce the uncertainty to a manageable level and gain actionable insights ahead of these rivals. Once this goal is within sight, a new set of questions about how to gain and hold a viable competitive position in the emerging opportunity space comes to the fore. The frameworks, methods, and best practices that are covered in this chapter can help illuminate these issues.[4]

Three Approaches

Useful assessments of future markets for emerging technologies, when uncertainties are intolerably high, are guided by the following premises:

1. *Diffusion and adoption.* Each emerging technology will diffuse at a different rate and pace into their prospective markets. Some markets leap ahead while others languish for years before gradually taking off.

Others never come close to realizing their potential before they are pushed aside by rival technologies. Each path is the outcome of the interplay between contending forces that inhibit or facilitate the rate of diffusion.

2. *Exploration and learning.* Advantage comes from informed anticipation. The emphasis should be on rapidly learning from a series of market probes with successively refined versions of the product, using the lessons from each probe to guide the subsequent stages in the development process, and anticipating critical inflection points in the market ahead of competitors. Winners are able to surface opportunities faster, invest in more attractive options, and shape the market to their benefit.

3. *Triangulation for insights.* The ability to absorb uncertainty and anticipate opportunities faster is enhanced by divergent thinking processes that surface and explore a wide range of possibilities, rather than convergent thinking that seeks a closure on a satisfactory answer. This need is best served by starting with diverse market research methods, with different assumptions, levels of analysis, and sources of data. Insights come from a process of triangulation that looks for convergence of conclusions across the different methods. A corollary to this premise is that (conventional) market research methods have limited utility because they were designed for other purposes. Different research approaches are needed when the customer requirements aren't known, usage situations can't be described, and prospective customers can't envision the product concept.

DIFFUSION AND ADOPTION OF REALLY NEW PRODUCTS

New product innovations take time to spread or diffuse into markets. Some innovations have a long gestation period and then grow explosively, while others penetrate their potential market very slowly and exhibit modest sales growth for many years. The diversity in patterns of growth can be largely explained by the following characteristics of the product:[5]

- The *perceived advantages* of the new product relative to the best available alternative. The value, set by the perceived relative benefits minus the perceived relative costs, must be sufficiently compelling to motivate the switch.

- The *risk* perceived by prospective buyers because of their uncertainty about performance, fears of economic losses, or concerns about standards changing.
- *Barriers to adoption* (such as a commitment to existing facilities, investment in the previous generation of technology or regulatory restrictions) which slow acceptance.
- *Opportunities to learn and try.* Not only must the new product be readily available (for trial, purchase, and servicing), but the buyer must also be informed of the benefits and persuaded to try it.

The main driver of the rate of diffusion is perceived relative advantage, but the other three factors can dampen or impede this rate.

The erratic history of videoconferencing shows the importance of the perceptions of relative advantage. Initially, the developers of these systems thought videoconferencing would be a substitute for travel. Meetings among people from different cities would be conducted by audio and visual transmissions between these locations, with great savings in time and expense compared to bringing people to one place. Increasingly, the prospective users of videoconferencing accept that significant travel savings are possible, although the immediate and high set-up costs are often more salient than the subsequent savings. The problem is that these prospective users discount the benefits because they don't believe that electronic meetings can deliver the subtlety and richness of face-to-face encounters. Thus, videoconferencing is now being perceived as a complement rather than a substitute for travel. Face-to-face meetings are needed to nurture relationships and build teams, while videoconferencing is used for the ongoing coordination needed to sustain these relationships between get-togethers. This change in perceived benefits has meant that videoconferencing has been growing in parallel with business travel.

Relative advantage depends on the performance inherent in the technology and the intensity of stimulative efforts by competitors offering the new technology. Not even the most promising technology will find a market unless the collective efforts and investments of the competitors to innovate, market, and reduce the cost of the technology can unlock the potential. These factors work together to determine how soon the trajectory of performance of the emerging technology will meet and then exceed the trajectory of market demand. The prospects for the electric car depend on when it will be able to (1) go 120 miles before needing a battery charge, (2) reach a top speed of 80 mph, (3) accelerate from 0 to 60 mph in under

10 seconds, while (4) being readily available at a competitive price. Unless and until these performance thresholds are crossed, the electric car will appeal only to a specialized segment.[6]

Stimulating Diffusion

While extrapolative models of technology evolution such as Moore's Law can help assess the rate of performance improvement, the stimulative effects of investments and price cutting by competitors are tougher to assess. It is an inherently dynamic process in which investment decisions hinge in expected growth which in turn is a cause and consequence of competitive activity. This iterative sequence is shown in Figure 6.1.

The process is triggered by pioneers who act on the belief that it is usually better to be a pioneer than a follower.[7] Consequently, the promise of an untapped or emergent market invariably attracts numerous aspirants. Each entrant is likely to make investments in technology development, facilities, and entry programs that may not fully account for other entrants

Figure 6.1
Drivers of Market Growth

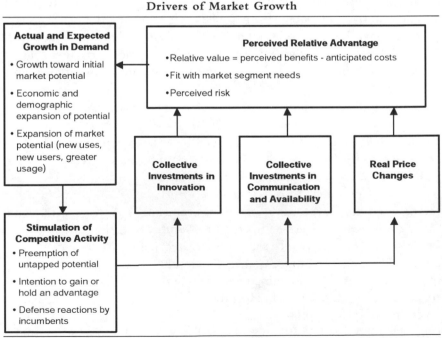

with similar plans. The intensifying competition also puts downward pressure on prices, as cumulative experience helps to lower costs. It is the combined impact of these investments and real price declines that stimulates market growth by increasing the market potential and/or accelerating the rate of growth toward that potential. Among the stimulants of more rapid diffusion are:

- *Innovation*. Progress in technology is largely driven by the competitive need to match what the rivals have already achieved while finding new edges that the rivals can't easily imitate.[8] The more intense the rivalry; the more will be spent on R&D, and the greater the urgency to bring the results to market. The growth of the facsimile transmission market illustrates how innovation drives market growth. The basic fax technology has existed since the early 1960s, although lack of speed and poor image quality precluded initial usage. The first machines were analog devices requiring four to six minutes to transmit a single page. It was the advent in the early 1980s of digital machines capable of higher resolution plus faster transmission speeds of 15 to 30 seconds per page that gave the fax a relative advantage over telex. These advances were accelerated by intense competition among 13 separate manufacturers in Japan. As competition shifted to making digital capabilities available at lower prices and more convenient formats, the fax machine became affordable for small business, home workers, and departments within large organizations. With more machines to communicate with the utility of those in place increased rapidly (what are called network externalities). As a result, fax sales and usage accelerated in 1987, with the number of fax machines installed in the United States reaching 2.5 million in 1989, up from 1 million in early 1988. The number of pages transmitted by fax grew at a compound annual rate of 37 percent in the early 1990s. This growth came at the expense of telex traffic, which decreased by 50 percent between 1984 and 1987 and has continued to decline.
- *Price*. The most important stimulus to growth is likely to be declining real prices relative to substitutes. The main reasons for the real price declines of technology-based products are: (1) experience effects, as a joint result of cumulative learning, economies of scale, and technological breakthroughs that result in productivity increases and cost declines, and (2) a persistent squeeze on the size of the margin between the prevailing prices and average total costs due to competitive

forces. The rate of decline in the relative price also has a direct impact on the expansion of market potential by increasing the number of new users who enter the market, and encouraging heavier usage among current users. To some extent, the rate of decline is also a self-fulfilling prophecy. As lower prices expand the market and stimulate sales, the faster increase in cumulated experience enables costs to be lowered, followed eventually by lower prices and the cycle continues.

- *Collective investments in education and access.* The acceptance of an innovation will be hampered if the target customers are not aware of it, do not fully understand the benefits, are not persuaded of its merits, or cannot find it. Investments in overcoming these barriers are critical in achieving the market's growth rate potential. The greater the levels of collective spending on advertising, personal selling, promotional support, and distribution coverage, the greater the impact on the perceived value of the product, which in turn accelerates market growth.

  This spending is better viewed as an investment with multiyear benefits. The purpose is to lead prospective customers through the stages of the adoption process: **awareness → knowledge → interest → evaluation → trial → adoption.** This is an education process, that is most effective with personal selling that enables a two-way interaction to identify needs and problems, and show how they can be overcome with tailored solutions. In the emergent stage of the market, individual firms make these investments both to grow the market and preempt other rivals. As growth accelerates and competition intensifies, the purpose shifts to gaining or sustaining an advantage and defend market share. However, it is the combined effect of all advertising messages, sales calls, and trade show programs that moves customers slowly or quickly through the response hierarchy. If expectations for the emerging technology are bright, then investments are heavy; conversely if expectations are modest, or confidence is lacking then collective investments are modest. In this respect, their collective behavior becomes a self-fulfilling prophecy.

Rate of Adoption

The speed of diffusion of an innovation into a market depends on the number of buyers who progress through the adoption process, when they start, and how quickly they make the decision to try. This was a crucial issue for strategists in the market for digital imaging technologies that enable images to be saved in a computer and sent over the Internet to be printed out by a digital mail box that uses ink-jet technology. Prospects for relative advantage depend on cutting costs (in 1998, a 4 × 6-inch silver halide print cost 8¢ versus 50¢ for a digital image) and improving quality (which meant increasing the number of pixels by a factor of 5 or 10 so images didn't look grainy). Meanwhile there were entrenched habits to overcome. Instead of dropping a roll of film off at a photo store, would consumers prefer to input them to a PC? Would they be willing to invest the time to manipulate photos in a computer? How valuable is the benefit of using the PC to store photos? Would they be willing to pay extra for a scanner and printer and put up with the headaches of hooking up the system? The answer was that some consumers would quickly see that the benefits outweigh the costs and inconveniences. Who are these early adopters and how can they be identified?

Prospective customers for a discontinuous innovation will self-select into segments based on degree of risk aversion and intensity of need. This leads to differences in time of adoption that can be represented as a bell-shaped curve when plotted over time. After a slow start, an increasing number of people adopt the innovation, this number reaches a peak, and then declines as fewer non-adopters remain, as illustrated in Figure 6.2. The adoption curve can be divided into segments, such that the early and late majority are one standard deviation away from the mean, while early adopters and laggards are at least two standard deviations away.

These five segments have distinct identities, behaviors, and requirements,[9] demanding different strategies:

1. *Innovators = technology enthusiasts.* These people are committed to the possibility that any new technology in their area of interest has promise, and are willing to take the time to master it. They are often "lead users" who have needs in advance of the rest of the market. They not only help to prove the new product but their endorsement

Figure 6.2
Adoption Curve

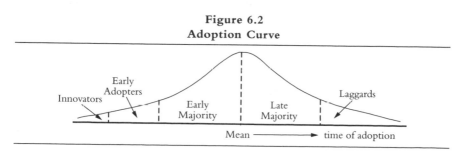

is key to acceptance by the other segments. Later in the chapter we will describe how to study this influential segment.

2. *Early adopters = visionaries.* These adopters see the opportunity presented by the new capability to change the rules of competition in their market. They help to publicize the new technology, but are costly to support because they require special adaptation to their requirements. Often these visionaries are in specialized niches, such as the businesses that are attracted to hybrid digital camera/cell phones that can take pictures in the field and instantly send them to a remote image printer.

3. *Early majority = pragmatists.* This is a large group that decides to adopt only when the benefits of the technology are well proven, and the risks are tolerable. They typically buy from the leading firm because in a technology market these vendors usually have the most reliable configuration and attract the largest number of third-party companies into the aftermarket.

4. *Late majority = conservatives.* This segment adopts an innovation only after a majority of people have tried it. They tend to be price-sensitive, skeptical of their ability to derive any value from the innovation, and very demanding. They have high needs for service support and assurance, but are usually not willing to pay much to have their demands met, which reinforces their doubts.

5. *Laggards = tradition bound.* These people are suspicious of changes, and are likely to adopt the innovation only when they have no choice or it takes on a measure of tradition itself.

The immediate implication of this model is that markets for discontinuous innovations should be developed by proceeding from one segment profile to the next. Once the visionaries are interested, make sure they are satisfied so they will be good references for the much larger group of pragmatists. At this point, the strategy shifts to trying to become market leader and set the de facto standard.

The compelling logic of this sequential market development strategy may be seriously flawed, however, because the early adopters often have almost nothing in common with the pragmatic early majority. Whereas the visionaries were risk-takers, intuitive in approach and motivated by future opportunities, the pragmatists are just that—analytic, more evolutionary than revolutionary, and motivated by solving present problems.[10] While the visionaries accept the bare product itself to get the superior performance or new functions, the pragmatists won't adopt until they have a complete product that meets all their requirements. Thus, to make the transition, high technology firms found they had to target specific segments within the mainstream market, and develop a fully augmented offering, rather than trying to diffuse their resources across many different end-use segments.

CONTINUOUS EXPLORATION OF MARKETS

Most successful discontinuous innovations follow a halting development path, marked by stop-and-go metamorphoses, before "emerging" from a series of market experiments with a feasible application. The trial-and-error learning that led to General Electric's digital X-ray, and the replacement of film with computerized imaging is typical. Basic research began in 1975 in the aerospace business. Sometimes the technology was aimed at industrial applications, and at other times, medical diagnostic imaging. After languishing in 1989, it was revived in 1993 when the Internet opened up the possibility of online medical consulting using digital images. This time the technology was ready and there was a strong champion to drive the project forward. The first machine was successfully shipped in 1996.

This iterative sequence has been termed "probe and learn" to denote a process of successive approximations and accumulating learning.[11] The path to market for fiber optics, cellular phones, and CT scanners was found to be guided by probes with immature versions of the product, learning from those probes and trying again in different market segments. This process has a great deal in common with the generic market sensing process that recycles as shown in Figure 6.3.

The process of market learning is typically sparked by an emerging problem or opportunity, a technological advance or a belief that further innovation requires deeper insights into latent needs. This begins the active collection and distribution of information from prospective customers about their problems and requirements, decision criteria and constraints, early reaction to experiences with prototypes in beta versions, as well as ongoing monitoring of secondary sources and competitive activity.

Framing the Inquiry

This critical step asks: What are we trying to learn about? What decisions have to be made and what alternatives should be considered? The market inquiry should be viewed as insurance against making bad decisions. It

Figure 6.3
Learning about Markets for Emerging Technologies

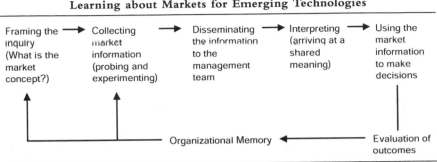

should not be done to satisfy curiosity or justify a decision that has already been made.

The inquiry needs to be especially alert to a variety of possible market concepts that precede the establishment of a dominant design. As late as 1994, there was considerable uncertainty about which concept for personal digital assistants (PDAs) would eventually prevail.[12] The possibilities included (1) palmtops configured like miniature PCs that could run PC software; (2) electronic organizers with a diary, address book, and calculator; (3) mobile phones with computer capabilities; or (4) pen-based computers without keyboards that could perform some of the above functions. Each was vying to become the industry standard. It was not until 1996 that the now ubiquitous Palm Pilot emerged as the early winner.

By contrast, once the dominant design has emerged and the market is already established, the market concept and product requirements can be quite tightly specified early in the development process. Indeed, this is one of the keys to successful product innovation in established high-technology markets.[13] A robust product definition that is well grounded in customer and user needs assessment is an essential guidance mechanism for the entire stage-gate process, enabling the development team to make trade-offs and design choices quickly.

Interpreting and Acting

Before the welter of conflicting, biased and incomplete information can be used, it has to be interpreted so patterns can be revealed and understood. These interpretations are guided by mental models that affect the information that is sought, selected, and simplified. Interpretations of market signals about nascent or emerging markets are especially difficult because the mental models of managers are incomplete and poorly structured, and prospective customers usually have difficulty envisioning the final version from their experience with the crude early version. Instead of relying on direct customer feedback, the interpretation must draw on contextual information about latent needs, persistent problems or trends in requirements. This is why a wide array of research approaches is needed.

The cumulative lessons are eventually lodged in the sprawling memory of the organization—perhaps to be retrieved when needed. Too often, however, there is collective amnesia about these lessons because of team turnover, inadequate repositories for the findings, or an unwillingness to treat interim failures as learning experiences.

This market-learning process can be subverted in many ways, which accounts for the wide variance across firms in their ability to learn about markets for emerging technologies, and anticipate when the time is right. Studies of the organizational impediments to a firm's ability to learn about markets find three persistent barriers.[14] When *acquiring information* there is a tendency to avoid ambiguity and presume greater familiarity with the market than is warranted. This means that user requirements do not matter as much as the "obvious" needs to improve performance. The enemy of *information dissemination* is compartmentalized thinking where each department or function focuses on its own goals, so information does not cross boundaries or is interpreted very differently be each group. *Usage of market information* is susceptible to inertia which means that the information will be used only when it conforms to prior expectations and market research methods and tools will be used only if they are deemed to be technically adequate. A related barrier to learning is skepticism about disconfirming information, which is subjected to much more criticism and scrutiny.

TRIANGULATION OF INSIGHTS: THE VALUE OF MULTIPLE METHODS

It has become conventional wisdom that methods such as concept tests, focus groups, surveys, conjoint analysis, and market simulation are inappropriate and even misleading when used in embryonic markets for disruptive/discontinuous innovations.[15] This is hardly surprising since these methods were designed to understand opportunities and strategies for incremental innovations in established markets. They are well suited to support formal stage-gate product development processes, where requirements are well-defined, but fall apart under the weight of uncertainty and proliferation of alternatives during the trial-and-error development of an emerging technology.

This does not mean that one cannot systematically learn about nascent markets with barely known requirements, applications, and attributes. Available methods have to be adapted and new approaches fashioned to accommodate endemic uncertainty. In deciding which methods to use, and how to use them two considerations must be weighed.

First, no single method will suffice, because all methods are flawed or limited in some important respect. Thus, analogies with markets for technologies with similar characteristics are suspect because the situations may not be comparable in critical but unknown respects. Similarly, surveys of experts using Delphi methods to assemble composite forecasts of demand may be no more than a pooling of collective ignorance. While a single method is limited, a combination of methods—each asking the same question in a different way and prone to different biases, with the various methods yielding conclusions that are directionally similar—deserves greater confidence. The process of triangulation of results looks for common themes and patterns after accounting for probable biases.

Second, it is a truism that prospective customers can't envision radically new products based on discontinuous innovations, and judge the early versions of the emerging technology from the standpoint of the refined versions of the established technology. However, they can be eloquent about their needs, problems, usage or application situations, and changing requirements that will dictate their eventual acceptances—but only if the right questions are asked.

This point seems to have escaped many commentators, judging from the following quotations:

> Customers are notoriously lacking in foresight. Ten or fifteen years ago, how many of us were asking for cellular telephones, fax machines, and copiers at home, MTV, 24 hour discount brokerages, cars with on-board navigation systems . . .[16]

> The familiar admonition to be customer-driven is of little value when it is not at all clear who the customer is—when the market has never experienced the features created by the new technology.[17]

By dismissing conventional methods that obtain direct feedback from prospective customers, they overlook rich possibilities for market insights from indirect methods of inquiry.

The story of Corning's early exploration of the most promising markets for fiber optics has been cited as an example of the shortcomings of conventional market research. A team of consultants proceeded to match the key attributes and benefits of optical fibers with potential market applications. The most promising candidate was judged to be local area networks (LANs) because they needed high capacity and cost was not a major stumbling block. This analysis pointed Corning *away* from the most significant opportunity—long distance telephone lines.[18] They could, however, have adopted an indirect approach similar to the one used by Xerox to get an early estimate of the market for fax machines (for details, see Chapter 2). Instead of getting responses to concept statements, they looked at the latent needs for the fax capability based on analysis of the frequency of customer needs to send messages. Corning could have learned more by undertaking detailed analyses of the future capacity requirements of long distance carriers to estimate the future demand.

While the centerpiece of an assessment of a future market for an emerging technology should be the results of probe-and-learn experiments, there are four specific methods that help to interpret and extrapolate these results. Lead user and latent needs analyses are especially useful while the market is still emerging and the product concept is still fluid. As the market moves toward take-off, more formal diffusion and information acceleration models are appropriate.

Learning from Lead Users

The guiding premise of lead user analysis is that some prospective consumers have pressing needs that may eventually be widespread in the market and face them ahead of the rest of the market.[19] Because they expect large benefits from finding a solution to those needs, they innovate on their own. These innovators and early adopters are often pioneers in their own markets or activities—such as developing biochips—but find their progress is being thwarted because they can't find processes, materials, or instrumentation that meets their novel requirements. In frustration they may try to solve their problems by making their own equipment.

The virtues of lead user analysis can be seen by contrasting it with established ways of identifying market trends and latent needs. Firms customarily go to users at the center of the market, using methods such as focus groups to get reactions to proposed concepts, site visits to observe users at work, queries to sales representatives in contact with customers, or customer evaluations of current products. The in-house development team uses these inputs to brainstorm their way to new ideas. By contrast, lead user analysis presumes that savvy users are already working on innovations in response to their pressing needs. The job of the development team is to find especially promising users and adapt their ideas to the business's needs.

There are three kinds of lead users to be found. Of immediate interest are those in the target application who have actually experimented with developing prototypes. Thus, an auto manufacturer looking for designs for innovative braking systems would talk to builders of race cars. Next are those in analogous markets with similar applications. A health care firm interested in antibacterial control products for humans might find a lead user in veterinary sciences. Third are lead users involved with important attributes of the general problem. Refrigerator makers could look at the supercomputer industry where cooling technologies are critical to the operation of the computers.

Lead users can be elusive. This is especially so when the emerging technology has many possible applications. This was the problem faced by developers of organic, light-emitting diodes that are light, bright, ultra thin and flexible, and easier to produce than most other types of flat screens for computers and television. Instead of conducting a survey of all prospective applications to uncover a few lead users, it is better to start with an underlying dimension these users desire. This would identify users who have needs for bright screens that are very lightweight, such as makers of jetliners seeking to reduce the weight of the current bulky ceiling light fixtures. Lead users are also highly dissatisfied with existing products and are actively searching for alternatives by participating in informal networks and user groups.

Because these lead users have early experience with the problems the emerging technology is trying to solve, they can provide rich and accurate feedback about needs, application requirements, and reactions to design concepts. They are also highly motivated to participate in beta tests, early market probes, and joint development activities because the pay-off is so great. They are also less likely to be deterred by the high initial prices of the early development versions. In short, they are leading indicators that are far more valuable than a random collection of prospects collected for a focus group.

Most lead user projects begin with a major trend that is changing the arena being explored. Thus, a team focusing on improvements in medical imaging was well aware of a trend toward the detection of ever smaller features such as early-stage tumors. They started by contacting experts in radiology to identify those working to solve the most challenging imaging problems. Their next step was to ask these lead users if there was anyone who was ahead in any aspect of the problem. These queries surfaced a separate community of specialists in the military who were enhancing the resolution of images with the aid of pattern recognition software. Eventually, specialists from these different areas were assembled for a three-day workshop to combine their technologies and experience to design product concepts that could meet the needs of the medical imaging company. During this process, the focus of the project shifted from the incremental improvement question, of how to create higher resolution images, to enhancing the radiologists ability to recognize medically significant patterns in images. This had profound strategic implications and led the firm to master some new software technologies that their rivals didn't understand.

Learning about Latent Needs

Sometimes the technology is at a point in its development when even lead users have not emerged or the most attractive markets may be different from those that first adopt the technology. Sometimes the technology can address needs that customers do not even know they have. How does one hear the unspoken voice of the market and identify these latent needs? One way to improve the targeting of market opportunities is to look for indirect evidence of market needs through an immersion into the customers' world. It takes a "prepared" mind to devise the right method for surfacing and understanding the latent needs that will be satisfied by the emerging technology. By defining latent needs as evident but not yet obvious, we are reminded that it will take energy, intuition, and informed judgments to extract useful lessons from the following methods.

Problem Identification. There is no better place to start than with the problems and frustrations customers have with the currently available solutions to their needs. The concept of relative advantage is based in the ability of the new approach to deliver more benefits at a lower cost. The attractiveness of biochips is their ability to address the doctor's difficulties in arriving at a diagnosis and then prescribing the right drug. The costs of the current trial-and-error approaches are highly visible; the most attractive diagnostic opportunities are those where the costs are greatest.

Problem detection approaches can be used throughout the development process to uncover barriers to acceptance of the emerging technology itself. For example, despite high early interest in solar heating systems, few systems were installed. As prospective customers learned more about total system costs, including maintenance over the lifetime of the system and the risks of fire damage due to system failure, their interest waned.

Story-Telling. Another kind of dialogue asks customers how they behave and how they truly feel. Kimberly-Clark listened over and over to stories from parents before they realized that parents viewed diapers as clothing that signals particular stages of development, not as waste-disposal fodder. Armed with this insight, they developed training pants that looked and fit like underwear, yet still kept accidents on the inside. Such finely detailed stories and case experiences help surface unanticipated purchase criteria. While these techniques have apparently not been applied to probing for underlying beliefs and motives about emerging technologies, there is no reason they couldn't be adapted.

Observation. The advantages of observation over direct inquiry are that it occurs in a natural setting and doesn't interrupt the usual flow of activity; second, that people give nonverbal cues of their feelings as well as spontaneous, unsolicited comments that are stimulated by an actual product or prototype; and third, trained observers with knowledge of technical possibilities can see solutions to unarticulated needs or problems which users could not conceive.[20] This is why firms like Sony and Sharp have set up "antennae shops" so they can watch prospective customers pick up and try to use their new products. The salespeople are trained to delve into the reasons for the observed reactions.

Anticipating Inflections

An inflection is a noticeable shift in the character of demand that presents opportunities to seize or lose advantage. An inflection point is encountered when the slope of a curve (mathematically, the second derivative) changes, as when it goes from concave to convex. In the ambiguous and uncertain market for an emerging technology, rewards come from anticipating these inflections ahead of everyone else.

The two inflection points that matter are (1) the take-off in market demand as the product starts to diffuse beyond the lead users and technology enthusiasts, and (2) the onset of aggressive competition aiming to capture the most attractive opportunities. Because market turning points are the product of the interplay between contending forces that facilitate or inhibit growth, there will be confusing signals and conflicting points of view. The anticipation of an inflection point is foremost a question of being able to read the pattern in the signs. This requires knowing which indicators to pay attention to, and an ability and willingness to separate the signals of transition from the background noise. This takes a combination of methodical guesswork, tracking of leading indicators, and diffusion modeling.

Methodical Guesswork. This in effect is what Forrester research did in 1999 to forecast a take-off in Internet advertising revenues from $550 million in 1997 to $33 billion worldwide by 2004.[21] They first charted ad spending per person in conventional media, and then compared each medium to the Internet. Key questions and assumptions were: Will Internet ad spending reach the level of newspapers? (No, because of the local presence of newspapers.) How does online advertising compare to TV advertising? (Inferior due to a lack of bandwidth.) But ad spending per person was expected to surpass that of magazines and radio. A key assumption was that more dollars would be drawn to the web because of new technologies that would improve the accountability of advertising in this medium. The total online ad-spending figure was multiplied by a forecast of Internet population growth, to reach the final numbers. The efficacy of this forecast procedure depends on having diverse sources of information and an intense dialogue and debate about major assumptions. What gets lost in the seemingly precise figure that emerges is any sense of the uncertainty surrounding the assumptions. It would be much more appropriate to provide a range of estimates.

Tracking Leading Indicators. Methodical guesswork needs to be complemented with careful tracking of early signals of the take-off in market demand as the product diffuses beyond the lead users and technology enthusiasts, and the onset of aggressive competition aiming to preempt attractive opportunities. Among the variables to watch are:

- The trajectory of the performance of the technology on key parameters compared to target customers expectations.

- Experience of lead users and other early adopters during market probes.
- Customers' perceptions of barriers to adoption and level of risk.
- Rate of competitive entry and collective investments in product availability and market access.
- Progress in building the infrastructure and resolving issues about standards, and complementary products.

Diffusion Modeling.   After a market for an emerging technology has taken off—and the technology is now emergent—the remaining uncertainty is over how long growth will continue. Will sales slow abruptly or continue to expand? These are questions well suited to a diffusion model known as the Bass model. The essence of this model is a forecast of the rate of adoption (or initial purchases) and a forecast of the inflection point when growth starts to slow. These two predictions are based on estimates of the eventual market potential, and two parameters corresponding to the propensity of buyers to innovate or imitate.[22]

The underlying structure of the diffusion model is shown in Figure 6.4. The number of new adopters per period is the same bell-shaped distribution we discussed earlier in the chapter. This distribution peaks at T* which corresponds to the point of inflection of the S-shaped curve that reflects cumulative adoption. Cumulative adoptions peak at a ceiling, which is the estimated potential demand.

In practice, this model is seldom used in the earliest stages of the emergence of a market but it is very useful for later stages. The major drawbacks are: (1) the model cannot be estimated without a few periods of actual sales data; (2) the forecasts are unstable when there is high uncertainty about the potential number of adopters, because the relative advantages over existing technologies (which are also improving) have not been established; (3) the model assumes there are no supply restrictions. If the infrastructure is not in place or the product is not readily available, then the excess unmet demand will presumably generate a waiting line of potential adopters. Here is where artful adaptation is needed. With some informed guesses about the rate of early sales and the eventual market potential, along with estimates of model parameters based on sales histories of analogous products a forecast can be derived that is useful for testing the feasibility of other forecasts using different approaches.

Information Acceleration.   This method literally "accelerates" potential consumers into an all-encompassing future environment. These consumers

Figure 6.4
Modeling the Diffusion Process

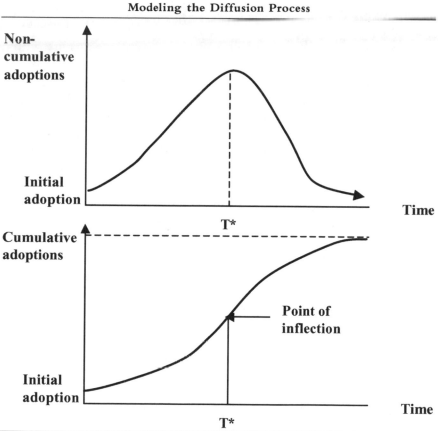

can experience an array of potential products and services that are simulated on an interactive multimedia workstation. They see and hear advertisements, product descriptions, simulated testimonials, sales presentations, and other communications. Once they are conditioned to the future, they are asked to choose among a variety of offerings at different price and performance points.

This method was applied in 1995 to assess the attractiveness of a wide array of multimedia products that were enabled by planned broadband networks that could send enormous amounts of digitalized information over fiber optic cables. Although the Internet had only penetrated 2 percent of households and broadband networks were just beginning to be built, there was already a consensus that (1) video-on-demand services would boom, (2) in the near-term, the Internet would not achieve penetration outside of sophisticated high-end users, and (3) video telephones would still not become a short- to medium-term opportunity.

To test these assumptions and identify the areas of opportunity within the broadband market, Mercer Consulting undertook an information acceleration study with 850 randomly chosen consumers.[23] These consumers were asked how they would use the new technology—as though they were in the richer, more "futuristic" environment of 2000 to 2005. They were introduced to services ranging from home banking to video-on-demand to time-shifted television and asked whether they would buy them at different price points or stick with the ones they currently had.

The results predicted that the conventional wisdom would be wrong and that investments based on these assumptions would be wrong. Although the study found there was demand for video-on-demand services, and that a consumer would pay up to $45 a month for them, this was not enough to justify an investment of as much as $1,600 per household to upgrade the telephone and cable networks to full two-way broadband capability. The research also predicted that consumer online services would be a $5 billion market by 1999 with 30 to 40 percent household penetration. Both predictions have been verified by events, which increases our confidence that consumers can evaluate emerging technologies when they are put into a realistic choice situation.

Conclusion: Informed Anticipation about Markets

The disruptive character of emerging technologies makes point forecasts and extrapolations of their eventual market prospects a futile exercise. It could even be counterproductive if misplaced precision leads to overconfidence and insensitivity to surprises. No single forecast can possibly absorb all the uncertainties about customer responses, competitive activity, and technological progress, or consider all the complex interactions, discontinuities, threshold effects, and other nonlinearities. The best that can be done in the early stages of the development of the technology is to demonstrate that the market is likely to be big enough to warrant a development project.

The proper emphasis of market assessment activities should be on learning from market probes and anticipating the critical inflection points ahead of the competition. However, tracking and probing are only the first steps in informed anticipation. The actual learning comes when the organization can make sense out of the data and resolve important uncertainties. The problem is that most managers find high levels of uncertainty so difficult to tolerate that they impose patterns where none exist. They may borrow or create seemingly logical rules of thumb to decide issues in the absence of discernable patterns. The result is that decisions are made on the slippery basis of unwarranted and untested assumptions about the market opportunity or the proper path of technological development. Two useful methods for combating premature closure are discovery-driven planning and scenario analysis[24] as discussed in Chapter 10. Both are designed to surface and challenge key assumptions and focus on the sources of uncertainty The need for validation of assumptions brings the learning process full circle, by directing specific market inquiries at the most critical areas of uncertainty.

CHAPTER 7

TECHNOLOGY STRATEGY IN LUMPY MARKET LANDSCAPES

IAN C. MacMILLAN
The Wharton School

RITA GUNTHER McGRATH
Columbia University Graduate School of Business

Which technologies should a company develop? At any given point in time, it may have the opportunity to push technology in a variety of different directions. Should R&D concentrate on advances that will reduce price or boost performance? The answer depends on what particular customer segments value. Since these markets are "lumpy," a relatively small push against one technology barrier can have a much greater impact than a larger movement of another barrier. This chapter presents a framework for examining these lumpy markets and identifying opportunities to deploy emerging technologies across lumpy market landscapes. Using this framework, the authors present three strategies for using new technology to shift the company's position the market. They also discuss approaches for identifying out-of-market applications for new technologies.

The leader of a company that produces laptop computers is faced with the challenge of setting the firm's technology strategy. Some customers will purchase a laptop computer that is more rugged rather than more portable, while others will select a laptop computer for greater portability over increased ruggedness. Should the company push for technology that will increase the portability of its machines or should it invest its resources in pushing the limits of ruggedness? Assuming the company has limited resources to devote to advancing its technology, which direction will give it the best results?

At any given time, a particular technology operates within a set of performance constraints or technology barriers.[1] These barriers define a technology envelope whose edges are determined by the current limits of the firm's (and sometimes of an industry's) capabilities.

The attribute set offered to customers is thus constrained by an envelope of technology barriers. This set of barriers is depicted two-dimensionally in Figure 7.1, a highly simplified illustration of two attributes and associated limits that are important for laptop computers. (In reality, the manager would have to consider many other dimensions but we have limited our discussion to two—portability and ruggedness—to simplify the presentation.)

For laptops, key technology barriers that limit portability and ruggedness might be the size of the components, the limits on energy storage, and the limits of the materials themselves:

- *Component size.* Even as the size of components necessary to make a working laptop has been steadily decreasing, the most state-of-the-art

Figure 7.1
The Technology Envelope for Laptop Computers

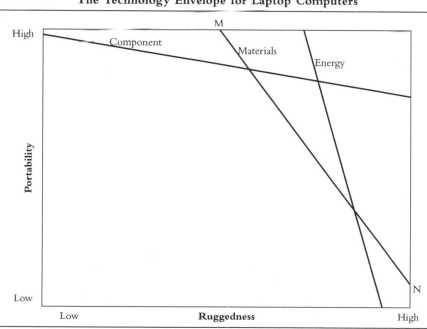

systems cannot be created using smaller components without compromising price or ruggedness to the point of rendering the system too expensive or fragile. Companies work to develop new component technology, including smaller chips, drives and modems to push this barrier.

- *Energy storage.* As customers consider battery life an important feature of a laptop, the size and weight of a battery limits how light and small the laptop may be. Companies work to develop new energy storage technology to push this barrier.

- *Materials technology.* Even the state-of-the-art plastics manufactured today have limits on ruggedness. Companies work to develop tougher plastics or apply new materials to push this barrier.

The reality that managers face is that every product or service is trapped within a complex multidimensional envelope of barriers. Possible configurations of attribute sets and their associated processes for production or delivery run up against a set of technological and other barriers faced by the industry. These barriers place limits on the extent the company can deliver attributes such as ruggedness or portability to customers. Particular firms may advance further on some barriers than others, so that different firms each find their attribute sets confined within not only an industry-specific but also a firm-specific technology envelope.

Given limited resources and constantly developing technology, it is impossible for any firm to maximize all the desired attributes of any product offering. Technology strategy comes down to choices about the allocation of resources aimed at pushing back selected barriers in the technology envelope. For instance, in the case illustrated by Figure 7.1, the firm may choose to focus on reducing component size, improving the weight and/or strength of the battery, developing stronger and lighter construction materials, or some combination.

The key question for the strategist is: What are the most important barriers in the envelope to push? Would the firm create more strategic value from developing and deploying an emerging technology or by pushing the barriers of a known, existing technology? This requires not only taking the technology into account, but also analyzing the market in which the technology will be introduced. We examine this question next, considering the preference structures of limited population markets.

THE "LUMPINESS" OF SEGMENTED MARKETS[2]

Technologies per se are not what a customer will pay for. No matter how exciting or innovative an emerging technology may be, from the customers' perspective its value stems only from creating an attribute set from which a customer actually derives satisfaction. It is these valued attributes, or *dimensions of merit*,[3] provided by the technology that are what customers respond to. Products and services offered by a firm can be characterized as translation devices. They are *transponders* that translate the technological capabilities of a firm into an *attribute set* that satisfies customer needs.

These valued attributes change over time. For example, as users become familiar with a set of features, technological developments improve those features and competitors offer alternative ways to satisfy the same needs. In the case of an emerging technology, the design challenge is to determine whether and where the technology can be deployed to offer more desired attributes or fewer undesired ones, while recognizing that the desirability of attributes will change as market needs shift.

Anticipating market preferences for emerging technologies is highly problematic, especially for firms facing high-velocity technological change. Studies show that traditional market research is less useful in this type of environment.[4]

A staple concept in the marketing literature is that customers, each making their own trade-offs among competing attribute sets, tend to cluster in segments around different attribute preferences. In evaluating the purchase of laptop computers for commercial use, for instance, customers make preference tradeoffs between attributes such as ruggedness, processor speed, weight, keyboard size, battery life, and price.

Such preferences create uneven concentrations of customers, each seeking a different attribute set. These submarkets for attributes tends to be "lumpy" rather than evenly distributed, with sometimes sharp distinctions between preferences in what might initially appear to be an evenly distributed market space.[5]

To illustrate market lumpiness, consider laptop users in a hypothetical industry in which competing firms install and service expensive building and plant equipment on the sites of their clients. In such an industry there might be three primary kinds of laptop users, each with very different needs: corporate executives, the salesforce, and the service technicians.

Assume that contending models of laptops are within acceptable ranges on most attributes for all three sets of users, with the exception of ruggedness and portability.

Corporate executives spend their travel time in transit lounges and on planes en route from one client headquarters to the next. Their key need is for a more conventional, small, light computer to handle basic word processing, spreadsheet duties, scheduling, communications, and other executive applications. Other things being equal, size and weight are the most important criteria for this group.

Salespeople, on the other hand, travel primarily by car and place less of a premium on portability. Ruggedness is much more important to them than to the corporate executives because they are always on the road and their equipment is more physically abused in multiple trips from car to site to plant and back.

For service technicians, ruggedness is a top priority because their laptops are used on-site to perform systems diagnostics, recall blueprints, obtain pricing information, and design drawings at sites that tend to be rough environments for computers. Hazards such as dust and dampness are common. Portability, in contrast, is relatively unimportant because the service technician sets the computer up on-site and leaves it there for the duration of the project.

For a given price, each segment would be prepared to make a very different set of tradeoffs of ruggedness and portability (Figure 7.2). At a specified price, members of the corporate executive group would be prepared to buy a very light and small but rather fragile model at point A' or a slightly heavier but sturdier model at point A" or any combination along curve AB. At the same price, the salesforce would only accept combinations on or above curve CD, reflecting the need for greater ruggedness and less concern about portability. Finally, service people would only accept combinations of attributes on or above curve EF, reflecting a high need for ruggedness and a relative lack of concern for portability.

Note that for a specified price, executives would not even consider many models that are absolutely required by the sales and service personnel. To maintain simplicity, we also assume that portability and ruggedness are the only dimensions that matter. As Figure 7.2 shows, at a specified price, each group of users will buy different combinations of features.

A key objective for the technology strategist is to develop technology that allows the continual improvement of the attribute set in ways that are profitable for the firm. Strategic success usually occurs when a given

Figure 7.2
Lumpy Market Segments for Laptop Computers

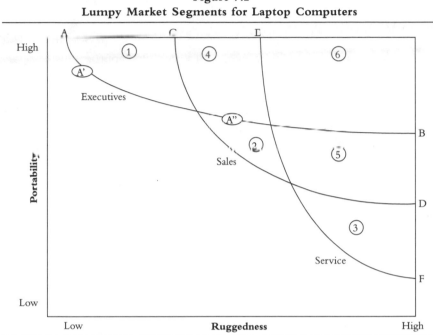

attribute set becomes the "most favored set" for a market, or segment of a market, thereby profitably capturing a controlling proportion of the transactions of that segment for that firm. (The most favored set is similar in concept to that of dominant design but emphasizes the fact that it is the set of attributes, positive and negative, that a given market selects, not the component technologies underlying the offering.)

The challenge for the strategist therefore is to decide ex ante whether the emerging technology might be developed to create a "most favored attribute set" that dominates the offerings of competitors. In a given segment, to do this requires an understanding of how the limits of the current technology impose constraints on the current attribute set. This assessment cannot be made without a rich understanding of both the technology barriers and the lumps in that market landscape. The answers are found in the interaction between the two.

PUSHING TECHNOLOGY BARRIERS IN LUMPY MARKETS

The manager responsible for setting technology strategy at our laptop firm might respond to this lumpiness in several ways. One approach is to decide to serve only a single target segment. With such a niche strategy, the firm would attempt to deeply understand the most desired attribute set for a target group of customers and then design products to precisely match this set. For instance, laptop computers priced at the reference price and positioned at points 1, 2, and 3 in Figure 7.2 offer a combination of weight and ruggedness attributes that fit what each target segment, but only that target segment, desires. A company might elect to design small, light computers with low levels of ruggedness for corporate executives, design bulkier but more durable machines for the salesforce, or offer even bulkier but highly durable laptops for the service group.

Focusing on a single segment can create substantial vulnerability for the niche player if a competitor can determine how to profitably develop a design to meet the requirements of more than one niche. For instance, a computer with attributes allowing it to be positioned at points 4, 5, or 6 could capture two or all three of the relevant segments. Since in this kind of market there are advantages to capturing the maximum number of customers, companies would ideally like to position their products and services to do so.

To see whether this is feasible, the manager needs to simultaneously consider the intersection between market segments and the technology barriers created by limits to current technology. Figure 7.3 offers a simplified illustration of how this might be done in the laptop computer case. We have kept the market landscape map, but overlaid the firm's current technology envelope from Figure 7.1, with its three technology barriers: component, energy and materials.

What becomes evident from this diagram is, given current barriers, some of the most attractive market opportunities may lie out of reach because of the intractability of current barriers to current technology. Using an emerging technology to move a currently intractable barrier can open up reconfiguration opportunities. Sometimes small shifts in a barrier can yield colossal benefits if they allow a firm to cross the appropriate barrier and access a new market segment that was blocked before.

Suppose that you are currently able to deliver design X in Figure 7.3. This design offers a highly portable light laptop computer with midlevel ruggedness. It appeals to executives (because of its portability) and to the salesforce

Figure 7.3
Overlay of Technology Barriers on Lumpy Markets

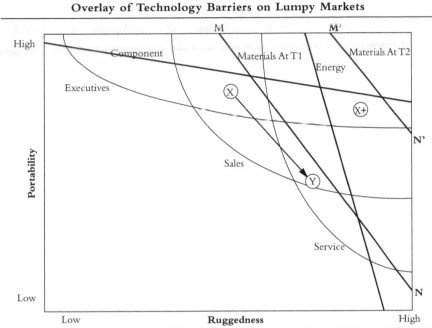

(because it is both portable and durable). The service force, however, would still regard it as insufficiently rugged. To access all three markets, you would need to improve ruggedness (in other words, to move design X to the right in the attribute space). The problem is that this cannot be done given the constraints imposed by the current state of materials technology. Additional ruggedness is feasible, by moving to design Y, for example. But this could be accomplished only by reducing the portability of the machine, but which would then lose the appeal to the executive portion of the market.

By imposing technological constraints on our hypothetical lumpy market, the investment choices become clearer. To begin with, there is a large area of technological development that, while technically feasible, will not result in combinations of attributes that will attract customers because they are below any isoprice line. Further, since the market landscape is lumpy, relatively small shifts in the attribute set offered can result in large changes in market access. Finally, the most tempting combination of market segments may be entirely out of reach of current technology.

Furthermore, there are some barriers that if disrupted will yield much better access to expanded lumpy markets than others. Clearly, a firm capable of currently delivering design X in Figure 7.3 would benefit most by moving the materials barrier MN to the right. This emphasizes the need for those involved with technology development to understand the lumpiness of markets in which the technologies eventually will be deployed.

This way of framing the technology/market problem also provides guidance on where the firm should *not* make investments in emerging technologies. A firm capable of offering a laptop with design X will be ill-advised to invest in emerging technologies that might improve the energy barrier, for instance, since this does little to provide the firm access to more attractive demand spaces.

If the firm offering design X could somehow deploy an emerging technology to move the materials barrier far enough to the right (say to M'N'), it might create a design (X+) that satisfies the needs of all three relevant niches, causing them to collapse into a single new lump. This process represents the tangible mechanism through which the Schumpeterian process of "waves of creative destruction."[6] When an emerging technology allows for new designs that can create new attribute sets, it becomes possible for firms to invade and dominate an entire attribute space or for new entrants to reconfigure the existing pattern of market demand.

This pattern of disruption and invasion has been observed in many industries.[7] Many "disruptive" technologies initially fail to meet the needs of the core market segments. They are introduced in niches below what existing competitors and major segments of the market consider to be attractive. Subsequent rates of performance improvement allow the attribute set associated with the emerging technology to rapidly move up, making it possible to invade and conquer previously inaccessible segments.

For example, an initially heavy laptop launched in the service market is made successively lighter and so can move into the executive market.

IDENTIFYING VALUABLE TECHNOLOGIES FOR LUMPY MARKETS

How is this framework applied to making decisions about investments in emerging technology? The following section walks through an approach for developing an understanding of market lumpiness and technology barriers. It then offers ways to use these insights to develop technology strategy.

Identifying opportunities calls first for a thorough understanding of the market landscape. To understand market lumps, strategists need to have insight into three sets of conditions.

1. Those attributes that meaningfully differentiate one offering from another.
2. How sets of attributes appeal to different market segments (including the size, purchase propensity and profitability of these segments).
3. How technology barriers influence the interaction between attributes and segments.

Identifying Market Lumpiness

We begin with a simple but powerful tool—the attribute matrix—to tease out the lumpiness of the market landscape prior to applying more sophisticated analysis. Attributes are organized in the matrix based on the nature of the customer reaction and the intensity of that reaction to the attribute sets. The nature of the customer response to a product attribute may be positive, negative, or neutral. We categorize the spectrum of response into three components:

1. *Basic.* When the market regards an attribute as basic, it is a taken-for-granted feature or component that all providers are expected to offer. Not offering it precludes the firm from the market, but offering it does not get the firm the order.
2. *Discriminators.* These distinguish among providers. These can be both positive and negative. The slow speed of Internet connections is an example of a negative discrimination attribute.[8]
3. *Energizing features.* These are discriminators that draw a sharp distinction between offerings. These are often new features and functionalities, such as auto-focusing in 35-mm cameras. Energizing attributes are features of the product that have a dramatic influence upon customers' propensity to purchase that product.

Companies can begin by examining the firm's current attributes to identify attributes that may be susceptible to reconfiguration using emerging technology. This is done first by conducting an internal assessment, using a team from within the company, ideally with rich representation from functions that come directly in contact with customers. This team is charged with developing a comprehensive list of attributes and categorizing them. Employees from different areas argue about how to categorize each attribute by customer type. Through this process, the company develops a separate attribute set for each customer segment.

After this internal discussion, we typically test the results with customer research. At its simplest, we ask customers in the segments to provide their own assessment of the importance of attributes. This is often an extremely enlightening step as firms discover that customers have very different views of the value of attributes than employees expect. Attributes that were thought to be positive actually turn out to be negative, such as airlines announcing on time departures and then circling the destination airport for up to an hour. Some companies take this a step further by interviewing competitors' customers. This often can yield considerable insight into attributes that are important to customers but are not currently offered by the firm.

Once the segments are identified, a technique such as conjoint analysis[9] can be used to clarify the attributes on which customers in each segment make decisions among competitive offerings. These attributes should reflect the customer needs satisfied for each segment. For instance, a baseball bat made of carbon-reinforced fiberglass is not purchased for *impact resistance;* it is purchased for durability—the bat lasts longer without cracking or chipping, and therefore needs less frequent replacement.

Once these matrices have been developed, it is then possible to begin to sketch out (at least in illustrative diagrams) the lumpiness of the market. The result would be a series of diagrams similar to those we examined for laptop computers. Since these can only deal with two dimensions at a time, managers may move to more sophisticated graphing or break the analysis into pairs of attributes and use a series of these diagrams.

Once managers understand the attributes that are valued in the present, the next challenge is to understand the attributes that will be most valued by segments in the future. Are laptop-toting executives becoming more concerned about connectivity to the Internet and voice mail? Using market forecasting techniques, managers need to identify the emerging differentiators and dissatisfiers, emerging or eroding needs, and population changes that affect the composition of the segments. These analyses provide the basis for making conjectures as to what future dimensions of merit might be emerging for each segment. Managers move from the current most-favored attribute set to the needs it satisfies to the trends in those needs and then back to estimating the changes in dimensions of merit that may be required in the future. Through this process, they begin to scope out potential directions for enhancing or augmenting their attribute sets.

Identifying Technology Constraints

Now that managers have a better understanding of the current and future attributes valued by customer segments, the next step is to identify the technology constraints that prevent the firm from delivering these attributes. These barriers represent the target of future technology development. The company works with technology specialists to tease out which current technology barriers might become obstructions in the future. Even though there are significant uncertainties involved in both the technology and markets, the backdrop is now in place for creating a "barrier register" to pinpoint specific technology barriers that could constrain the firm's ability to enhance or augment the attribute set along the hypothesized future dimensions of merit. With the help of the firm's technology forecasting specialists (or outside consultants), managers can identify places where the emerging technology could allow the firm to reconfigure the future attributes. As each opportunity to deploy the emerging technology to a segment is identified, add it to a register or list of opportunities.

This comprehensive homework provides a foundation for decisions about whether and how to explore an emerging technology. The more applications are identified for the technology, and the larger the markets affected, the more attractive it is to begin pursuing the emerging technology.

Some technology may come into an industry out of the blue. While these types of unexpected applications seem to defy careful analysis, they can be explored systematically using the framework presented above. Once the technology barrier that needs to be pushed is identified, managers can begin to scour the planet for potential technologies that might achieve this result. By looking outside the firm, they might be able to make the creative leaps to a new technology that had not been applied to the problem before. While many of these creative applications are the result of serendipity and inspiration, there is no reason they cannot be the result of the "99 percent perspiration" such as Thomas Edison used in identifying materials for the electric light bulb. (On the other hand, managers sometimes have a new technology in search of applications. This challenge and a process for dimensional search are discussed later in the chapter.)

Investing in Options

The static analysis of market lumpiness and technology barriers offers a useful framework, but it is far too simplistic. Virtually every element of the above process is uncertain. Customers' needs may change in ways that were not anticipated. Technology barriers may prove unexpectedly resistant (or unexpectedly flimsy). New barriers may become evident as development proceeds. Competitors are not standing still—particularly in highly dynamic environments, the entire landscape may shift rapidly as a consequence of competitor behavior. Technologies that are radically new to the market can appear and shake up the entire industry.

Given this level of uncertainty, even if the evidence is strong that the emerging technology will deliver future dimensions of merit, the search for ways to deploy the emerging technology needs to be guided by options reasoning.[10] As discussed by William Hamilton in Chapter 12, investments in emerging technology can be viewed as real options. Small investments at the outset give companies options to make more extensive investments in further development or commercialization in the future. The dynamic future value determined by real options frameworks is in contrast to the more static view of net present value and discounted cash flows used in traditional financial analysis. One thing remains the same: With real options, as in any investment, the company wants to maximize the value of its investments.

How can an understanding of lumpy market landscapes and technology barriers contribute to these decisions about investments in technology? The value of an option on a given technology is affected by the potential size of the market the technology might open. The more the technology can contribute to an attribute that differentiates the firm's product or service from that of rivals, allowing the company to dominate a particular segment, the greater the value of the option on that technology. By combining an analysis of the potential of the technology to change a given attribute with an assessment of the importance of that attribute in capturing a particular customer segment, the company gains a far clearer view of the potential value of that option.

Managers can approach this assessment by looking at the potential for *positioning* the firm in the market or by *scouting* with a potentially valuable technology:

- *Positioning options.* In some cases, the company will examine its market landscape to identify ways to push the technology barriers (as discussed in the laptop example), and then go out and seek technologies that will help achieve that positioning. It will invest in technology options that it expects to help improve its positioning. We examine several strategies used to create these positioning options based on the analysis of market lumpiness and technology barriers described above.

- *Scouting options.* In other cases, the company has a technology in search of a market. The company then looks for ways to apply this technology to create attributes that are of value to particular customer segments. We examine a process of dimensional search that can be used in this scouting process.

IDENTIFYING POSITIONING OPTIONS TO EXPLOIT LUMPY MARKETS

Given this picture of overlapping lumpy markets and technology barriers, there are number of ways companies can use technology development to move into new markets. These strategies range from capturing or consolidating a single niche to disrupting the entire industry structure. The choice of strategies depends upon how the market lumps and barriers are configured and the firm's level of ambitiousness and appetite for change. Three primary strategic alternatives are:

1. Single niche domination.
2. Niche fusion.
3. Creating a new technology envelope.

Single Niche Domination

The first strategy is "single niche domination," in which one firm moves a technology barrier in a way that allows it to offer a superior product in a specific niche. As shown in Figure 7.4, Firms X and Y are competing for a single niche with designs X and Y that lie on the same isoprice curve, so that customers are currently indifferent to the two designs. The design envelope within which both firms compete is bounded by barrier MN, which constrains the range of possible attributes either firm can introduce. Firm X makes investments in technology options with the potential of shifting barrier MN. If Firm X is able to shift this barrier in a way that Firm Y cannot copy, it creates the opportunity to move to a design such as X', which dominates Firm Y's product. Firm X captures the niche from Y because its design is now superior.

An example of this kind of strategy is seen in Citibank's pursuit of new sources of differentiation for its consumer credit card. The bank experimented with many new technologies, from smart card recognition and identification systems to intelligent links to automotive and shopping

Figure 7.4
Single Niche Domination Strategy

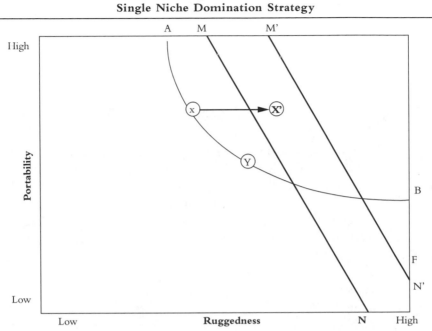

services, before finding the technology making it possible to personalize credit cards with a photograph of the cardholder. This allowed it to offer a powerfully enhanced attribute set for its target market by adding a fraud-prevention claim. As it turns out, the credit-card-with-photo also changed the nature of the landscape in other ways. It is commonplace for a drivers' license to serve as official identification for many purposes, such as admission to nightclubs, purchase of alcohol, and check cashing. Non-drivers thus were at a disadvantage. A photo-bearing credit card became accepted as adequate identification in many venues, creating an entirely new category of needs the card could fulfill.

A single niche strategy is appropriate when competitors for a fairly uniform and attractive segment cannot deliver improved attributes at a reasonable price because of technology barriers. The firm that moves first to remove these barriers can dominate the niche, and is generally able to extract significant price premiums as well. The kind of options that are appropriate here are limited-scope investments in technologies that have a direct connection either to enhancing the attractiveness of the attribute set or to reducing the cost and/or asset structures of the offering

Niche Fusion

A second way to deploy emerging technology is to pursue technology that will precipitate the fusion and domination of one or more segments through disrupting one or more technology barriers. As shown in Figure 7.5, Firm X with design X cannot compete against firm Y because it is trapped behind barrier MN, while Y for whatever reason is not. If by deploying an emerging technology, Firm X uncovers a way to move Barrier MN to M'N', allowing it to offer design X+ it will be able to dominate Firm Y in both segments. This strategy essentially fuses two or more niches together and allows the firm to dominate this larger "niche." Such a strategy often is suitable for technologies with a rate of performance improvement that is steep relative to alternatives.

Figure 7.5
Niche Fusion Strategy

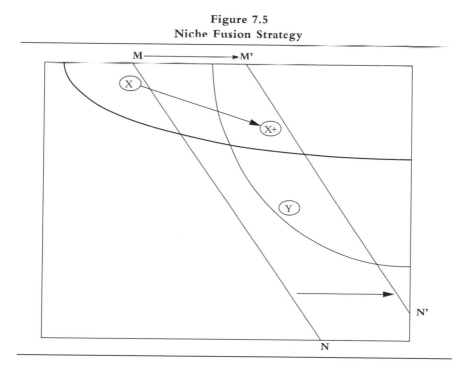

This strategy can be seen in the invasion of incumbent steel manufacturers' markets by steel minimills.[11] Initially, the state of the art of the existing technology precluded minimill manufacturers from entering markets requiring high quality finished products, and confined them to niches in which these attributes were unimportant. But the minimills continued to push technology barriers that allowed them to offer acceptable products to customers in increasingly demanding markets. This allowed them to finally square off against the integrated steel mills nearly across the board, including competing in their core, profitable markets for hot-rolled steel.

To seek opportunities to fuse several segments, the first step is to identify the dimensions of merit that differentiate the firm's current attribute set from those anticipated in each of the target segments under consideration for attack. The next step is to identify potential technology development projects that if successful will put the firm in a position of delivering a most favored attribute set that dominates all the segments under consideration. There will be fewer opportunities than for segment invasion strategies, and the costs and risks will be greater.

Creating a New Technology Envelope

The most challenging strategy for deploying emerging technologies occurs when a firm discovers the opportunity to radically change the configuration of attribute sets for virtually all existing configurations, typically by introducing an entirely new technology, as illustrated in Figure 7.6. This radical discontinuity, which completely changes the rules of the game and shifts the configuration of attributes desired by customers, often results in the demise of incumbent firms.

Such radical shifts in underlying technologies and the attributes they allow are becoming a fact of everyday life in an increasing number of industries. For instance, film-based photography is being challenged by rapid technological improvement in digital imaging, traditional means of powering automobiles are being challenged by a variety of electricity-based power sources and telecommunication through the Internet is putting conventional telephone network operators on edge.

As discussed by Day and Schoemaker in Chapter 2, incumbent players—especially those with considerable investments in fixed plant and capacity (or with considerable investment in established methods, such as long-lived service firms)—have difficulty adapting. They find themselves locked into a given technology envelope. When the boundaries of this envelope are

Figure 7.6
Revolutionary Strategy

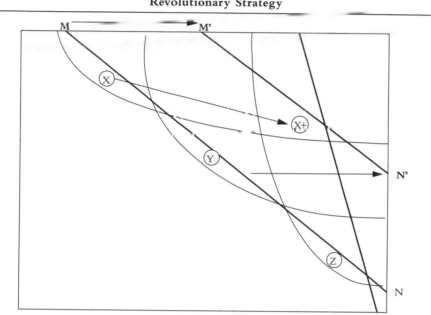

completely disrupted, segment demands may shift radically, leaving few or no customers in the previous technology space and creating a new lower performance bound for all (or most) segments. Established technologies are simply unable to offer the new functionality, despite the tendency of incumbents to invest in improving them.

In exploring opportunities for industry reconfiguration (or at least anticipating the danger of being reconfigured), the initial step is to identify whether the emerging technology may be capable of delivering attributes of the current technology (or similar ones) expected to dominate in the future. Identify places where the emerging technology could enhance or augment attributes, which could lead to major shifts in favor of the emerging technology.

The costs and risks of reconfiguration can be enormous, implying that no one firm may want to bear all the risk. Joint ventures, research consortia or investments in startups may be used as alternatives to going alone. If the emerging technology shows promise for delivering future dimensions of merit, identifying the fewest number of projects to pursue those dimensions of merit will allow firms to begin exploring the emerging technology as a vehicle for future competitiveness.

IDENTIFYING SCOUTING OPTIONS FOR APPLYING NEW TECHNOLOGIES

A similar, but reverse, analysis must be done by firms that have or are pursuing a new technology for its own sake and want to find valuable market applications. Some technology development projects find unexpected applications in other areas. DuPont's invention of Kevlar for instance, has fundamentally transformed materials technology by permitting the substitution of a polymer-based material for applications that previously could only use materials such as steel. However, DuPont had to explore a multitude of applications, in many different markets, before the final applications were determined. Or consider carbon-reinforced fiberglass, whose lightness, strength, impact resistance, and resistance to cosmic radiation were deemed attractive for satellite housings The final killer applications were in sports equipment—such as baseball bats, hockey sticks, bowling balls—where every attribute except resistance to cosmic radiation rendered the material attractive. Thus one of the key challenges for emerging technologies is identifying the killer applications as early as possible. The challenge is to see potential killer applications when both the technology and market applications are uncertain.

The key in this case is to look for potential attributes that can be created by the emerging technology and then identify market needs that can be served by those attributes. The tool we recommend for this is what we call "dimensional search." It works like this: The final outcome of an emerging technology is some type of ability to manipulate a property. For instance, the development of carbon-reinforced fiberglass resulted in a material that had properties such as a low specific gravity, high tensile strength, high impact resistance, low corrodability and high radiation resistance. The development of the Internet resulted in a communication network that had properties of high reach and high interpersonal connectivity, among others. Each of these properties can be measured (e.g., tensile strength in the case of fiberglass) or at least described (e.g., connectivity in the case of the Internet) along one or more dimensions.

If the company determines that the emerging technology has the potential to change the dimensions of a specific property, the challenge then is to find market segments in which that change is valuable. What segment would value an increase in the tensile strength of fiberglass? What segment would value the increased connectivity of the Internet? If we can find offerings or applications in which a certain dimension is important, and the emerging technology shows promise of changing this dimension, there is a possibility the new technology could be deployed to reconfigure the offering to make it more attractive.

How does the company find such products that could benefit from its technology? Those attributes that are truly important usually end up in performance specifications, standards or descriptions of the offering, and for the attributes to be specified they must be dimensioned (measured or at least described). This provides the basis for dimensional searching. First use a multifaceted panel to brainstorm and systematically identify and list all the dimensions that the emerging technology has the potential to change. Push the panel to identify measurement changes rather than descriptive changes, but use descriptive changes if necessary.

Then, create a search for all specifications, standards, patents or any other formal descriptions of product or service offerings. (One might set up a search-bot to scan Internet directories and other record depositories that list dimensions or descriptions, or search the U.S. patent database.) Search for mentions of all the dimensions that the emerging technology is expected to affect. Separately list all offerings where one, two, three or more dimensions are mentioned in the description. The offerings in these lists constitute the opportunity set for application of the technology. For a specific offering, the more dimensions that are flagged the more promising the emerging technology may be for that offering.

In particular, flag all cases in which the dimension is part of a standard or specification. In the attribute set, negatives are usually seen in terms of some maximum or minimum (it must be better than X, or it can't be worse than Y). If it is possible to deploy the emerging technology to remove or reduce negatives to create a new most favored attribute set, then it is a double win because it also puts the old offering at a disadvantage to the new offering. Also, positives are usually specified at some level above the best alternative—so if you can use emerging technology to exceed this level you may have a chance of using the emerging technology for building a new most favored set.

If you have a long list of potential applications and limited resources, you might ask multifunctional groups of technical and business people to help select the most promising applications. Identify customers in each of the potential applications and have them complete the attribute set as they see the current product they are purchasing without revealing the dimensions you may be seeking to change. From these sets, identify those attributes that would be changed if the emerging technology is successful. Interview potential customers to identify the extent to which the attribute in question would have to be enhanced for them to switch business, pay more, buy more, recommend more, or otherwise justify your developing the emerging technology to deliver an enhanced set. This should provide a good idea of how much of a challenge it would be to push back barriers with the emerging technology and provoke enough adoption to justify the investment.

Next, identify what the firm will have to do to deliver the other attributes in the current most favored attribute set, and determine whether these can be delivered, especially the non-negotiables. For all applications where customers indicate a willingness to support a product and the firm is confident it can deliver all attributes, add the applications to a register of "semi-validated" applications. Each product on this list of semi-validated applications becomes a potential scouting option, which if pursued can be used to more fully validate the application. The more scouting options identified, the more attractive it is to pursue the emerging technology for its extrinsic opportunity value.

CONCLUSIONS

We in no way suggest that the future applications for new technologies will ever be a predictable or orderly affair. Disrupting a technology barrier can be very costly and highly uncertain. Even if barriers are disrupted there is no guarantee that the market will embrace the result. There may be other attributes that are more important to customer segments. For example, despite a twofold jump in computing speed, Reduced Instruction Set Chips (RISC) failed to make much progress against Intel's Complex Instruction Set Computing Chip (CISC). The most significant barrier to adopting the RISC design is that software designed for the Intel chip will not work on the RISC chips. An insufficient number of computer buyers (other than Macintosh users) were willing to leave their existing software investments (and associated investments in training, customization and learning) to adopt the new technology. In this case, the attribute of compatibility was more important to most customer segments than raw performance of the chip.

Finding applications for new technologies is a messy process, but a significant part of this "messiness" is a result of the lumpiness of markets. By better understanding this lumpiness and the market impact of pushing technology barriers, managers can develop technology strategies that use their technology resources for the greatest impact.

This approach helps forge tighter connections between emerging technological potential and strategic market opportunities by perceiving products and services as attribute sets that translate technological capabilities into need satisfiers. We can, for example, begin to identify dimensions of merit that differentiate market segments on the basis of responsiveness to specific need sets and their associated attributes, creating a more fine-grained way to think about markets than conventional segmentation (for instance, by age, size, or location).

A second set of ideas introduced here is the notion that opportunities can be discovered through an options-driven, experimental approach. This emphasizes first understanding how market landscapes are configured, then identifying places where low-cost experimentation can be conducted to use emerging technology to push back barriers to open attractive opportunity spaces. Firms can take on emerging technology projects that either position the firm to take advantage of resolved uncertainties or to scout for and uncover market opportunities.

The concept of creating a technology register gives the strategist an additional framework for identifying which technology barriers inhibit access to specific segments along identifiable dimensions of merit. The firm can use the register to identify and undertake technology development that positions it to use an emerging technology to occupy that segment. If this technology development is successful, the firm can use the new technology to move into new market spaces using one of the three strategies we described: single niche domination, niche fusion, and new envelope creation.

In conclusion, we believe that the destiny of a firm is intimately linked to the evolution of attribute sets available to various customer segments. We suggest that decisions regarding deployment of an emerging technology cannot be undertaken without an intimate understanding of the relation between the technology trajectories and barriers and the opportunity spaces created by the lumpiness of the firm's market.

CHAPTER 8

COMMERCIALIZING EMERGING TECHNOLOGIES THROUGH COMPLEMENTARY ASSETS

MARY TRIPSAS

Harvard Business School

In commercializing new technologies, companies often focus on the technological challenges alone. But successful commercialization requires more than mastering the technology. Managers must also understand and build complementary assets, address new market needs and meet the challenges of new rivals. This chapter presents a framework for analyzing these commercialization challenges. The author discusses how these three dimensions allowed typesetting leader Mergenthaler Linotype to retain its market leadership across more than a century, despite three periods of radical technological change. The author then uses this framework to analyze how these issues shape the challenge of commercializing digital photography for established firms.

In 1886, a company called Mergenthaler Linotype introduced an automatic typesetter machine called the Linotype. This machine revolutionized the world of printing—representing the first major technological advance since Gutenberg invented moveable type in the 1400s. The word Linotype became synonymous with typesetting and the machine dominated the market for over 60 years. Even more extraordinary, in 1990, despite three waves of radical technological change and the entry of over 40 new competitors, Linotype-Hell, the successor to the original company, dominated the market for what were now called imagesetters—sophisticated color digital image processors. How was this company able to stay on top through these turbulent waves of technological change?[1]

As discussed by George Day and Paul Schoemaker in Chapter 2, incumbents often are at a disadvantage when new technologies emerge. In the transitions from vacuum tubes to semiconductors, electromechanical to electronic calculators, turboprop to jet engines, mechanical to quartz watches, and countless others, leading firms were devastated by the introduction of radically new technology.[2] What did Mergenthaler Linotype do differently? What can managers in established firms facing a radically new technology learn from Mergenthaler about how to successfully launch a new technology?

Incumbent firms such as Mergenthaler often have advantages that allow them to persist and lead through periods of technological change. Because of these advantages, even when they do not win the battle to create the first and best technology, they can still go on to win the competitive war to commercialize it successfully. For instance, if existing distribution channels remain important under the new technological regime, then an incumbent's distribution relationships can help it to survive.

Mergenthaler's success was not a result of having the first or best technology. As in many industries, some of the earliest or best technology was created by new entrants. For Mergenthaler, it was not technology itself but the complementary assets of its proprietary typefaces and its existing customer relationships that were a big part of its success. Companies that can identify and manage these assets and relationships have a much better chance of carrying their firms across these life-threatening technological chasms.

MERGENTHALER'S STORY: FROM HOT METAL TYPESETTERS TO LASER IMAGESETTERS

From its founding, Mergenthaler's Linotype architecture became the dominant standard for the typesetter industry and by 1903, the company controlled over half of industry sales. When key patents expired in 1911, four new companies entered the industry, but only one of them survived, and after that, no one dared to challenge Mergenthaler's dominance in hot metal typesetting until the machines ceased to be sold in the 1970s.

Ever since Gutenberg's creation of moveable type, printers had set type by hand. Mergenthaler's machine allowed the metal type to be put together into words and documents using a typewriter-like keyboard. An operator used the keyboard to send molds of letters into rows. The machine then made an impression of the entire "line of type" by injecting molten lead into the molds. These lead "slugs" were arranged in frames for letterpress printing.

The typesetter industry, which was based on this mechanical hot-metal technology, was subsequently disrupted by the advent of analog photo-typesetting in 1949, digital CRT phototypesetting in 1965, and laser im-agesetting in 1976. All three generations were competence destroying, making the technological skills and routines of incumbents obsolete.

The shift to analog phototypesetters changed both the key knowledge and technology of the industry and also opened up new markets for in-house publishing. Some 90 percent of the key skills of the typesetter de-velopment team were lost in the transition. Companies had to bring in new skills from lens designers, optical engineers, and electrical engineers to mas-ter the new phototypesetting technology.

The transition to analog phototypesetters attracted 17 new entrants to Mergenthaler's cozy market. Still, the three dominant incumbent hot metal firms (Mergenthaler, Intertype, and Monotype) were ahead of the curve in adopting the new technology. Incumbents brought out their first analog pho-totypesetting machines an average of 12 years before new entrants (1955 compared to 1967). With such a long head start, it would appear to have been difficult for entrants to steal the show. But the first phototypesetters de-veloped by hot metal incumbents had significantly inferior performance. They tried to adapt their existing routines and architectures to the new tech-nology. For example, when hot-metal firm Intertype launched the world's first phototypesetting machine in 1949, it did not operate any faster than machines based on the old technology. In fact, the basic architecture wasn't much different from the company's traditional typesetters.

While the incumbents were aware of the new technology and realized its potential, they had trouble breaking out of the mindset of the old tech-nology. One manager who left a hot-metal firm in frustration over the lack of interest in electronics later explained that he felt like "a one-eyed man in a blind kingdom." And one of the phototypesetting machines cre-ated by an incumbent hot-metal firm was described as "an odd amalgam of mechanics, fluidics, and electronics . . . with anachronistic technical con-cepts which ran counter to prevailing development trends."[3]

By 1961, Intertype's share of the phototypesetter market had plummeted from 100 percent to just 12 percent. New entrants created machines with significantly different architectures and far superior performance. Looking at one important dimension of performance—the average speed of the ma-chines in setting type—even the more refined products from incumbents performed far more slowly than the first machines of new entrants. In the shift to analog phototypesetters, incumbent's first machines processed an av-erage of 14 newspaper lines per minute while new entrant machines blazed through 41 lines in same time. Even the third and subsequent machines of incumbents only achieved an average speed of 26 lines while the second and later machines of new entrants were more than twice as fast. Despite initial experiments with the technology for decades before new entrants arrived, incumbents were behind in performance at the start and never caught up.

The same patterns are found in the next generations of technology. The first incumbent digital CRT phototypesetters could set 399 lines per minute, compared to 974 for the first machines of new entrants. In the move to laser imagesetting, incumbents' first machines only set 381 words per minute compared to 648 for entrants. But incumbents fared even better during these shifts. In each case, based on technology alone, the incumbents would not have appeared to stand a chance of success. How did they survive and flourish? The answer lies beyond the technology.

THREE CHALLENGES OF COMMERCIALIZATION

Emerging technologies do more than change the technological skills needed to succeed. They often change the relevant complementary assets, relevant competitors, and relevant customers. Unfortunately, established firms are often so focused on mastering the new technology itself that they fail to understand the implications of the technology for these other areas. Proactive management of these other areas, in addition to developing new technical capabilities, is crucial to survival. If these other factors are not changed radically, incumbents have a far better chance of succeeding across a technological shift by making ongoing investments in existing nontechnical assets. If these other factors do change significantly, the challenge is much greater. The incumbent is much more likely to fail unless it aggressively develops new complementary assets, understands new customers, and responds to new competitors. The technological change should thus be only one of the factors shaping overall commercialization strategy, as illustrated in Figure 8.1. The technological change should thus be only one of the factors shaping overall commercialization strategy, as illustrated in Figure 8.1.

Change in Complementary Assets

To benefit commercially from an innovation, a firm must possess additional assets that enable it to uniquely bring that innovation to market. These complementary assets[4] include resources such as access to distribution, service capability, customer relationships, supplier relationships, and

Figure 8.1
Forces Shaping Commercialization Strategy

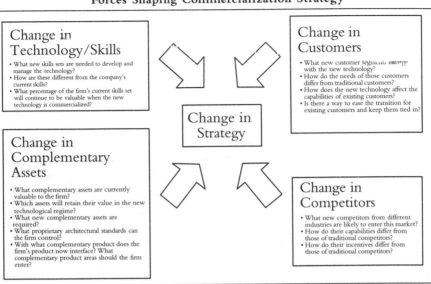

complementary products. When complementary assets are difficult for other firms to acquire or replicate, an innovating firm is much more likely to benefit commercially from its innovation.

Sometimes, even when the technology shifts, these complementary assets continue to give established firms an advantage over rivals. For example, the shift to CAT scans and MRI in medical diagnostic imaging were much less devastating for traditional X-ray competitors than might be expected because their existing market and customer linkages retained their value.[5] The sales and service relationships of the existing players were difficult for new entrants to replicate and provided a buffer for established firms.

Among the most powerful secrets of Mergenthaler's longevity were its proprietary typefaces. Font libraries affect the value of the firm's typesetter machine in much the same way the variety of software available for a computer affects its value to a buyer. The more fonts available, the better. Interviews with customers revealed that firms needed at least 500 typefaces to be considered viable and they were still at a disadvantage relative to companies with a larger library. By 1923, Mergenthaler had developed a font library with 2,000 faces. Despite large investments, it took Mergenthaler about a year to develop 100 new typefaces. At that rate, it would take an entrant more than 20 years to duplicate Mergenthaler's 1923 library and five years just to reach the minimum level of 500 fonts.

New entrants either had to license, design, or copy new typefaces. Established firms were unwilling to license their typefaces. Designing new faces was very time-consuming. Even copying typefaces was more involved than originally anticipated. The most successful second-generation entrant, Compugraphic, took 10 years and approximately $23.8 million to acquire 1,000 typefaces. Most of the similar typefaces were considered inferior to the originals and Mergenthaler had trademarked many of its names. The true "Helvetica" was preferred to some "cheap imitation." As one commercial typographer wrote to a new phototypesetting firm:

> Naturally, I understand that you have many typefaces available . . . My problem, however, is to perfectly match our present typestyles— all of which are Mergenthaler faces. We are in the middle of converting from hot to cold type and have many cautious customers who want the same quality that hot type produces. I'm sure you can see our predicament.[6]

Mergenthaler made explicit investments in maintaining its valuable complementary assets, in particular its proprietary font library. Established firms can also proactively influence what complementary assets will become important, particularly in the arena of architectural standards. When technologies are emerging, there are generally a number of competing models for implementing the technology or competing standards. Established firms can strategically influence the evolution of the industry by pushing their particular implementation or standard, as Microsoft and Intel have done in personal computing.[7] This standard is an asset that continues to give the firm an advantage in the market, even as the technology continues to evolve.

For instance, Adobe Systems' adept management of relationships with applications developers, printer and computer manufacturers, and font makers helped it succeed in establishing its Postscript page description language product as a standard. Ownership of the Postscript standard has served as a strong complementary asset for Adobe as the underlying technology continues to evolve.

Postscript software provides an interface between a computer program and a printer. When Adobe launched Postscript technology in 1985, the firm took a number of measures that helped it to become a standard.[8] First,

Adobe made major parts of the technology "open." Applications developers had free access to the Postscript language, and in fact Adobe helped them to write applications that were compatible with Postscript printers. In contrast to Xerox and its competing page description language Interpress, Adobe enabled multiple printer manufacturers to offer products that incorporated a Postscript interpreter. In fact, Adobe offered a boilerplate controller design to printer manufacturers to accelerate the development of Postscript printers. Adobe then collected a royalty fee for every Postscript printer sold.

Adobe also invested in developing its own Postscript-compatible products, in particular, an extensive Postscript font library, and sophisticated desktop publishing software such as Adobe Illustrator and Adobe Photoshop. These products served as complementary assets by helping to increase the demand for Postscript printers. Finally, Adobe formed a strategic alliance to offer a complete desktop publishing system when Postscript was first announced. Apple provided the Macintosh computer and the Laserwriter printer, Aldus provided Pagemaker software, and Linotype-Hell provided access to typeset quality fonts. In combination, these firms' products provided customers a complete desktop solution.

Interestingly, Adobe's adept use of complementary assets ultimately eroded the competitive position of Linotype-Hell and other typesetter makers. With the growth of desktop publishing and the availability of fonts in Postscript format, typesetter firms lost proprietary control over an important complementary asset, their font libraries. It then became much easier for other firms to introduce new technology, in the form of desktop publishing. After over 100 years of strength, Linotype-Hell's performance suffered in the mid-1990s, and the firm was acquired by Heidelberg Press, a German printing press company, in 1996.

Thus, while some new technologies may make the firm's complementary assets more valuable, others can make obsolete the existing complementary assets needed for success. Another example is found in the calculator industry. When calculators were electromechanical, an important complementary asset was a large, well-trained sales and service organization. Leading competitors in this industry had over 1,500 individuals dedicated to this activity.[9] With the shift to more reliable electronic calculators, however, these large service organizations were no longer required. In fact, office equipment distributors became a viable substitute for a salesforce/service organization. The value of the established firm's complementary asset all but disappeared and made it difficult for them to compete in the new generation of technology.

The task for established firms in evaluating a technological transition is to identify which complementary assets are currently important, and which will be important in the new technological regime. If certain assets—for instance a strong brand name—retain their value, then in addition to investing in technology, the firm needs to invest in those assets. For instance, it may need to invest in ongoing advertisement to maintain the strength of the brand. Alternatively, if new complementary assets are needed, then in addition to developing new technological capability, the firm needs to develop or acquire those new complementary assets.

Change in Customers

Emerging technologies are a revolution not only for the established firms seeking to commercialize them, but also for their customers. The next key issue in developing a commercialization strategy is to carefully examine the technology's impact on customers. New technologies often create new customer segments with new sets of needs. When evaluating a new technology, many established firms fail to invest in a new technological domain because it is only valuable to such a new, emerging market segment and not to the firm's existing customers.[10]

Although typesetter technology changed radically with analog photo typesetting, the customers purchasing typesetters—newspapers, typographers, and commercial printers—remained largely the same. Only one new segment emerged, the "office" or in-house publishing segment, but the needs of this segment and those of existing segments were not significantly different, despite the radical shift in technology. In this sense, Mergenthaler was lucky. But the emergence of the office market encouraged a number of rivals with connections to this market to give Mergenthaler a run for its money.

Effectively serving the needs of an emerging market segment is extremely difficult. Customer needs evolve along with the technology, and traditional market research methods are not effective. When customers don't have experience with a product or technology, it is difficult for them to articulate preferences. George Day discusses strategies for assessing these new needs and markets in Chapter 6.

Established firms also can ease the transition to the new technology for their customers. By understanding how a new technology affects the capabilities of their customers, companies can develop strategies to help their customers preserve their existing capabilities and build new ones. For instance, word processors required typists to develop an understanding of computers that they didn't need when using typewriters. A company commercializing a word processor can ease the transition for customers through training. Building bridges from the old technology to the new can also help companies preserve their market position. For instance, in selling to the workstation market, CISC chip firms that introduced upwardly compatible RISC chips outperformed those that didn't.[11]

Changes in Competition

The third factor that affects commercialization strategy is changes in competition. Emerging technologies reshape the competitive landscape. Start-up firms are likely to enter the market and diversifying firms also find attractive opportunities in the emerging technology. Each wave of technological change in the typesetting business attracted new competitors.

For firms accustomed to competing with a stable, predictable set of competitors, this new environment is particularly uncomfortable. These firms must resist the temptation to focus just on competitors that were relevant in the old technology and instead identify and track the activities of a much broader range of potential rivals. Beyond simply recognizing these competitors, managers need to understand what makes them tick. How do the capabilities and incentives of these new competitors differ from those of traditional competitors? What strategies does the firm need to change to compete against these new rivals?

One way to gain a deeper understanding of these new competitors is to become embedded in the emerging social networks of the new technological field. As discussed by Lori Rosenkopf in Chapter 15, informal interaction with competing firms through meetings, technical conferences, and trade shows, and even e-mail can give a firm a better understanding of its competitors' activities and strategic knowledge of the industry.

While each of the shifts in typesetting represented a significant technological change, they had varying effects in changing complementary assets, customers, and competitors, as summarized in Table 8.1. By looking beyond the technology at these other factors, we better understand incumbent performance across these generations of technology. While Mergenthaler and other hot metal incumbents lagged in technological performance across all three new generations of typesetter technology, they only suffered significant market share losses in the first shift from hot metal to analog phototypesetting. This shift resulted in significant changes to complementary assets and

Table 8.1
The Impact of Technological Changes on Mergenthaler's Business

Transition / Dimension	Hot Metal to Analog Phototypesetters	Analog to Digital CRT Phototypesetters	Digital CRT to Laser Imagesetters
Change in complementary assets	**Significant** • Specialized manufacturing capability no longer valuable • Extension of sales/service network needed for new office market segment. • Proprietary font library still valuable	**Not significant** • Sales/service network still valuable • Proprietary font library still valuable	**Not significant** • Sales/service network still valuable • Proprietary font library still valuable
Change in customers	**Moderate** • New customer segment: offices However office segment needs were the same as for existing customer segments	**Not significant** • Same customer segments with the same basic needs	**Moderate** • Same customer segments • Additional need: integration of text and graphics
Change in competitors	**Significant** • A large number of strong, capable new entrants	**Moderate** • A great deal of new entry, however, few strong new competitors	**Moderate** • A great deal of new entry, however, few strong new competitors

competitors as well as moderate changes in customers. In contrast, the subsequent two shifts resulted in substantially less change to nontechnological factors, and incumbent typesetter firms performed quite well in the market without major strategic changes.

APPLYING THE FRAMEWORK: DIGITAL IMAGING

Mergenthaler was fortunate that major technological shifts did not lead to significant shifts in complementary assets, customers, and competitors.

Table 8.2
Challenges for Photography Firms

Transition/ Dimension	Transition from Silver Halide Photography to Digital Imaging
Change in complementary assets	**Significant** • New distribution channels: computer stores vs. camera shops • Film manufacturing capability has less value • New complementary products, software are needed
Change in customers	**Significant** • New customers: Internet users • New uses/needs: e.g., electronic transmission of images; posting of images on web pages, editing of images, compatibility with computers/printers
Change in competitors	**Significant** • New, strong competitors from consumer electronics, computers/software, graphic arts as well as a number of startups • "Internet time" provides a different competitive dynamic

Established photography firms, however, may not be so lucky. How could an incumbent photographic firm use this framework to analyze the challenges of shifting from old chemical-based technology to emerging digital imaging technology?

The technological challenges are clear. Digital imaging requires fundamentally new technical knowledge on the part of traditional photography firms. While their technical understanding of lenses and optics is still valuable, they must develop new knowledge in semiconductors, electronics, and software and then translate that knowledge into competitive products.

The technological challenge is, however, modest in comparison to the commercial challenge facing photography firms, as summarized in Table 8.2. To succeed, established photography firms must rethink their strategies, business models, and basic identities. Incremental change will not suffice in this new world.

Analyzing each of these factors provides deeper insights into the competitive challenges facing incumbent firms than examining technology alone:

- *Complementary assets.* The complementary assets required to capture the value of digital cameras are quite different from those in traditional photography. While distribution through specialized camera shops mattered in traditional photography, digital cameras are often sold through computer stores as computer peripherals. Photography firms generally have no relationship and no pull with this new channel. The chemical-based film that accompanies a traditional camera provides a strong source of revenue for traditional photography players. With digital cameras, display of the images occurs through printing on film as well as printing with alternative technologies (e.g., ink jet) or not printing at all, but just electronically displaying the images. Finally, imaging software is an important complementary asset for a digital camera, but chemical photography firms have no particular expertise in this area.
- *Customers.* The relevant customer segments and customer needs also change. When computer users buy a digital camera to transmit and display images via the Internet, their purchase criteria differ significantly from traditional camera users. The limitations of a computer CRT make lower quality, lower resolution pictures more acceptable. Factors such as how fast the camera can download pictures and how efficiently those images are stored are new criteria that didn't matter before. In addition, as they use digital cameras, customers are discovering new uses and needs. Products therefore must be continually evolving at a much more rapid pace than traditional 35-mm cameras have historically evolved.
- *Competitors.* Finally, competition is fundamentally different. Rather than competing with a clearly defined set of camera and film makers, traditional photography firms are now competing with firms from diverse backgrounds. Firms from consumer electronics, graphic arts, computer hardware, and computer software all have some claim on the digital imaging space. It is extremely difficult for established photography firms to understand and evaluate such a wide range of competitors.

These competitors are vying for control of the architectural standards that will define the digital imaging landscape. Intel, for instance, has proposed a digital camera architecture that it is willing to license to any takers. Flashpoint, a start-up firm, has proposed a digital camera operating system that it hopes will serve as a platform for future innovation. The value of any given firm's technology will be substantially lower if someone else's standard wins out. This competition at a system and architectural level, as opposed to product-based competition, is also new for established photography firms.

Based on the above analysis, one can easily understand why, despite credible investments in digital imaging technology and development of award winning products, firms such as Polaroid and Kodak are not leading the digital imaging market. In fact, Sony, a consumer electronics firm has taken an early lead. Since the digital imaging market is still evolving, this race is far from over. This analysis would lead one to believe, however, that unless traditional photography firms develop strategies to address the challenges of complementary assets, customers, and rivals, they will continue to be at a significant disadvantage in turning their new technological capability into products and services that are successfully commercialized.

THE THREE HURDLES OF EMERGING TECHNOLOGIES

Developing emerging technologies can be envisioned as a race with three hurdles. The first hurdle is the decision of whether to invest in developing the new technology. The second hurdle is the organizational challenge of using that investment to effectively develop or acquire a new technological capability. Assuming the firm makes it through this step and actually has the technological capability to compete, it now faces the third and highest hurdle: the challenge of commercializing that technology.

Many established firms stop at the first hurdle. They are reluctant to invest in new technologies for fear of cannibalizing their existing sales.[12] Even if there is no threat of cannibalization, firms may feel that market for the new technology is too small or does not appeal to their existing customers, so they don't invest.[13]

Assuming they make it over this first jump, they then face the challenge of developing the new technology. In making this leap, established firms often are weighed down by the strong organizational routines and procedures that make them exceedingly efficient at developing products based on existing technology.[14] This strength in the context of the existing technology becomes a weakness in developing products using radically new technology. Therefore, despite their superior resources and experience in the old technology, initial products developed by established firms are often inferior to those developed by newcomers. For instance, during periods of significant innovation in photolithography, the products of established firms did not perform as well technically as those of new entrants.[15] This technical inferiority often translates into an inferior market position as firms approach the final hurdle, commercializing the technology.

The race is far from over, however, after the second hurdle, so firms with an inferior technical position still have the opportunity to perform quite well. Similarly, firms with a technological lead can still fail when it comes to bringing the product to market. The obvious implication is that even if an established firm does manage to develop strong, technologically competitive products, it can still lose the race. On the other hand, if it stumbles over the first two hurdles, as long as it is still in the race, it may still have a chance to win in commercialization, as Mergenthaler did.

A careful study of the Mergenthaler Linotype case shows that developing new technical capabilities through timely investments and appropriate organizational structures is an important first step in managing emerging technologies. But it is not the whole story of Mergenthaler's survival and success. It is often the manner in which the firm commercializes the technology that determines whether a firm succeeds or fails. Too often firms are so focused on the difficult challenge of developing radically new technical capability that they miss the broader picture. For instance, they assume that once technologically advanced products are developed, they can rely on existing strategies to bring them to market when, in fact, fundamentally different business models with different resource requirements are needed. Established firms must develop new complementary resources, address new market needs, and compete effectively against a new set of rivals.

The outcome of commercialization is not decided in the laboratory alone. Companies that successfully win the battle for technology development thus sometimes end up losing the war for commercialization. The framework presented earlier can help managers of incumbent firms take a broader perspective on the challenges of commercializing new technologies. Systematically examining relevant complementary assets, customers and competitors under old and new technological regimes, enables firms to develop consistent, integrated strategies for competing in the new technological environment.

PART III

MAKING STRATEGY

3M has more than 100 technologies that it uses to create more than 50,000 products, and "our lab people are constantly exploring—and creating—others." These inventions range from heart-lung machines to yellow Post-It notes. These technologies include 30 technology platforms that can be applied to multiple products or multiple markets. But Senior Vice President William E. Coyne notes that "last year's technology can quickly become this year's technology platform, and last year's platform might be superseded by new developments in our labs . . . our technology base ebbs and flows."[1] With 30 percent of its revenues coming from products introduced in the past four years, the company has to constantly redesign its strategy to develop a rapidly changing universe of technologies and decide which ones to move toward commercialization.

While most companies may not face challenges on the scale of 3M's, any firm developing emerging technologies faces a similar challenge of developing strategy in an environment that is complex, risky, and uncertain. Most firms are competent at competing in established technologies, where trajectories and strategies are well known and understood. In emerging technologies, the technological, strategic, and organizational factors are largely unknown.

PLANNING UNDER UNCERTAINTY

Formal strategic planning needs to be rethought in an environment of rapid change and high uncertainty characteristic of emerging technologies. Long cycles and the rigidity of formal planning can be a burden.

Companies that go on retreats every few years to come up with the "big thought" will often find they come down from the mountain with ideas that are already obsolete. Managers traditionally have moved between the extremes of discipline and imagination. In Chapter 9, Gabriel Szulanski and Kruti Amin offer insights into the strengths and weaknesses of discipline and imagination, and offer methods for combining the benefits of both in developing strategy.

Scenario planning offers a powerful tool in developing strategies to address disruptive new technologies. In Chapter 10, Paul Schoemaker and Michael Mavaddat examine the dilemma facing traditional newspaper companies under siege from the Internet. The authors describe their work with various newspapers and present four different scenarios about the future. The chapter also offers pointers on how to develop and conduct a productive scenario exercise in your own organization.

Even if companies successfully develop technologies and find markets for the resulting products and services—in other words, create value from the process—they still are not guaranteed that they can capture this value. Often rivals will move into the market with imitations or similar products. Other times, buyers or suppliers will appropriate the gains from the innovation. Protecting and capturing these gains is a key strategic issue, discussed by Sid Winter in Chapter 11. He points out that while much attention has been focused on patents and legal protections, these are just one approach—and often not the most effective way to appropriate gains from innovation. He also explores the use of secrecy, control of complementary assets, and lead times to hold onto the value created.

Strategy making in disruptive environments is always a dynamic process. Like mountain climbing, each new advance offers a fresh set of vistas and challenges. But as managers ascend these new peaks and make leaps into entirely new areas, these frameworks and perspectives can provide the ropes and anchors to guide this progress into the unknown.

CHAPTER 9

DISCIPLINED IMAGINATION:
STRATEGY MAKING IN
UNCERTAIN ENVIRONMENTS

GABRIEL SZULANSKI
with KRUTI AMIN
The Wharton School

The accelerating pace of emerging technologies is shrinking the window in which any given strategy, however well thought out, remains viable. For this reason, an elaborate but mechanical attempt to plan or a three-day retreat in the woods may no longer qualify as strategy making. Companies, big or small, new or established, are paying increasing attention to how fast and how well they are able to create new strategies and migrate to them. This is a capability that develops over time. Advice for strategy making has oscillated between a single-minded emphasis on discipline and an equally single-minded emphasis on imagination. Neither one alone, however, is as effective as both are together. The authors review the strengths and limitations of both discipline and imagination. They conclude by discussing ways in which companies can develop a capability for strategy making that combines both ingredients.

Geneial Instruments (GI), a manufacturer of set-top cable boxes, had a rude awakening. Protected by regulation, company engineers grew accustomed to setting the technical standards for their industry. GI's competitors—a handful of other hardware manufacturers—simply followed. When the cable TV industry began to deregulate, however, the competitive field started to widen. Suddenly, satellite, fiber-optic lines or software could render set-top cable boxes obsolete.

GI's strategic planning process, a thinly disguised yearly budgetary ritual, provided little help in coping with these changes. As the pace of change in the industry accelerated, the thick 100-page reports, which took months to prepare, were irrelevant and obsolete before they could be read These reports were thus mostly ignored. It was rumored that someone had placed a $100 bill inside a Phase One report at the beginning of the year only to find the bill at exactly the same place a year later. The gears of the strategic planning process continued to turn at GI, meaninglessly.

GI and other companies are finding that as the competitive environment accelerates, the window of opportunity in which a strategy, however well thought out, can remain effective, is steadily shrinking.[1] The market today constantly demands new solutions to new problems.[2] Elaborate but mechanical planning no longer works. It is becoming gradually and—for some—rather painfully apparent that strategy-making efforts that rely primarily on enforcing rigorous planning have precious little, if any, impact on an organization's ability to produce new wealth,[3] nor do the proverbial three-day retreats in the woods where, every five years or so, senior management is summoned to construct a "big thought."

Emerging technologies, with their associated high uncertainty and rapid change, create a particular need to adjust strategies quickly and to create new ones continually. Downes and Mui characterize this shift toward what they call "digital strategy" as a more democratic, intuitive, nonlinear mode of strategy making where strategies are reformulated every 12 to 18 months, instead of the customary three to five years, to respond quickly to technological threats.[4] Similarly, in a study of the fast-paced computer industry, Brown and Eisenhardt argue that organizations must compete "on the edge," where effective strategy making yields a "relentless flow of competitive advantages."[5]

The complexity and uncertainty of the environment overwhelms standard formulas for strategy making. A frantic search for ideas or responses often gives way to chaos, laissez-faire, and resignation. These, however, are not options. It is very dangerous to give up on discipline completely. Strategy making must not only be imaginative but also disciplined. A new kind of discipline is required, "disciplined imagination."

THE ART OF STRATEGY MAKING

Strategy making is an art, not a science. Mintzberg compares making strategy to creating pottery.[6] In his view, strategy, like pottery, is crafted.

Uncertainty is endemic in strategy formulation. Ex-ante a firm cannot know with certainty its relevant environment, the reach and extent of its capabilities, the lessons from its past, or what the future will bring. Thus, the quality of a strategy cannot be fully assessed until it is actually tried.

Strategy can be thought of as a theory of success that has not yet been tested. Peter Drucker claims that to be successful, every organization must work out such a theory of the business, which he defines as a set of assumptions about markets, technology, and its dynamics, about a company's strengths and weaknesses, and about what the company gets paid for.[7] The strategist cannot know whether such a strategy would be successful, but can only guess.

In this respect, the problem of the strategist is similar to the problem of the theory builder in that neither a theory nor a strategy can be known to work until it is actually tried and tested. Like theorists, strategists must choose one among many competing strategies for the actual test. Strategists cannot validate all possible strategies by implementing each one of them, so they must resort instead to other criteria for selecting strategies that will suggest, ex-ante, whether the chosen strategy is likely to succeed.

Although no amount of planning can fully guarantee the success of a strategy, the odds of success can certainly be improved. Strategy making, thus, can be thought of as an organizational capability, where different approaches are generated and considered and where past successful approaches are just an option for the future among many. Lessons from the art of theorizing are particularly valuable for strategy making.

Seen in this light, the process of strategy making can be thought of as a set of mental experiments in the strategist's mind, whereby many strategies are created and one is chosen for implementation. In his path-breaking work, Karl Weick argues that although one can never guarantee the success of these mental experiments, one can increase the chances for their success through the use of *disciplined imagination*.[8] In brief, disciplined imagination means "deliberate diversity" in the formulation of the problem, the generation of alternatives, and the variety of rules that are used to evaluate those alternatives. The discipline is reflected in the degree of consistency by which those rules are applied to evaluate each one of these alternatives.

This line of reasoning suggests that the quality of the organization's ability to make strategy will ultimately impact the quality of the strategy and the results. Thus, the quality of the strategy process is based on the degree to which it exhibits disciplined imagination.

TWIN PILLARS OF STRATEGY

Both discipline and imagination are consistent themes running through several decades of strategy research and practice. Advice for strategy making appears to swing back and forth between discipline and imagination, causing some confusion among managers as to the "right" way to make strategy.

In the early 1960s, strategy formulation revered the imagination, the creative insight of the CEO. The leader was a brilliant visionary, able to intuitively form a coherent strategy for the future of the company by matching the strengths and weaknesses of the company to the threats and opportunities posed by the company's environment. The weakness of this approach was that it reflected only one perspective, so it failed to incorporate diverse strategic insights from across the organization. The strategy "process" occurred for the most part inside the mind of the CEO.

Soon, however, discipline asserted itself and the process of deciding the future of the enterprise was entrusted to the professional planner, following the example of firms such as General Electric—a pioneer of formal, centralized strategic planning. This process rode a wave of popularity throughout the 1970s when many companies developed specialized planning departments featuring elaborate systems to create strategy.

In the 1980s, a new kind of discipline emerged. Many strategic planning departments were disbanded. Corporations began to focus on operational improvements as the keys to success, using concepts such as Total Quality Management, re-engineering, and benchmarking. Some went as far as to claim that they didn't need strategy.[9]

In 1994, C.K. Prahalad and Gary Hamel carried forward the banner of imagination in their influential book *Competing for the Future*.[10] They pointed out that there is a limit to the extent to which improvements in operations can sustain growth. They advocated a more democratic strategy-making process that enhances corporate imagination by involving more people and focusing on creating the future.

Thus, with the benefit of hindsight, trends in strategy appear to alternate emphasis between discipline and imagination. Proponents of either approach advocate their prescription passionately. However, both approaches have value. The limitations of either discipline or imagination alone seem to suggest that both discipline and imagination are essential components of a high quality strategy-making effort. At any given time, one may be in the foreground while the other serves as background, however, neither can fully supplant the other.[11]

In the following sections, we discuss the meaning of discipline and imagination for strategy making and why the quality of a strategy-making process might hinge on disciplined imagination. We then define disciplined imagination in terms of what it means for the tasks of problem finding, alternative creation, and evaluation, and conclude by sketching a few practical considerations for sustaining, establishing, or re-establishing disciplined imagination.

Discipline

Discipline is the consistent application of rules to evaluate a set of alternatives. Discipline has been informed by the "rational actor model,"[12] which assumes that the problem to be solved and all of the alternatives available for its solution are known in advance. The actor or strategist makes a choice by collecting information, developing alternatives, and selecting the one that "maximizes value."[13]

Discipline is exemplified by consistency, which does not necessarily require a formal planning process and is feasible both in stable and in chaotic environments. While strategic planning is a disciplined approach, discipline can certainly be achieved through other means. For example, it is possible to identify patterns in the evolution of the market, and to compare these trends to similar markets (if they exist and information on them is available). Once these factors have been dealt with, the manager in an uncertain environment can move forward with a clear framework, without getting lost in the seeming chaos of constant change.[14] There can be discipline without a formal strategic planning process.

However, the strategic planning movement was the first broad effort to instill discipline in strategy making. Its rallying cry was to bring method to the strategy-making efforts of senior decision makers with a formal process that would allow them to consider issues consistently and systematically. Also, formal tools such as the planning cycle and the capital budgeting procedure were used to help organizations in cross-referencing and integrating decision making at different levels of the company.[15]

The advocates of discipline pointed to its many benefits. Discipline was linked to greater consensus among top managers (associated with higher levels of success).[16] It was seen as a way for managers to be more judicious in their decision making, to identify and correct "avoidable errors."[17] Discipline also contributed to a more exhaustive or inclusive process for making and integrating strategic decisions, which was associated with improved performance in stable environments.[18]

As this type of discipline fell out of favor in increasingly turbulent competitive environments, a new form of discipline began to assert itself. In the mid-1990s, McKinsey started a research program with the express intent of "bringing discipline to strategy."[19] They suggested that even at the highest levels of uncertainty, such as those faced by emerging technologies, there are systematic ways to evaluate the competitive environment that can lead to better strategy making. They argued that "the secret of devising successful strategies lies in ascertaining just how uncertain the environment really is, and tailoring strategy [making activities] to that degree of uncertainty."[20]

In a related article, three other McKinsey executives explicitly illustrated a method for coping with uncertainty in strategy making. They proposed four basic levels of uncertainty, and offered specific tools for each level. They admonished firms that allow themselves to be carried by the tide of chaos, relinquishing all discipline in their strategy making. The central message from the McKinsey program was an emphasis on the importance of maintaining focus and control throughout the process of strategy formulation.[21]

Imagination

There has been increasing recognition of the importance of imagination and the associated concepts of synthesis, vision, foresight, creativity, and intuition within the realm of strategy making. For example, studies have found that a deliberate effort to generate and evaluate more options in strategy making appears to be related to a higher chance of success.[22]

A strategy-making process is said to exhibit imagination when there is "deliberate diversity" in the way problems are defined and alternative solutions are generated and selected. This means that a large variety of distinct options are generated in response to each formulation of the problem.[23] This notion of deliberate diversity applies not only to the number of alternatives generated but also to how different each is from the other. Basically, alternatives must be varied and distinct rather than mere variations on the same theme. There also should be diversity in the statement of the problem at hand, by using multiple cognitive frames to examine and define the problem.[24] Finally, there must be deliberate diversity in the number of rules used to select among alternatives. The plausibility that an option will turn out to be superior increases with the number of different selection rules or screens that point out that alternative as the most plausible and desirable.

There have been many calls for imagination in the strategy literature. Russell Ackoff urges managers to be "future-oriented" to imagine the direction of the company and work backward from that future.[25] Because, it takes time to generate alternatives, Peter Williamson recommends that companies maintain a portfolio of options, which he describes as diversified "launching pads," to quickly adapt and change direction in response to rapid market change.[26] Likewise, there have been calls for adaptive thinking and a somewhat vague but persistent emphasis on creatively "thinking out-of-the-box."[27] Gary Hamel and C.K. Prahalad exhort explicitly for imagination in every aspect of the business.[28] They call for creating new rules, not just breaking existing ones.

Hamel argues that pursuing imagination requires that we abandon the "hierarchy of experience" in favor of a "hierarchy of imagination."[29] The strategy-making process should be led not by those with the most experience but by those with the greatest ability to envision the future. These new voices in the process can include young people at the bottom, newcomers, and satellite offices that haven't yet been indoctrinated in the company culture, who are more likely to see things with a fresh outlook. A person who ranks high in the hierarchy of experience qualifies as futurist only if the future mirrors the past.

PECO Energy, a $12 billion Fortune 500 utility company in Philadelphia, put Hamel's ideas to practice to stimulate imagination in the organization. Operating in a regulated environment, the company had relied on an essentially reactive planning process to cope with the whims of industry regulators, provide adequate service, and grow the company.

However, when the industry began to deregulate, competition became less predictable and therefore mastering new technologies and new markets for energy distribution became critical for long-term success. PECO launched Project Leapfrog, using a process called "Journey" to involve employees from all levels of the company. Together they explored a variety of approaches to leverage PECO's capabilities, some of which were radical departures from how PECO was doing business. Instead of strictly looking to the past for a vision of the future, they began to foster active imagination through inclusiveness in the strategy-making process. Most of the ideas that resulted from the project broke the mold of a regulated utility company. The alternatives considered by the Leapfrog teams included developing an electric car, forming a utility consulting company, or creating an energy theme park. In fact, an employee in the theme park team spent a significant portion of his personal time researching the economics of roller coasters.

For PECO and other companies, a process for encouraging creativity has been a critical part of the strategy-making process. This approach sometimes leads to wild and impractical ideas, but can also create powerful new strategies that break with the past and create enormous value.

LIMITATIONS OF THE PRESCRIPTIONS

Prescriptions for both discipline and imagination have limits that detract from their usefulness. Discipline, by sticking to tried-and-true paths, rarely generates original insights and creative alternatives. It overemphasizes analysis rather than synthesis and selection at the expense of generation of ideas. It also assumes that the past will be similar to the present, often a dangerous assumption when it comes to emerging technologies. On the other hand, imagination can lead to chaos, losing touch with reality, and undervaluing the past. It also can dilute creativity and slow the process. Alone, neither discipline nor imagination is enough.

Limitations of Discipline

Strategic planning illustrates the strengths of discipline as well as its limitations. A common problem for companies that rely on strategic planning processes is that the routine nature of the process tends to atrophy strategic thinking. This happens because strategic planning stimulates:

- *Analysis rather than synthesis.* With the regularity and predictability of each planning cycle, the company can fall into a rut of automatic thinking. Mintzberg goes further to say that strategic planning was never meant to foster strategic thinking: "So-called strategic planning must be recognized for what it is: a means, not to create strategy, but to program a strategy already created."[30] He calls this the "grand fallacy" of strategic planning: "Because analysis is not synthesis, strategic planning has never been strategy making . . . analysis cannot substitute for synthesis."[31]

- *Selection at the expense of generation.* As Nobel Laureate Herbert Simon points out, classical decision theory—which informs strategic planning—is useful to evaluate strategies already created, but has paid little attention to the framing of problems that these strategies are meant to solve or the generation of the different alternatives available.[32] Yet, alternatives are rarely given.[33] Where do these options come from?

- *Extrapolation from the past.* As scenario planning expert Paul Schoemaker argues, an extreme emphasis on discipline is inherently limited because it presumes that history can be extrapolated to novel situations, which, to the extent that they are truly novel, have no history by definition. Particularly in emerging technologies, extrapolating from the past could seriously mislead strategy makers.

- *Overconfidence in the power of analysis.* Analysis is predicated on the availability of information, but in reality, information is often too limited for meaningful analysis. (For example, the use of "profit pools"[34] to analyze opportunities, while appealing in theory, is often very hard to put into practice because of limited information.) In addition, the plan resulting from detailed analysis may not offer a compelling motivation to inspire employees to implement it. This is especially true for emerging technologies, in which the quality of information is often insufficient to make compelling forecasts that will mobilize the organization.

Given the high levels of uncertainty endemic in strategy making, strategic decisions frequently resist analytical treatment. In complex systems, it is futile to orchestrate strategy formulation and implementation beyond a certain level of detail. Complex systems are inherently indeterministic, and thus detailed formal plans, however carefully prepared, are likely to leave room for surprise and possibly disappointment. This is not due to lack of insight into the market, but rather because there is a dim link between cause and effect in complex systems. As complexity theorists claim, a complex system can be disturbed, but not determined. It is impossible to control or predict the future perfectly, and any method that relies solely on the ability to do so is likely to be ineffective.

Professor Spiros Makridakis, an authority on business forecasting, claims that when analysis fails ". . . intuition, gut feelings, and creativity become the important elements of success in planning, while the role of planning models is restricted to providing an analytical framework for formalizing the planning process."[35] The creation of a rich set of options that are the fodder for analytical processes hinges on imagination, something that no amount of discipline can provide.

Limitations of Imagination

Although imagination is a wellspring of creative strategies, it is not a universal panacea. Taken to an extreme, imagination may lead to:

- *Chaos.* Imagination often requires the broad inclusion of diverse players in the strategy process. As noted, Hamel advocates including more outsiders (new recruits, peripheral employees) in discussions, finding new ways of thinking (gaining a new perspective), and igniting more passion (involving more people, so they have a proprietary interest in the success of the strategy).[36] Yet, this kind of participatory process that encourages imagination creates new problems. If everyone participates in the process, then who makes the final decision? Unchecked imagination could create a sense of chaos in the organization.
- *Losing touch with reality.* The pull of new opportunities that may dominate the firm's future activities may distract it from attending to its present ones.[37] And it is in the current activities that value is created or destroyed.
- *Undervaluing the past.* Organizations that are always focused on reinventing themselves for the future may lose track of their own past. This can mean there is no foundation for learning and self-improvement or building on their strengths. Companies that forget their pasts (both their own experiences and those of others) are more likely to repeat past mistakes. Also, sometimes the greatest strategic gains are achieved by consciously examining and disproving past success stories.
- *Diluting individual creativity.* Many heads are not always better than one. Involving more people does not necessarily generate more creative suggestions. Many heads often reduce creativity by producing group-think. Also, new people are not necessarily better people for the task; if they bring a narrow view into the strategic conversation, they are far more likely to retard, rather than enhance, the process. In contrast, some well-performing companies have built their success on employees that embody a specific company culture, such as "Andersen

Androids." They have purposely suppressed genetic diversity and there's no evidence they would do better emphasizing it.[38]

- *Slowing the process.* In general, involving more people in the strategy process means taking more time, in meetings, memos, and sometimes in building consensus around a particular strategy. This involvement can slow down the process.[39] As discussed, the traditional disciplined processes can also be very slow, for different reasons. In emerging technologies, time is already a scarce resource. Opportunities are often lost when no agreement is reached on the best thing to do.[40]

Some companies rely too heavily on imagination in their strategy-making process. Many of these firms compete in fast-paced industries, and they feel that the creativity of visionaries is the only way to out-innovate competitors. Far from the orderly thinking of the disciplined organizations, these companies assume that since there is no way for them to precisely define the environment, or the future, then there is no reason to set any boundaries. However, unchecked imagination can have the same paralyzing effects as too much discipline.

DISCIPLINED IMAGINATION

As we have seen, the themes of discipline and imagination have been prominent in strategy making. The limitations of both suggest that the combination of disciplined imagination, rather than an extreme emphasis on either, could be a sensible way to capitalize on their strengths while avoiding their weaknesses.

Strategy making is an artistic process. Thus, one way to understand the ways discipline and imagination can be combined in the strategy process is by examining the interplay of discipline and imagination in the arts. Disciplined imagination is a hallmark for many forms of visual and performing art. For example, movie-maker George Lucas told to a television reporter, "Movies are a make-believe business that feeds on imagination. When you're creating them, you live in a fantasy world. Without some kind of discipline, however, it is too easy to get distracted." Likewise, commercial songwriters must satisfy their inner creative needs while simultaneously obeying some form of rhyme and rhythm.

Even in those areas of music where discipline is not readily apparent, such as in jazz improvisation, a distinction is often made between uncontrolled and controlled improvisation. Effective improvisation must be controlled. Zinn notes, "in evolving to controlled improvisation the musician acquires both knowledge and technical abilities, and along with them the ability to transform them into 'logical musical conclusions.' "[41]

In addition to parallels with the visual and performing arts, the artistic processes used in science and mathematics to construct new theories offer insight into the functioning of disciplined imagination. Scientific break-throughs certainly depend on creativity but they still must follow some sort of rules. Physicist Murray Gell-Man, the discoverer of quarks, once said:

> Any art that's worth the name has some kind of discipline associated with it. Some kind of rule—maybe it's not the rule of a sonnet, or a symphony, or a classical painting, but even the most liberated con-temporary art . . . has some kind of rule. And the object is to get across what you're trying to get across, while sticking to the rules.[42]

Disciplined Imagination in Strategy Making

Examples of disciplined imagination also can be found in strategy-making processes. Scenario planning expert Paul Schoemaker (see Chapter 10) ex-plains why he prefers such a balance in scenario planning by distinguishing among *intuitive, statistical,* and *heuristic* approaches to generating scenar-ios. He says:

> I like neither intuitive nor statistical. The first can be very imagina-tive but lacks discipline. They are, however, always very insightful, and I use them to get ideas but they are rarely precise enough. That can be seen, for example, by looking back at how few times science fiction writers actually got their predictions right. In contrast, the statistical method is too mechanical. It relies on clustering techniques that require little use of imagination. So I like a balanced technique I call heuristic. It has some structure to it, and it gets you going.[43]

In general, imagination generates diverse options and encourages cre-ative thinking. Discipline grounds the process in reality and ensures that these options are rigorously evaluated and systematically developed and implemented.

A major telecommunications company found the discipline of its well-oiled planning process derailed by deregulation. Employees reacted to the sudden need to respond to the multiple pulls of the opportunities and threats of a deregulated environment by generating and pursuing a large number of ideas. They found themselves unable to evaluate those ideas, however, because of the burden imposed by the intricate financial screens mandated by the formal planning process of the company, which had proven so valuable in times of stability.

Ironically, the very screens developed to bring order to the planning process fostered *less* discipline; managers abandoned orderly evaluation because they could not follow the intricate process and respond rapidly enough to the fast pace of change. When the pace of change outstripped the ability of this cumbersome process to keep up, the system broke down. The dam of the strategic planning process was overtopped by the flood of new strategic initiatives and too many half-baked ones were being implemented.

A small group called the Opportunities Discovery Department (ODD) spontaneously emerged from the grassroots of the organization and started to bring a new kind of discipline to this process, by applying scenario-planning techniques in division after division. Scenario planning techniques provided them with practical rules to evaluate alternatives, consistently. A member of the ODD described the situation as follows:

> Traditionally, [our company] was unable to drop an existing strategic option. We had to create a huge amount of paper, showing . . . many reasons why it wouldn't work. People over-analyzed each option, and it led to slow decision making. This method of strategy making was fine when [our industry] was regulated, but not in a fast-changing environment.

For every available strategic alternative, ODD did first some basic back-of-the-envelope calculations. In many cases, these rough estimates provided sufficient grounds for dropping an option, thus avoiding the need to fully implement the customary intricate financial metrics. It was an eye-opening experience for the departments they helped. Soon, upper-level managers in charge of Corporate Strategy began to notice ODD's impact on strategy formation, they noticed a new kind of discipline. This was a discipline that was less rigid than the old process and one that encouraged rather than stifled innovation.

In general, disciplined imagination combines a process for generating diverse options with another for evaluating them consistently:

- *Generate imaginative options.* The diversity in problem statement
  entails examining reality from a variety of perspectives, applying multiple frames. The statement of the problem should be broad enough to
  accommodate many alternatives yet keep the number manageable. To
  be meaningful, these alternatives must be distinct, as opposed to being
  variations on the same theme. An example of such a process was seen
  at PECO Energy, in which the employees tried to find new ways for
  PECO to respond to deregulation while remembering the organization's capabilities in utilities.
- *Evaluate the options consistently.* Discipline is introduced by consistently applying rules to select alternatives. A strategy is more likely to
  be plausible if it survives the filter of a large number of diverse rules.
  The discipline part of the process manifests itself in the degree of
  consistency by which these rules are applied to each alternative. For
  instance, a women's apparel retailer relied on a set of rules to "filter"
  new product-line introductions, as shown in Figure 9.1. The company went beyond fashion considerations looking also at how each
  option fits with their organizational capabilities, their brand reputation, their culture, and their financial benchmarks. Each option passed
  through the same filter to ensure consistency, with a final review of
  the process by the CEO.

Figure 9.1
Multiple Filters for Strategic Options

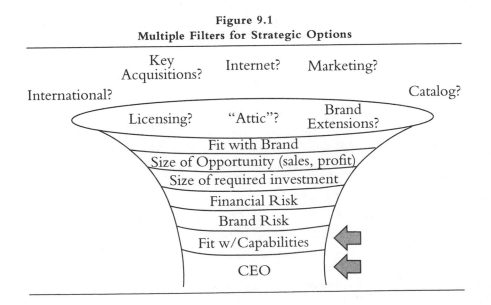

Practicing Disciplined Imagination

Strategy-making sophistication develops with practice. The hallmark of such sophistication is the ability to suit strategy making to the characteristics of specific situations. A study of 285 top managers found that firms that were capable of making strategy in multiple modes (such as centralized and participatory) in response to different situations outperformed those that used only one model or were less process-capable.[44] This finding supports Mintzberg's long-standing observation that there is "no one best way" to make strategy.[45] General Instruments, discussed at the opening of the chapter, was well served by its formal planning process when the cable business was regulated, but this was no longer true when the industry was deregulated. Discipline may suffice in some cases; in others, imaginative prowess may do. In some, however, discipline and imagination may both be required. When the environment accelerates, the organizational capability to make strategies is put to the test. The capability to deliver discipline and the capability to deliver imagination must both exist and be readily deployable.

There are a variety of ways to combine and organize these twin capabilities for discipline and imagination. Some companies develop and maintain these capabilities by alternating between periods of idea generation using imagination and periods of disciplined idea evaluation, just like PECO did during Project Leapfrog. During the initial phase, they explored divergent themes, which converged slowly into three central ones. Later, based on these themes, they generated specific ideas for the business—such as the energy theme park—and then screened them by applying a set of rules consistently, settling on those that could truly be implemented within the company.

Other companies preserve discipline and imagination in different parts of the organization and combine them when needed. For example, the Walt Disney Company cultivates imagination in the Walt Disney Imagineers (WDI); a think tank, not a profit center, responsible for maintaining the vitality of the corporation. Its "only job is to come up with ideas to build on—and then build on those ideas."[46] Among other things, Walt Disney Imagineers develop many of Walt Disney's theme park attractions, including everything from rides to restaurants. The Imagineers is a team of about 1,000 individuals from diverse backgrounds, including artists, engineers, architects, songwriters, and accountants, each member contributing his or her own expertise to imagine and make real Disney's imaginary world. Disney's corporate office, in many cases Michael Eisner personally, brings financial discipline by exploring and questioning the feasibility and commercial sense of an Imagineer's project. In this way, the company exploits the special skill sets of the different types of employees maintaining them more efficiently, thus implementing concurrently both discipline and imagination in the process.

The Complex Route to Disciplined Imagination

Discipline and imagination, however, are interdependent processes. Separation across time or the organization is not always possible or desirable. The best path to disciplined imagination is not always a straight line. As organizations with highly formal disciplined processes try to infuse them with imagination, they find often that their efforts have the same effect as throwing sand into the gears of an elaborate machine. Either the existing process kicks out the irritant, or the irritant stalls the machine. Organizations often find that they must take a more complex route to disciplined imagination. They must first dismount their present formal system, use imagination to create a new way of making strategy, and only then bring discipline back to the new process.

For example, when PECO began retooling its disciplined strategy-making process for deregulation, the company hired several consultants to infuse creativity into their existing process. These attempts had little impact. Hamel's suggestion, however, was to build a new process from scratch, seeking to create new space or new rules by encouraging the company to seek new lenses, challenge the orthodoxies of the industry and evaluate new trends. The purpose of the process was to energize the organization to find out its future and to stimulate collective imagination. The new ventures that were generated through the Leapfrog initiative were then carefully evaluated and selectively funded. PECO managed to inspire employees by creating strategy through an off-line process that exhibited disciplined imagination.

CONCLUSION

Perhaps few would question that both discipline and imagination are important ingredients for strategy making. Strategy making can be an art, but in situations of uncertainty, strategy making *must* be an art. It should not be surprising that discipline and imagination have surfaced as central themes in strategy-making advice. Disciplined imagination seems to be a common feature of the process by which artists, scientists, theorists, and strategists venture into the unknown.

The rapidly changing reality of emerging technologies strains established approaches to strategy. It exposes paralyzing excesses of discipline as well as chaotic creative efforts. Strategy making in fast-moving environments requires the development of an increasingly sophisticated organizational capability, of a different kind of discipline that allows for a greater involvement of the imagination. While organizations in fast-paced environments need to develop capabilities for both discipline and imagination, they may not always need to deploy both simultaneously at any given time. But those capabilities take time to develop and should be deployable when a particular challenge calls for that.

At first glance, like jazz music or abstract art, creative new approaches to strategy making may seem less disciplined, more chaotic than classical approaches. However, with time, new promising approaches grow their own kind of discipline. Like the potter, songwriter, scientist, jazz musician, mathematician, and movie producer, successful strategists in emerging technologies will find their way to develop and sustain disciplined imagination.

CHAPTER 10

Scenario Planning for Disruptive Technologies

PAUL J.H. SCHOEMAKER
The Wharton School

V. MICHAEL MAVADDAT
Decision Strategies International, Inc.

Planning for emerging technologies may seem like an oxymoron. Uncertainty and complexity undermine traditional planning approaches. How will a given technology develop? Will customers accept it? What if regulators or competitors change the environment? If you don't invest in this technology, will you be left behind? Scenario planning offers a framework designed to address complex and highly volatile environments by revealing and organizing the underlying uncertainties. This chapter describes the 10 steps involved in scenario construction and shows how this approach was used by major newspapers to address the emerging challenge of the Internet. The authors draw upon their extensive research and consulting experience in strategic planning to explore specific ways in which scenario planning techniques can be applied to the very turbulent environment of emerging technologies.

In the 1980s, videotext was hailed as the future of newspaper publishing. Knight-Ridder, along with Times Mirror, Dow Jones, and other major newspaper publishers, rushed to embrace this new technology that would deliver news to the television monitors via cable rather than traditional newsprint. But the technology proved to be disappointing and the market virtually nonexistent. After investing $50 million and nearly three years in its Viewtron system, Knight-Ridder pulled the plug in 1986. At the time, James Batten, then president of Knight-Ridder Newspapers, was asked by investors what the company had learned from the failed experiment. "What did we learn?" he replied. "Sometimes pioneers get rewarded with arrows in the back."[1]

Nearly a decade later, information technology had receded from the center of the issues facing new Knight-Ridder Chief Executive P. Anthony Ridder when he took charge of the $2.8 billion newspaper empire founded by his great-grandfather. In 1995, newspaper readership was declining nationwide, ad revenues were falling and moving toward less-profitable preprint inserts, and there was intense competition for news and advertising spending from radio, television, and other media. He was focused on cutting costs, so he abandoned a research lab working on a flat-panel display that would allow newspapers to be delivered electronically. He also cut loose the company's financial information business, ceding the territory to Dow Jones, Bloomberg, and others.

The cost cutting was necessary to boost margins, but where would the company's growth come from? At this point, the Internet was only a small blip on the radar screen. The company was working on plans to issue its papers electronically through America Online and the Internet, but still derived 85 percent of its revenues from its traditional print business of 31 major daily newspapers.[2]

About two years later, when we visited Knight-Ridder's Philadelphia Newspapers, Inc., electronic commerce had moved front and center. The most pressing strategic issues on the minds of its management team in 1997 were:

- To what extent will electronic commerce gain market share?
- To what extent will key advertisers experience profound changes in their industries and markets because of e-commerce?
- To what extent will the government regulate the Internet?
- Will advertising and news continue to be bundled and delivered to consumers in the form of a newspaper?

There were deep doubts. Was the Internet the future of publishing or the next "videotext" sinkhole? How might the emergence of the Internet and other related e-commerce technologies transform the newspaper industry's long-standing business model? How could the company move aggressively forward without taking another arrow in its back?

In a span of a few years, the mental landscape of a typical newspaper executive suddenly became more complicated. An industry that had been blessed with relatively incremental technological innovations since the time of Gutenberg (bigger, better, faster presses to put ink on paper) now faced a fundamental challenge to its business model. Managers knew they had to do something. But what? They had to place huge bets on an uncertain and rapidly changing future. They knew from firsthand experience how difficult it was to predict the future. It was clear the wrong bets could be disastrous.

On the other hand, they could not stand still. Established firms often become paralyzed in the face of discontinuous or radical change.[3] Standing still could mean watching its business be slowly eroded by fast-moving new rivals as the Internet rapidly advanced on the news business. This danger was amply demonstrated by technological shifts in other industries. Integrated circuits went from 20 percent to 80 percent of the electronic component market in six years; radial tires gained 50 market share points in 18 months; electromechanical cash registers went from 90 percent of the market to 10 percent from 1972 to 1976; and the Encyclopedia Britannica lost 50 percent of its revenue between 1990 and 1995 due to inroads made by CD ROM technology. No manager wishes to experience such a dramatic decline.

Newspapers outlived the rumors of their imminent demise with the rise of radio, TV, and magazines. But with the Internet, as Intel's Andy Grove put it succinctly "newspapers are under attack from both sides." Internet auction and transaction sites are attacking on the advertising front while online news services are attacking the editorial side. How should newspaper managers respond to these emerging technology issues? They found themselves thrust upon the horns of a profound dilemma. On the one hand, they didn't want to repeat the videotext failure. On the other, they didn't want to become the next Encyclopedia Britannica. How could managers think through the challenges of the Internet, let alone make the decisions about next steps? Which future should they prepare for?

An Uncertain Future

The question, "Which future should they prepare for?" cannot be answered. Moreover, it is probably the wrong question. It implies that managers can know what the future holds and then prepare for it. This is a fundamental assumption underlying linear strategic planning and forecasting. Figure out where the world is headed, or where you can take it, and then execute. There is only one little problem: The future is highly uncertain. As Winston Churchill once said of Russia, the future is for the most part "a riddle wrapped in a mystery inside an enigma." And when it comes to emerging technologies, that uncertainty goes off the chart. So the question should not be "which future" but rather "which set of multiple futures" might be likely and how can the company best prepare for all of them?

Sometimes managers think if they find experts in the technology, they can more accurately determine where the industry is headed. They attribute missteps to relying on the wrong technical advice. But most of the time, no one really knows exactly where the technology or the market is headed. Even the most seasoned experts on the technology have more often than not been dead wrong about what the future holds. Consider a few classic examples:

The phonograph . . . is not of any commercial value.

Thomas Alva Edison
Inventor of the phonograph, c. 1880

Heavier-than-air flying machines are impossible.

Lord Kelvin
*British mathematician, physicist, and
president of the British Royal Society, c. 1895*

I think there is a world market for about five computers.

Thomas J. Watson
Chairman of IBM, 1943

There is no reason for any individual to have a computer in their home.

Ken Olson
President of Digital Equipment Corporation, 1977

In hindsight, these massive mental blunders seem foolhardy, but these are not isolated instances. Rather than making definitive statements about an inherently uncertain future, these experts might have been better served by preparing and considering multiple scenarios for the future of these emerging technologies. The challenge is not to draw a detailed picture of a single future but rather to sketch a vision of many futures. Then, the company can develop the best strategy for preparing for this portfolio of futures and adjusting that strategy as the path becomes clearer.

Another reason why understanding the technology trajectory is not enough is that often known technologies are recombined to create radical innovations (see Chapters 1 and 2). Discontinuous or radical change comes about through the confluence of seemingly disparate and imperceptible changes in different disciplines such as economics, technology, politics, social sciences, and so on. Clunky initial prototypes become cheap and practical. As Alfred Marshall wrote in *Principles of Economics* (1890), "A new discovery is seldom fully effective for practical purposes till many minor improvements and subsidiary discoveries have gathered themselves around it."

The reason that most firms stand paralyzed in the face of technological change is usually a failure of imagination and an inability to make sense of weak signals that don't fit the traditional frames of mind. It also can be a failure of leadership, since senior executives should serve above all as the organization's primary sense makers. However, the mind can only see what it is prepared to see. Scenario planning helps prepare the corporate mind so that it will recognize opportunities faster than rivals, and can move more quickly, with more resolve. Good scenarios help overcome strategic myopia and frame blindness by forcing organizations to scan beyond the boundaries of their current core businesses, in a systematic and purposeful manner.

STOP THE PRESSES?

The daily newspaper offers a good test case for demonstrating the value of scenario planning. Its core business (that of delivering news and other timely information) should not be tied to any one technology per se, so newspapers have to consciously and carefully consider the impact of new technologies that are largely being shaped outside the firm or even industry. As an incumbent, Knight-Ridder faced a fundamentally different challenge in wrestling with new technology than the challenges facing an Internet or biotech start-up. These companies in many cases have no choice but to gamble on a commitment to a specific technology; they will have a great business if the technology works and may have no business if it fails. An established company such as Knight-Ridder, on the other hand, already has an existing successful business. It would prefer not to gamble the whole business on a single unproven technology (although this has occasionally been done by incumbents). The company can't stop its presses, but has to retool on the fly. It has to rebuild a corporate culture and mindset to transition from one world to another while still tending to its current business.

Every day, there are hard decisions to be made that are largely shaped by managers' mental models and key assumptions about their business and their industry. Circa 1993, Knight-Ridder's Philadelphia Newspapers, Inc. (publisher of *The Philadelphia Inquirer* and *The Daily News*) installed nine

state-of-the art printing presses in its facility just outside Philadelphia cost-
ing about $300 million. This investment was needed to stay competitive in
a stagnant market besieged by suburban papers, pre-print advertising, and
supermarket shoppers. This long-term investment decision itself was made
before the explosion of the World Wide Web was evident (i.e., pre-1993),
under the assumption that in decades to come newspapers would continue
to deliver news, advertising and relevant information to households on a
daily basis in printed form. The mental landscape of these managers lacked
a vivid image of the World Wide Web and its impact on the newspaper
business. Yet in hindsight, the Web has become the most significant threat
to newspapers as we know them. Will we still have newspapers 10 or 20
years from now? And if you were the publisher of a large urban newspaper, would you continue to invest in new printing presses, each of which
might cost you $30 million or so?

The Power of Scenario Planning

Scenario planning was designed to address the kind of complex, uncertain
challenges facing Knight-Ridder as it developed a strategy for the Internet.
It has been used to grapple with technological, political, demographic, and
other shifts in diverse markets. Royal Dutch/Shell has used scenarios since
the early 1970s as part of a process for generating and evaluating its strategic options.[4] In addition to energy scenarios, we have used scenario planning to examine how new technologies might change the rules of the game
in diverse industries such as healthcare, print publishing, consumer electronics, insurance, agriculture and food, financial services, engineering,
and higher education. Scenario planning has proved to be a powerful tool
to change people's thinking, as well as that of entire organizations.

Scenarios address three challenges that are inherent in emerging technologies and often befuddle other planning or strategy techniques:

1. *Uncertainty.* Unlike most other tools, scenario planning embraces
   uncertainty as the central element in its process. It could not do without uncertainties, which we distinguish from risks that can be quantified using objective probabilities.
2. *Complexity.* Scenarios focus on the confluence of a diverse set of
   forces—from social to technological to economic—and explores how
   they combine, intermingle, and dynamically influence each other over
   time as a complex system.

3. *Paradigm shift.* Scenarios aim to challenge the prevailing mindset, surface core assumptions, and create intellectual turmoil to see things anew by amplifying weak signals that would otherwise remain unnoticed among the pressing day-to-day issues of the firm.[5]

How does scenario planning differ from traditional planning? Suppose you are going to climb a mountain. Corporate planning of the past would provide you with a detailed map, describing the predetermined and constant elements of the terrain. This traditional planning tool is very valuable and, indeed, indispensable in this case. Just as geographical mapping is an honored art and science, so corporate mapping can be very useful. However, it is incomplete. First, the map is not the territory, but an incomplete and distorted representation (e.g., any two-dimensional map distorts the earth's surface). Second, it ignores the variable elements, such as weather, landslides, floods, animals, and other hikers. The most important of these uncertainties is probably the weather, and one option is to gather detailed meteorological data of past seasons, perhaps using computer simulations.

Scenario planning goes one step further. It simplifies the avalanche of data into a limited number of possible states or scenarios. Each scenario tells a story of how the various elements might interact under a variety of different assumptions. Ultimately, each scenario becomes a richly textured, internally consistent, plausible narrative description of a possible future. It is important to construct scenarios that are internally consistent and plausible. For example, high visibility and heavy snowdrifts are an implausible combination. Although the boundaries of scenarios might at times be fuzzy, a detailed and realistic narrative may direct your attention to aspects you would otherwise overlook. Thus, a vivid snowdrift scenario (with low visibility) may highlight the need for skin protection, goggles, food supplies, radio, shelter, on so on. Where relationships between elements can be formalized, quantitative models can be developed to further explore the implications of each scenario for the chosen strategies.

Scenario planning is different from planning methods such as contingency planning, sensitivity analysis and computer simulations. *Contingency planning* examines only one key uncertainty, such as, "What if we don't get the patent?" It presents a base case and an exception or contingency. Scenarios explore the joint impact of various key uncertainties, which stand side by side as equals. *Sensitivity analysis* examines the effect of a change in one variable, keeping all other variables constant. Moving one variable at a time makes sense locally, for small changes. For instance, we might ask what will happen to the demand for high definition televisions if prices drop by just a fraction of a percent, keeping everything else constant. However, if the price drop is much larger, other variables (such as programming, ancillary devices, consumers' behavior, etc.) will not stay constant. Scenarios change multiple variables at the same time, in quantum ways, without keeping others constant. They try to capture the new states that will develop after major shocks or deviations in key variables occur.

Last, scenarios are more than just the output of a complex *simulation model* where the focus is on computational complexity. Instead, scenario planners attempt to discover patterns and clusters among the millions of possible outcomes a computer simulation might generate. They often include elements that were not or cannot be formally modeled, such as new regulations, value shifts, or radical innovations. Hence, scenarios go beyond objective analyses, entailing subjective interpretations. In short, scenario planning attempts to capture the richness and range of possibilities, stimulating decision makers to consider changes they would otherwise ignore. At the same time, it organizes those possibilities into narratives that are easier to grasp and use than great volumes of data.

A special challenge of scenario planning is that the scenarios must challenge managerial beliefs. A scenario that merely confirms conventional wisdom is of little use. On the other hand, the scenarios have to be credible. A scenario that challenges deeply held beliefs, and therefore seems to have little chance of actual occurrence, might simply be dismissed as "off the wall." The dilemma is that the future is often "off the wall." Hence, a balance must be struck between what the future may really bring and what the organization is ready to contemplate or must consider for its survival. It's best to start with an intellectually honest and wide-ranging set of views and then rein in the scenarios—reluctantly and carefully—to accommodate an organization's legitimate political or emotional concerns. How do you know whether you're breaking out of the old paradigm? Watch people's reactions. Do they exhibit denial, confusion, discomfort, or outright anger? Do the scenarios stimulate vigorous debate and deep dialog? If so, you are probably challenging their fundamental beliefs in a healthy way.

In sum, scenario planning differs from many other planning techniques in its goal of a paradigm shift. It does so by painting concrete and vivid narratives of the future that hinge on key uncertainties whose outcomes will shape the future environment. The act of jointly developing stories about the future, in a disciplined but imaginative way, enhances both the learning about and the acceptance of these futures—in contrast to developing a bare-bones set of bullet points.[6] To see how scenarios are developed, let's first briefly review the steps involved in scenario construction in general and then later apply this to newspapers and the Internet.

CONSTRUCTING SCENARIOS

The basic steps in the scenario planning process can be characterized as follows:[7]

1. Define the issues you wish to understand better in terms of time frame, scope, and decision variables (e.g., prices of natural gas over the next five years in the Far East or the extent of inroads made by e-commerce). Make sure the scope of your scenarios is broader than the industry, product segments, customer groups, and technologies that currently define your business. Also, you may wish to review the past to get a better feel for degrees of uncertainty and volatility that your industry has already witnessed and use this information as a way of calibrating your time frame and scope. If much change has occurred in the past decade(s), don't develop tunnel-vision scenarios but inject that magnitude of change into your future projections.

2. Identify the major stakeholders or actors who would have an interest in these issues, both those who may be affected by it and those who could influence matters appreciably. Identify their current roles, interests and power positions. These stakeholders could be both internal or external. For example, in the case of environmental scenarios, judges, journalists and special interest groups have become much more powerful than a few decades ago. Consequently, their interests and points of influence should be understood.

3. Identify and study the main forces that are shaping the future within the scope of the issues established in step one. Your forces should cover the social, technological, economic, environmental, and political domains, and perhaps such subdomains as legal, medical, or scientific. The overall aim in this step is to gather relevant information about those forces that may change or perhaps shape your future. The process of identification and study is necessarily iterative. As more is learned about various forces, through reading, seminars or internal speakers series, new issues will emerge. At this

stage, the firm should try to consult with a diverse group of internal and external sources to identify the forces shaping the future, even those deemed to be too remote or weak today.

4. Identify trends or predetermined elements that will affect the issues of interest from the list of main forces. As described later, a survey of forces may be scored (when large numbers of people participate) or a workshop can be organized to identify those forces that are pre-ordained to happen. A simple example of such a force is the aging population in the United States, which is a statistical given, barring extreme shifts in immigration or an unusual illness that disproportionately kills the elderly. Briefly study each trend to understand at a deeper level how and why it will continue to exert an influence on the future. Constructing a diagram may be especially helpful to show inter-linkages and key causal relationships, thus avoiding the mistake of just extrapolating an existing trend that will in effect end or reverse itself. Remember, of course, that no trend lasts forever. By examining the deeper relationships among the trends, you can surface those fundamental drivers that affect multiple trends.

5. Identify key uncertainties (forces deemed important whose outcomes are not very predictable) from the list of main forces. Examples of key uncertainties could include the outcome of political elections, the granting of a patent, or the ultimate impact of the Internet on our society. The key uncertainties can be identified through surveys or simply by asking senior managers to select the top three external questions that they would like to know. Once key uncertainties have been surfaced, it is useful to project a range of possible outcomes for each uncertainty. For example, if one key uncertainty were to be future GNP growth, the future level could be specified as being anywhere between say—1 percent growth and say 6 percent growth. This range, which can be thought of as a subjective confidence range, would become wider the farther we look into the future. Once the key uncertainties have been defined, it is important to explain why these uncertain events matter most, and to what extent they interrelate. For example, a correlation matrix could be constructed showing the extent to which each uncertainty is correlated with every other key uncertainty. Such a matrix also allows for a consistency test about people's underlying beliefs, since certain correlation patterns are statistically probable.[8]

6. Select the two most important key uncertainties. One way to do this is to have senior managers vote for the top two from the full list of uncertainties. Once the top two have been selected, you can simply cross their postulated outcomes in a two-by-two matrix (as shown in Table 10.2). Each cell of this matrix will represent the nucleus of a possible scenario. To develop a particular cell into a full-fledged scenario, you must add suitable outcomes from other key uncertainties to that cell, resulting in a master blue print for that scenario, as shown in Table 10.3. Furthermore, to complete the scenario's blueprint, the trends and predetermined elements must be added to all the scenarios. The two-by-two technique is of course a heuristic approach to developing scenarios. Some may favor a less structured approach and rely more on intuition,[9] whereas others might prefer a more statistical approach using some multivariate technique such as cluster analysis. This latter method is appropriate if many key uncertainties are about equally important, such that no natural two-by-two matrix emerges. In that case, either intuition or a multivariate approach is called for.[10]

7. Assess the internal consistency and plausibility of the initial learning scenarios. It is important that the logic of each scenario is internally consistent. For example, most economists would deem an economic scenario that postulates full employment and zero or negative economic growth not to be internally consistent. You can test for internal consistency in at least three ways. For each scenario, ask:

- Are the main future trends all mutually consistent with each other?
- Can the outcomes postulated for the various key uncertainties all co-exist?
- Are the presumed actions of stakeholders compatible with their interests?

Eliminate combinations that are not credible or impossible, and create new scenarios (two or more) until you have achieved internal consistency. Make sure these new scenarios bracket a wide range of possible future outcomes.

8. Assess the revised scenarios in terms of how the key stakeholders might behave in them. Where appropriate, conduct role-playing exercises or consult outsiders. You may wish to share these learning scenarios with customers, suppliers, strategic partners, regulators,

consultants, academics, or others whose opinions you respect. The purpose of scenario planning is to create a framework within which deep dialog[11] can occur which in turn may bring about deeper strategic insight. The nature of scenarios should evolve as the strategic conversation within the organization evolves as well.[12] Based on discussions with various stakeholder groups, identify topics for further study that would provide stronger support for your scenarios, or might lead to revisions of these learning scenarios.

9. After completing additional research, reexamine the internal consistencies of the learning scenarios and assess whether some of the more complex interactions should be formalized via a quantitative model. At first, it may be wise to portray each scenario's basic logic via an influence diagram, which highlights, with circles and arrows, the basic cause and effect relationships that characterize a scenario's dynamic nature. Importantly, feedback loops should be acknowledged, since most scenarios are not just linear progressions of cause and effect, but rather represent more complex systems of interacting forces. To capture these interactions quantitatively, techniques from system dynamic modeling can be used.[13] The real purpose of such modeling efforts is not to produce predictive models (like econometric forecasting models) but rather to help surface the mental maps that underlie managers' perceptions of reality. In the act of surfacing these maps, two things may happen. First, managers will appreciate more deeply that their maps are not the best representation of the environment. Second, new maps may take root as the scenarios draw new connections among key constructs while weakening or perhaps severing old ones.

10. Finally, when all initial scenario work is said and done, it is important to reassess the uncertainty ranges of the main variables of interest, and express more quantitatively how each variable looks under different scenarios. Also, managers may want to retrace Steps 1 through 9 to see if anything should be changed. If not, they have arrived at decision scenarios that can be used to help make strategic choices. The final scenarios might be presented in booklet form or a set of slides, and given to others to enhance their decision making under uncertainty. In principle, scenarios can be used in a variety of ways, from helping challenge the managers' mental models and key assumptions about the industry to better risk analysis in specific projects. Above all, they should be used to test the robustness of the

existing strategies and to create better strategies to cope with the full range of scenarios. Each strategy entails a certain level of commitment (i.e., how far to stick out one's neck) and flexibility (i.e., being able to change mid-course in view of new information). Scenario planning can be used to calibrate both the nature and the extent of commitment a firm should make in pursuing a particular set of technologies, products, and markets.[14]

A Case Study: Newspapers Meet the Internet

To see how scenario planning is applied, we return to the challenge facing our newspaper managers as they examine the impact of the Internet. The following discussion examines the scenarios facing a large, U.S.-based newspaper company such as the *Miami Herald, LA Times, Philadelphia Inquirer,* or *Chicago Tribune.* These papers, which traditionally derived about 70 percent to 80 percent of their revenue from advertising, face perhaps a radical change to their business model. Managers want to protect their newspaper franchises, while investing in emerging technologies such as the Internet and e-commerce. To understand all the forces at work, we must look more broadly than just newspapers, and construct scenarios about the future of the entertainment and media industries, as well as the future of information gathering and transaction facilitation. These are four key areas in which newspapers play a role today.

Various technologies over the last 100 years or so have gradually eroded the near monopoly that newspapers once enjoyed in delivering news, information, and advertising to the public. The advent of radio and then television created new conduits for delivering information to households. The reduction in printing costs and postal rates has created numerous substitutes from pre-print advertising and direct mail to super-market flyers. And with the potential of Internet and e-commerce, the nature of information delivery into homes and businesses may undergo further radical changes. These changes have put at risk the newspaper industry's $15 billion plus in U.S. classified advertising revenues, its franchise as a reliable source of news and information, and its overall financial performance. Will we still have newspapers 10 years from now, assuming that we define a newspaper to be a mass-printed product, delivered daily to the home, while deriving the lion share of its revenue from advertising?

The major threat to the newspapers' business model is the emergence of Internet-based classifieds-type products and services being offered by the likes of CarPoint, HomeAdvisor, or Career 2000 as well as alternative electronic information providers such as Sidewalk and Digital Cities. Today, individuals with an Internet connection can get all the information they need to make decisions about purchasing cars, mortgages, clothing and electronic products, etc. when and where they want. In some cases, they can even conduct the transaction over the Internet (for example, purchasing a customized PC from Dell Computer Corporation). All this is possible without ever having to read a daily newspaper or other printed material. The widespread availability of high-speed, constant-connection, high-bandwidth Internet access that cable modems and other technologies will provide over the next five years will generate even greater consumer demand for this mode of transacting business and gathering information.

However, as radical as some of the emerging technologies in electronic business sound, there remain unresolved issues about people's ultimate acceptance of these new technologies, which in turn will greatly affect the speed at which they might proliferate. As a result, the newspaper industry is faced with a dilemma: on one hand newspapers need to embrace these new technologies to remain viable in the future, and on the other hand they need to protect their existing franchises in case these emerging technologies do not deliver on their promise in the near future.

Scenario planning offers an effective method for dealing with such uncertainties in a systematic and holistic way. We shall present here a stylized case description based on several consulting engagements with large and small newspaper companies. (To respect the confidentiality of the actual work conducted in each case, we present a collage of the various issues these newspaper companies faced and paint a generic profile of the organizational issues encountered.) We believe that this synthetic example captures the key challenges facing most U.S. newspaper companies circa 1997, and indeed many other organizations that have a well-established business model, high fixed costs with stranded assets, and a loyal customer following. In a sense, the newspaper case presents one of the more extreme examples of what kind of impact emerging technologies can have on an established firm.

Organizational Issues

Before we can begin the scenario development process with any organization, we must understand the organizational context in which the scenarios are to be developed and used. In our discussions with several newspapers, we found out that management was often unable to make critical decisions about the firm's future products, investments in new technologies, the type and caliber of future key employees, and the overall strategic direction of the firm. Uncertainties about the Internet further complicated endless internal debates that had paralyzed several newspaper companies over the last few years. In view of these cultural challenges, the scenarios had not only to address emerging technologies but also needed to help the organization overcome paralysis in the face of potentially serious threats to its existence. Typically, companies faced with disruptive technologies pursue the "ostrich policy" of sticking their heads in the sand to avoid the threats. In such cases, one needs to develop powerful scenarios that challenge the existing business model at its core, while being credible enough not to be dismissed as off-the-wall views or rationalized away as manageable in a business-as-usual mode.

Our first step had nothing to do with technology or even scenarios. It was to gain the political support and involvement of senior management early on, because without them it would be nearly impossible to break the various intellectual and ideological stalemates paralyzing the organization. The ideological debates concerned the very mission of a newspaper: profit versus community service. Especially in the newsroom, the traditional ethos was one of being society's watchdog at almost any cost. Investigative journalism ranks high on the value scale of editors and reporters, while profit was often viewed as a necessary evil. Although these ideological conflicts may seem somewhat unique to newspapers, which is the only industry mentioned in the U.S. Constitution (freedom of the press), we have encountered it in many other industries from healthcare to credit unions to educational institutions. Indeed, many firms have a reigning ideology that transcends profit making, and which defines its culture or occasionally its "religion." In such cases, scenarios must challenge the collective mental model of senior management. Otherwise, there can be no significant change in strategy at the end of the process. Hence, the key decision makers must become the primary stakeholders in the scenario development process.

We also made sure the scenarios would address the needs of the organization as a whole and not those of a select few, such as the champions who invited us in. We always try to refocus people's attention on the external world, asking the group to ignore for now differences of opinion about the firm but instead concentrate on understanding how the world is changing around them. Many strategy exercises start with the question "so, what do you want to be when you grow up." We think this is a dangerous question to start with, although it needs to be answered eventually. We favor an outside-in approach, where we start with external changes and then address the internal issues in a structured and disciplined way, such that intellectual honesty will prevail rather than politics and hidden agendas. Doing so, allows a shared sense of challenge to emerge around external issues, which often helps senior management move past unproductive internal debates. Had we failed to consider the organizational issues, we would have produced technically correct and interesting scenarios but ones that would perhaps not address the real needs of the organization.

Identifying Forces

Once we addressed the organizational issues, we were ready to construct a set of richly textured archetypal scenarios about the future of media, entertainment, information and transaction processing industries. We started by identifying relevant forces that would or might shape the future of these industries over the next 10 years. We focused on identifying fundamental drivers, rather than derivative issues, covering a broad range of social, political, economic, and technological forces. In identifying these drivers, we interviewed newspaper industry insiders, advertisers, readers, and new media experts. We gathered and developed a master list of 74 fundamental forces that would likely shape the future of the industry.

Once the master list of forces was completed, managers rated each force in terms of its importance and predictability by considering the following two questions:

- How important is a particular force (relative to all the other forces) in shaping the future of the target industry or issues of interest?
- How predictable is this force in terms of its overall direction and impact within the time frame considered (the next 10 years in this case)?

Those important forces deemed highly predictable were labeled *Trends* and those deemed unpredictable were labeled *Key Uncertainties*. Where appropriate, we combined two or more forces based on the similarity of their content and meaning. The aim was to reduce the set of forces into no more than about 10 key uncertainties and perhaps 15 or so trends. The survey results were depicted graphically as well as in a tabular form as shown in Figure 10.1 and Table 10.1.

Building the Scenarios

From this list of uncertainties, we then selected two central uncertainties: one about changes in the business model and another about how customers

Figure 10.1
Classifying Forces into Trends and Uncertainties

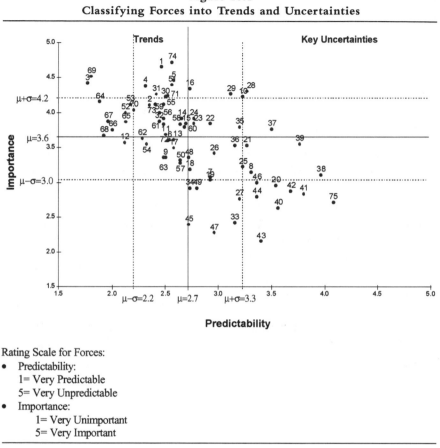

Rating Scale for Forces:
- Predictability:
  1= Very Predictable
  5= Very Unpredictable
- Importance:
    1= Very Unimportant
    5= Very Important

will use information (U_1 and U_7) to construct four newspaper industry scenarios, as shown in Table 10.2. The business model (U_1) could either continue to be based primarily on advertising revenue or it could move to a more radical model in which the sale of information is unbundled from the sale of advertising. And consumers (U_7) could either make minor changes in how and where they use information or make more radical shifts. The other uncertainties are not set aside, but are incorporated into the consideration of each scenario.

Each cell of the matrix becomes the nucleus for starting the scenario construction using a heuristic approach.[15] We also developed a catchy and memorable title for each of the four scenarios and placed them in the

Table 10.1
Key Uncertainties and Trends

U_1 (F10+F20). How will the future media companies make money: from selling content and advertising to transactions and commerce?

U_2 (F23+F24). How will the emergence of highly targeted, interactive and measurable media affect advertising strategies?

U_3 (F28). To what extent will suppliers eliminate intermediaries such as real estate brokers, car dealers, etc.?

U_4 (F29). What new intermediaries will emerge?

U_5 (F35). To what extent will privacy of individuals be protected?

U_6 (37). To what extent will media cross-ownership rules be relaxed?

U_7 How will people prefer to access and use information?

U_8 Who will be the future providers of information?

U_9 (F14,15,16,17,18). How will the newspaper industry's key classified and retail advertising categories be affected by technological changes?

U_{10} To what extent will new revenue sources emerge?

U_{11} (F36). To what extent will environmental regulation of newsprint change?

U_{12} (F38). To what extent will anti-trust regulations and legislation change?

U_{13} (F39). To what extent will postal de-regulation occur?

Trends

T_1 (F1). Technological change is heating up competition in news and information industries.

T_J (F3). Information is becoming a commodity product.

T_6 (F30). One to one marketing will continue to proliferate aided by an increase use of data.

T_7 (F31). Businesses are placing products in the hands of consumers through new and different channels.

T_{10} (F52). Consuming On-the-Go: Consumer multi-tasking, looking for short-cuts, consuming on-the-go.

T_{12} (F54). Telecommuting increasing, which may change reading habits and shopping.

T_{15} (F59). Privacy is becoming a major issue for people, as technology advances have led to increasing incursions into domain of personal information.

T_{20} (F67). Minority populations are growing and concentrating in certain geographic areas.

T_{23} (F70). Technology has enabled a "global" economy, where information crosses country boundaries, as well as regional boundaries, and can be accessed instantaneously anywhere at any time.

Table 10.2
Scenario Framework

U_1: Business Model

		Traditional (1)	New (2)
U_7: Information Use by Consumers	Minor Change (1)	Scenario A Business as usual ... with a twist	Scenario B Unbundling of information and advertising
	Radical Change (2)	Scenario C Consumers in control	Scenario D Cybermedia

appropriate cell of the matrix. The trends and other remaining uncertainties were then combined to be consistent with the postulated outcomes of the two chosen uncertainties. The result was a blueprint for each scenario that the scenario writers then used to create a different narrative story line for each worldview, as shown in Table 10.3.

Table 10.3
Scenario Blueprint

	A. Business as Usual ... with a Twist	B. Unbundling of Information and Advertising	C. Consumers in Control	D. Cybermedia
U_1	Traditional	New	Traditional	New
U_7	Minor change	Minor change	Radical	Radical
U_2	Minor	Major	Minor	Major
U_3	Low	High	Low	High
U_4	Few	Some	Many	Many
U_5	No change	Less	More	Less
U_6	No change	No change	No change	More flexible
U_9	Little	Dramatic	Dramatic	Dramatic
U_{10}	Some	Some	Many	Many
U_{11}	Little change	Little change	Little change	Highly restricted

The scenario writers consisted of teams of company personnel and out-side advisors. It is in the writing process that much of the learning and de-bate occurs, very much the way a movie director must take a script and make it come alive with a team of specialists. The act of creating multiple stories of the future serves both a learning and a communication objective, as 3M has found once it moved away from number-oriented planning to more of a story telling approach.[16]

Developing the Story.   Each of the scenarios was written in a logical and internally consistent format from the perspective of a historian in the year 2007 looking back in time over the last 10 years and chronicling the events that shaped the world leading to the year 2007. Once completed, each sce-nario, addressed the following issues (under rather different assumptions as postulated by the scenario blueprint):

1. A snapshot of what the world would be like in 2007 under that sce-nario.
2. A description of the key events and the linkages among them that led to the evolution of that particular world over the 10-year period of 1997–2007.
3. A discussion about the strategic implications of the scenarios for the newspaper business model of 1997
4. An illustrative vignette describing a typical "day in the life" of a key advertiser who needs to make decisions about advertising strategy or a typical consumer needing information.

To provide some flavor of each scenario's content, a brief narrative sum-mary of each scenario is provided below. Table 10.4 also summarizes some of the key themes of each scenario. The actual scenarios are usually more detailed than the thumbnail descriptions below. They would include the illustrative vignettes and strategic implications for the current business model. Our aim here is to convey only the gist of each scenario, not its rich detail.

Scenario A in 2007: Business as Usual . . . with a Twist.   The world has evolved as many industry observers predicted back in 1997. New telecommunications and networking technologies, including the Internet, have continued their evolutionary progress to-date, but not quite as quickly as Silicon Valley enthusiasts had hoped. As a result, newspapers have held

Table 10.4
Scenario Themes

	Scenario A	Scenario B	Scenario C	Scenario D
Consumer Markets	Consumers have multiple options across media to access information	Consumers are willing to pay for products that meet unique needs	Consumers are highly dependent on affinity groups, buying clubs, etc.	Consumers have numerous choices across multiple mediums
	Consumers won't pay for content	Highly targeted marketing proliferates	Information sources radically change; new commercial entities emerge	Strong demand for local and tailored information
Technology	Changes the nature of advertising—even in traditional media	Bandwidth challenges resolved	Internet takes-off rapidly	New printers allow home printing of custom newspapers
	Bandwidth challenges continue	Intelligent agents abound	Customized delivery dominates—both door-step and electronic	Inexpensive portable Internet appliances abound
		Electronic commerce proliferates	Electronic commerce takes off	
Industry Players	Content providers create fierce competition	Niche marketing is a large opportunity	Product companies compete with news providers to create information/entertainment that customers demand	Hypercompetition is the norm
	Industry consolidation		Custom publishing is an attractive business	

	Scenario A	Scenario B	Scenario C	Scenario D
Business Models	Heightened economic pressure due to competition	High quality journalism is important	Media companies have multiple revenue streams	Print media are struggling
	Media displace old ones where they're more cost effective	New business models emerge with less focus on advertising	Disintermediation wipes away much of classifieds ad base	Classified revenue is gravely threatened
Legal Issues	Privacy concerns continue; no major changes	Environmental movement progresses	Privacy issues remain important, but unresolved	
	Antitrust and cross ownership issues still important; no major changes	Privacy regulations pose distribution challenges		
		Libel and slander charges increasing against media companies		

227

their ground as an effective mass medium, but they're facing increasing competition—from new as well as the traditional media. Classified advertising revenue has continued its gradual decline over the last 10 years, putting pressure on the newspaper industry business model. In response, some newspapers have launched specialty publications for unique market segments and niches; others have become aggressive in attracting "image" advertisers. Still others are failing because they've been unable to adapt their organization to match the market realities of today brought about by the evolutionary changes of the last 10 years.

Scenario B in 2007: Unbundling of Information and Advertising. Internet and communications technologies have been slowly maturing, allowing ample opportunities for the newspapers to change their business model accordingly and experiment with new sources of revenues. As a result, over 80 percent of newspaper revenue streams are coming from entirely new products that barely existed a few years ago. The Internet's success as a means of building virtual communities is apparent, but some of the greatest benefits for media companies, especially newspapers, have come from applying lessons from direct-Internet marketing to the print medium. In fact, newspapers have developed sophisticated database marketing capability that allow them to achieve unprecedented levels of customer intimacy. For the first time, customers who are willing to pay the price can subscribe to customized newspapers delivered to their doorsteps. Other customers who prefer not to pay as much have less valuable and less digested information sent to them courtesy of eager marketers who want to reach select market segments or even individuals with targeted promotional messages and offerings. Most media and newspaper companies had to chose between becoming providers of premium, but expensive, content versus assisting marketers in delivering targeted messages through innovative customized publishing offerings.

Scenario C in 2007: Consumers in Control. Rapid technology progression has fundamentally changed the way consumers use and access information. Media companies have—for the most part—held onto the advertising based revenue stream, but have also been successful in incorporating new business models, such as making money from facilitating transactions between consumers and vendors. Consumers demand their news in real time and through multiple forms of media. In response, media companies have changed their internal operations to be able to deliver content over different media at the request of their "subscribers." Fortunately for most media companies, advertisers are willing to pay more to reach these "picky" consumers, because they're able to reap great efficiencies in reaching their customers through highly targeted advertising campaigns. As a result, media companies have come to rely heavily on the ever-changing world of the Internet and e-commerce. The most successful media companies aren't necessarily the best journalists. Rather, they're the ones who most readily anticipate their customers' needs and adapt their product offering to fit these changing needs on a regular basis, using the best technologies available.

Scenario D in 2007: Cybermedia. Technology has progressed rapidly as predicted by Silicon Valley pundits; and the ways in which consumers use and access information have changed fundamentally. Most consumers either have customized newspapers printed at their homes or access their news through high-tech Internet appliances. The lines between newspaper and television station as well as other media channels have blurred, as multimedia presentations of textual and visual information and news proliferate. The business models have changed too: newspapers derive revenue from national advertisers (who have few other locally focused mass-market distribution channels), subscriptions, transaction services, and new businesses such as being an intermediary for high-end purchases, high-technology classifieds, and customer profiling. With the rise in electronic distribution, newspapers are struggling to sell their antiquated printing presses—generally at rock-bottom prices.

Analyzing the Scenarios

Each of these scenarios poses serious challenges for newspapers. Scenario A is the most hospitable for the existing newspaper business model. Although the Internet as an emerging technology has made inroads in this scenario, any widespread acceptance of the new technology by consumers has not yet materialized by the year 2007. As a result, most newspapers did not have to radically alter their way of doing business, other than to compete with various niche publications. In contrast, Scenario D has a devastating effect on newspaper companies as we know them today. The asset base of these firms (printing presses) as well as their logistics capability for delivering newspapers have become devalued and obsolete. To remain in the game, these firms need to develop sophisticated information technology capabilities similar to those of AOL. They need to leverage this capability by bringing their brand of hard-hitting journalism to the right segments of the market. If these firms succeed in maintaining a relatively loyal customer base, they might augment their revenues through transaction fees and other marketing oriented services provided to product manufacturers. All this would mean that newspapers have developed truly different organizational capabilities than the ones they posses today. And, they would probably no longer be referred to as papers.

In a fully developed scenario process, managers would then use these four stories as starting points for discussion of best strategies. They would examine strategies for each scenario and the best overall strategies given the uncertainties about which one of these four scenarios will be closest to reality in 2007. As this discussion proceeds, they will be able to identify critical competencies that must be developed and strategies that work under any of the scenarios. The latter are generally much less risky to pursue. They will also identify paths that only make sense under one or two of the scenarios. Given the uncertainty involved, these are usually more risky, so managers need to assess the likelihood of the scenario coming to pass and the resources required to prepare for it. At a minimum, they will probably develop strategic options for pursuing these paths or strategies. Understanding the dynamics of alternative worldviews allows managers to lead change by investing in those capabilities that confer the most flexibility and staying power to their organizations.

Sample Analysis of One Scenario.   To illustrate this process, let's examine Scenario C in more depth. This world would be characterized as follows:

- *Internet changes the way business is done.* American business perfected an industrial business model around distribution channels and dealer networks. This worked well until now. The emergence of Internet and e-commerce has spun a new business model where the length of the distribution channel has been drastically reduced. The result is disintermediation (the process of cutting out the middleman). The new net companies are far more virtual with their customers and suppliers than the traditional brick and mortar companies and because of this they don't need traditional sales and distribution. As a result, they need far less working capital to run their operations.

- *Internet wealth attracts talent.* The ability to hire and retain top-flight talent is critical to the success of any enterprise, and net companies have no problem attracting the best and the brightest. When the executives of brick and mortar companies see the rate at which net companies have proliferated and managed to take market share away from long standing incumbents, they have no choice but to wonder if their own future survival is not at risk. They are afraid of becoming dinosaurs. The wealth creation ability of net companies adds to their sense of anxiety as they watch the rate at which the net company employees are amassing wealth. Viewing the Internet and e-commerce as the major growth medium and industry for the foreseeable future, most executives cannot help but to get involved in these new companies. Not to be a part of it appears to them as a great missed opportunity.

- *There is wealth creation out there but how do you get capital?* Net companies are rewarded by their stockholders for plowing every penny of cash back into product and market development, even if it means great losses. The brick-and-mortar companies on the other hand are expected to show earnings growth quarter after quarter. So while their net brethren are rewarded with access to seemingly unlimited supply of capital at low costs, the brick-and-mortar companies are found struggling to raise expensive capital. Under these circumstances, very few executives of the brick-and-mortar companies have the fortitude to take away cash from moneymaking businesses and plow it into new Internet ventures with their on-going losses and uncertain futures. Confounding the problems is the fact that no one really knows what the return on investments in the Internet and e-Commerce would be and when they will materialize. Because of their past successes, the brick-and-mortar companies encounter significant limitations in how much they can try to emulate the net companies.

- *You have time to decide but it is in Internet minutes.* Making matters worse for the executives of brick-and-mortar companies, the markets don't give you much time to decide what the organization needs to do. The Internet does not tolerate caution. For a brick-and-mortar company that means cannibalizing existing sales through established channels in favor of the nascent Internet strategy. This is not only a gut wrenching decision for most executives, it is also a difficult decision to implement on Internet time. However, emphasized in Chapter 2, delay has dire consequences.

Solutions. Given just these features of Scenario C, let's turn to identifying some possible solutions and strategies. To win in this world, media companies need to develop a set of organizational capabilities such as the ability to:

- Form and manage alliances.
- Develop a deep and intricate knowledge of the customers' (both advertisers and consumers) information, entertainment and transaction needs.
- Develop compelling content and distribute through as many media (e.g., TV, radio, print, Web) as possible.

Having developed these capabilities, a media company can expect to achieve the following:

- Generate multiple revenue streams off of the same content.
- Consolidate and dominate advertising sales in multiple media.
- Achieve largest share of advertising spending.
- Access the broadest possible demographics.
- Provide the broadest possible array of advertising vehicles.

In Scenario C, incumbent companies facing the uncertainties of emerging technologies need to assess their investments in projects and strategic initiatives using an options approach instead of the more traditional NPV methods. The greater the underlying uncertainty associated with the emerging technology initiative, the greater the value of having an option for that future (as Chapter 12 by Bill Hamilton explains further).

To illustrate, consider some of the competitive moves some newspaper companies have made in the area of building and managing strategic alliances.

Classified Ventures was formed in December, 1997 by The Times Mirror Company, Tribune Company, and The Washington Post Company. The partnership's goal is to help newspapers expand their position, using Internet technologies, as the nation's leading suppliers of classified advertising. During 1998, Knight Ridder, Gannett, Central Newspapers, The McClatchey Company, and The New York Times Company joined the consortium. By participating, the newspapers are buying an option for their future.

The consortium's strategy is to build on newspapers' existing inventory of classified listings and respected local editorial content as well as its long-standing relationships with auto dealers, home builders, real estate professionals, property managers, and providers of related information and services for consumers. Classified Ventures currently operates the national Apartments.com web site and is developing its cars.com automotive site. Bob Ingle, president of Knight Ridder's New Media unit, provided the following rationale for Knight Ridder's participation in Classified Ventures: "Classified Ventures recognizes that the best way to win nationally is to win locally. The unification of so many publishers with strong local brands behind a common national brand gives us a tremendous advantage against competitors from outside the newspaper industry."

CareerPath, the online job site, is another example of a collaborative effort. By banding together, the newspapers have created an option for themselves in competing with net competitors, hoping to stall their advance.

As with financial options, real options don't always pay off. So far, in both these cases, the newspapers have found the going tough. In real estate and jobs, the sites operated by the newspaper consortia rank third in 1999 behind such fast-moving newcomers as Monsterboard and AOL's Work-Place Channel in recruitment, and Realtor.com in real estate. In the car marketplace, the newspaper-operated site ranks a distant fifth behind such newcomers as Microsoft's CarPoint.com at this time.

How Should Newspapers Compete?   In Scenario C, nothing short of drastic changes in the way newspapers gather and disseminate information is needed. Newspapers by their very nature as a mass product provide too much information; an average reader consumers less than 5 percent of what is printed in a typical newspaper. In Scenario C, intermediate depth in say news is no longer valued because it is ubiquitous. Consumers especially desire more depth is such areas as jobs, entertainment, arts, food. And, the Internet offers the best technological solution by allowing content providers to modulate the depth and the amount of each subject's coverage according to the consumer's taste. A daily printed product delivered to one's doorstep hardly can meet the demands and needs of this new market. They are a mile wide and an inch deep, and people increasingly want to reverse.

The one area where newspapers currently enjoy some competitive advantage is local coverage. Newspapers have developed the scale of operations to effectively cover and analyze local events (they have feet on the street). The new net companies will have a difficult time replicating this infrastructure quickly in Scenario C. The Internet, as a global medium, is not the best suited to local coverage. The Internet has succeeded in bringing people together based on their affinities and interests, not geography. As long as people within a community have a desire to learn about their locality, newspapers can carve out a somewhat defensible niche for themselves (as a novelty product providing deep local coverage).

Newspapers certainly have the ability to cover local news but the question is: Are they willing to and would it be profitable? In the end, a newspaper company may not have a choice but to dominate this niche, unless it resolves to re-invent itself. Some newspaper companies have followed such a niche strategy. For example, the Thomson Newspaper Group of Stamford, Connecticut, started hiring local residents as reporters on the beat. By doing so, the Thomson Group is trying to make its daily newspapers (56 of them in 13 states) more useful and relevant to their readers. In Scenario C, those companies who follow the local niche strategy will have to provide their brand with distinctive local content in multiple formats. Increasingly, they will have to produce and deliver targeted editions of their newspapers in print and on the Web.

Other newspaper and publishing companies have pursued a re-invention strategy. For example, Bertlesmann has followed a horizontal integration strategy where the Internet becomes the distribution channel for its existing core media and has made acquisitions where it lacks presence. This strategy is also pursued by other media companies such as Time Warner and News Corp.[17] Those who follow this strategy will be well positioned to become one of the few major players to dominate the mass news business through multiple distribution channels (TV, Print, and Internet). In Scenario C, the demand for well-researched, well-written news and analysis will not disappear but you don't have to be a "paper" to deliver it.

In a full-blown scenario exercise, managers would create similar views for each of the potential futures, and the analysis might be taken to even finer levels of detail. As illustrated by this example, although the scenarios may describe future worlds at a fairly high level of abstraction, the real insight comes when linking the scenarios to the likely business models that will emerge and the kind of products and services that consumers will demand. This requires an in-depth knowledge of the existing features of the product offering, and informed projections about what a new technology may permit. Table 10.5 offers a preliminary analysis of the features that print and electronic newspapers share and the features that might discriminate them.[18] Whether the additional features of the new technology will

Table 10.5
Traditional versus Online Newspapers

	Shared Attributes	Discriminating Attributes
Traditional Newspaper	• Timely information • Classified ads • Convenient distribution • Low cost to consumer • Variable subscriber base	• Portable and mobil • Local connection • In-depth coverage/leisure time • Established forum for classifieds • Integrity of the mass medium • Time tested/trustworthy
Online Newspaper		• Customizable/download only • What customer wants • Continuous updates/links • Free distribution (ad supported) • No geographical constraints • Measurable ad exposure • No litter or mess • Novelty appeal (at first)

compensate for the loss of the old depends much on which group of consumers one considers. By segmenting the market based on the trade-offs consumers are willing to make among these various features, firms can focus on likely adopters and growth segments, while continuing to serve those segments that value the old more than the new. Taste and preference in any market segment may change as standards are set and aspirations change over time, underscoring the importance of timely market research and sound tracking models for the migration of value. A strong market focus can also create lead-users, who can help shape the product or offering toward broader market acceptance.[19]

Scenario Planning Traps to Avoid. Because scenario planning is unfamiliar territory for many managers, there are many pitfalls that may not be apparent to those trained in traditional planning processes. A few of these are major pitfalls that can jeopardize the whole project and are often hard to see ahead of time. We list below potential missteps that we deem especially treacherous for those studying emerging technologies:[20]

- *Failing to gain top management support early on.* The first step in any scenario process should be to secure the political support and involvement of senior executives, because without them there can be no significant change in strategy at the end of the process. Back in 1969, the first generation of scenarios at Shell was brilliant, but their insights were not incorporated into the thinking of decision makers. Since the scenarios were the product of staff managers, Shell's senior executives felt no pride of ownership in them. Those 1969 Shell scenarios actually anticipated the 1973 oil embargo by OPEC, but this warning was neither believed nor acted upon by management. The solution: Make decision makers stakeholders in the scenario process.

- *Lack of diverse inputs.* Outside inputs should be actively sought. Sometimes it pays to put outside experts on the team because even managers who are recognized as technological experts or astute students of specific markets may feel uncomfortable addressing issues beyond the familiar boundaries of their industry. At the outset, the scenario learning initiative addresses only the external part of the world, and not proprietary firm strategies or market information. As a result, there should be no problem inviting customers, suppliers, regulators, analysts, academics, or other thought leaders into the process at that stage. With the advent of the Internet, these outside experts can be easily and economically invited into an electronic scenario conference.

- *Failure to stimulate new strategic options.* The ultimate payoff of scenario planning occurs when the organization embarks on successful new strategic initiatives. But too often, the scenario process fails to create legitimate breakthrough options that the firm can accept. One problem is that alternatives that are in fact truly breakthroughs may not look attractive when evaluated through a traditional net-present-value lens by people who haven't participated in the assessment of the organization's long-term environment. As noted, managers can use scenarios to convince leaders to view breakthrough insights about the future as real options, with value beyond what can be seen with NPV analysis. Ideally, the company should develop an internal market place where ideas, talent, and funding can co-mingle the way they do in Silicon Valley, with little top-down steering or bureaucratic interference.[21]

- *Not tracking the scenarios via signposts.* Even when good scenarios and appropriate strategies have been developed, the task is still not complete. Scenarios provide coordinates that help managers better understand the world they're in now, and the futures they might be heading toward. However, scenarios initially look at the world from the perspective of an orbiting satellite. The day-to-day issues have to be managed on the ground. Thus, the scenarios should be made specific and tracked by developing concrete signposts or markers (see Chapter 6 by George Day). To add more detail to the scenario story, managers can write imaginary newspaper headlines to characterize the events and driving forces of each scenario. For example, these headlines might postulate provocative mergers, new patents or products, bold competitive moves, law suits, customer defections, and regulatory changes. To make the link between the scenarios and the day-to-day world of managers, you will have to reset the resolution of your mental binoculars a few times. Zoom in and out, until the whole panorama is revealed and understood: from earth orbit to ground zero.

A Good Tool for Emerging Technologies

One of the benefits of scenarios planning is that it allows the organization to examine the interaction between technology and market that shapes the emergence of technologies. It also allows participants to visualize the impact of technological discontinuities on the organization's existing business models. Without the benefit of scenarios, this would be difficult in most organizations because defenders of "old-technologies" tend to disparage and dismiss the new competitive technologies. New technologies such as the Internet are often crude when they first appear and are seldom superior to the old technology such as printing in its prime. One of the problems with videotext was that the technology was still not fully developed, so the service was slow. Once the new technology benefits from a major innovation that shifts its performance parameters and highlights its advantages (such as high bandwidth and persistent connection to the Internet), the defenders of the old technology find themselves faced with a period of intense substitution. By this time, it is often too late for firms based on the old technology, since they are overwhelmed by the rate at which their markets erode.

Scenarios also can help enhance the budgeting and resource allocation process within firms. The budget allocation process in most firms is based on some measure of return on investment that by its nature favors situations where few resources are used to gain substantial returns. In this frame, the more substantial the required returns, the larger the markets supporting the investments need to be. In the case of emerging technologies, the resource investments are often larger compared to the returns just because the markets arc in embryonic stage and hence still small. As a result, the established firms may dismiss the new technologies as "mere toys" not worth investing in. This perception is further reinforced by the fact that the early adopters of new technologies tend to fit the stereotype of "hobbyist." Against the concrete returns offered by existing technologies, the returns promised by emerging technologies seem to be either very small or pie in the sky.

By painting a vivid, highly textured picture of the future, scenarios help managers better understand the commercial potential of emerging technologies and use this understanding to enhance their resource allocation process. This emphasis can be reinforced by using real options analysis as discussed in more detail in Chapter 12. A proper options analysis requires a full understanding of the uncertainties surrounding a new investment (akin to estimating the variance of an underlying asset in financial options theory). Scenarios can help estimate the range of uncertainty in a variety of the different dimensions that give the option its value.

Another related benefit of scenario planning is that it allows the organization to better understand in which technologies it should invest its limited resources to gain competitive advantage. It makes little sense for companies to further invest in developing and perfecting base technologies that are merely tickets to entry (such as printing presses for newspapers or web sites for e-commerce). Firms should maintain their investments in key technologies that currently offer a basis for differentiation and sustainable competitive advantage. Finally, companies need to increase their investment in developing pacing technologies that have the potential to overturn the existing competitive structure. Scenarios can help managers identify which technologies will be base, key or pacing and thus help them make wiser, staged investment decisions.[22]

Amplifying and Analyzing Weak Signals

Scenarios can amplify weak signals. Emerging technologies differ from other threats or opportunities in their highly textured nature. As discussed by Adner and Levinthal in Chapter 3, seemingly minor speciation events may later produce profound changes in an industry's development. Scenario planning tries to sniff out such speciation events as weak signals or

turning points, and tries to take these preliminary ideas to their logical conclusion if various other developments (either market or technology related)
were to happen. The confluence of forces, and the high degree of path dependence that often typify emerging technologies, are well-suited to the
scenario planning methodology which seeks to weave multiple coherent
stories from the various threads available (trends and uncertainties) that will
likely create the fabric of the future. Thus, scenarios can help track events
and make sense of new events as they emerge. After doing scenario planning, many managers tell us that they process events and information very
differently from before, since they now have multiple lenses through which
to view them.

To illustrate the enhanced sense-making function of scenario planning,
consider the following weak signal that was announced in 1999, and contemplate its potential relevance to newspaper companies:

> Xerox now offers a new service, called Pressline, that delivers news-
> papers electronically to hotels and other locales. This service allows
> customized printing in the quantity needed at the designated loca-
> tion of delivery. Xerox especially recommends this new service for
> travels to foreign places, so that readers can get their local newspaper
> delivered in a timely manner or simply be able to read a national
> newspaper of their choice in their native language.

The relevance of this announcement can be interpreted in multiple ways,
depending on the scenario lens adopted. In scenario A, it represents a niche
market (the traveler's market) and an alternative channel of distribution
besides the physical delivery of newspapers. Indeed, under this scenario
Pressline offers an expansion opportunity for newspapers beyond their
natural urban domain as well as a means of enhancing customer loyalty. In
scenario D, this could be the start of a customized home printing trend of
newspapers, a development that could undermine the very asset base newspapers depend on today (their printing presses and physical distribution
network). Furthermore, as the market segments into finer clusters, the mass
publication model of newspapers—imposed upon them by the constraints
of printing presses—may give way to a new world of truly individualized
newspapers in which the terms "reach" or "penetration" have lost their
traditional meaning. The implications of these changes for the way newspapers sell and price advertising, manage the newsroom, produce and package their product, and distribute it are indeed profound and uncertain. In

this world, newspaper presses, delivery trucks or broad-band mass advertising messages are no longer needed nor valued.

Creating a "Surrogate Crisis"

Most established companies have no choice but to respond to the challenges of emerging technologies such as the Internet. The only choice they have perhaps is whether to face this crisis now or later. We can think of scenario planning as creating "surrogate crises," since a real crisis is often very expensive and sometimes comes too late to still survive. Such imagined crises often elicit similar responses to a real crisis: They may at first engender denial or anger, but can eventually lead to bold action. Emerging technologies often have the potential to create havoc in certain industries or markets, as well as new opportunities. Good scenarios will depict these threats and opportunities in vivid detail. Their aim should be to push people past their comfort zone, and to inspire managers to stop following the pack. This is why scenario planning requires an artful balancing of the known and the unknown. Scenario planning can help companies to avoid being either ostriches who keep their heads in the sand or chickens running around without their heads.

This resistance to new ideas can be reduced if the scenarios are considered to be *learning* opportunities.[23] Learning scenarios are first presented as tentative hypotheses to be tested and validated through further discussion and research. It is in the act of learning that emotional and intellectual acceptance of the scenario themes occurs and alternative mental models may take hold. But this process takes time and requires the involvement of the key decision makers. These learning scenarios are in contrast to scenarios that have been accepted and ratified by senior managers and which will be used to determine future strategies.

Once the learning scenarios are revised and accepted, they can then serve as decision scenarios. Strategies can be tested against these decision scenarios to determine which ones will be winners in only one possible future and which will serve the organization well in a number of futures. How current projects would fare in various scenarios can be quantified using Monte Carlo simulation, options analysis, or other tools for strategic risk analysis. The most important use of scenario planning, however, is to generate new insights about the future environment. From such foresight organizations can craft more effective strategies and plans.

Scenario planning can provide a powerful framework for analyzing the complex and uncertain challenges facing a company such as Knight Ridder. It can help provide a means for managers to examine the technological, organizational, and strategic impact of important technologies whose exact evolution and significance remains uncertain. It prepares the organization not for one future but for many. It can shake up managers and precipitate a virtual crisis without waiting for customers or rivals to create a real one. Above all, it offers managers a way to think about and discuss the future so they can more effectively prepare for it.

CHAPTER 11

APPROPRIATING THE GAINS
FROM INNOVATION

SIDNEY G. WINTER
The Wharton School

*No matter how great the technology or how big the market for it, there is
no guarantee that the value from a new technology will go to the innovator.
The strategy for appropriating the gains from innovation is critical in taking a successful technological innovation to the bank. Will patents do the
job? This chapter points out that managers often place too much emphasis
on intellectual property law in protecting the gains from innovations. The
author identifies some of the inherent limitations of patent protection and
examines other mechanisms for appropriating the gains, including secrecy,
control of complementary assets, and lead time.*

The path to extraordinary profitability
runs through innovation, but to reach this goal, innovative efforts must
surmount three hurdles: First, the technology must be successful—that is,
the "wonder widget" must work. Second, it must create value. It cannot
be something that only a widget engineer could love, but something for
which buyers are willing to lay out a purchase price that exceeds the cost
of production. If there are enough such buyers, the cost of the innovative
effort will be covered, and the prospect arises that someone will benefit
substantially from this innovation.

This brings us to the third hurdle: appropriating those gains. The "someone" who benefits may not be the innovator. Most of the gains may go to
rivals who have either imitated the innovation or reached similar results
on their own. The value could also be captured by buyers or suppliers of

The author would like to acknowledge the very helpful comments on an earlier draft provided by
the editors, Wes Cohen, and Robert Merges. The errors and limitations that remain are my own.

242

other resources involved in the processes of production and use. The innovator hopes to find a way to appropriate enough of the total gains to make the whole thing worthwhile—or, preferably, *very much* worthwhile.

This chapter is about the management challenges at the third hurdle, value appropriation. Many discussions of this subject overemphasize the importance of intellectual property law, and patent law in particular, in securing the returns from an innovation. A number of considerations contribute to this overemphasis, the most important being that there is an intellectual property "industry." Like most industries, it promotes its wares. Also, there is a public policy dimension to intellectual property law, and the presence of such a dimension generally creates a higher level of visibility. This overemphasis on intellectual property obscures other mechanisms for appropriating gains from innovations and leads to a focus on the *duration* of the period for which gains are secured. This is probably because the patent law promises more in terms of the duration of protection than rival mechanisms can deliver (also more than patents typically deliver).

Four major classes of appropriability mechanisms are examined here: (1) patents and related legal protections, (2) secrecy, (3) control of complementary assets, and (4) lead time. These four mechanisms are not mutually exclusive (nor are they fully exhaustive). They often are complementary approaches. There are significant interactions among them and opportunities for using one to leverage another. The chapter also provides a framework for examining appropriability issues in emerging technologies and explores the role of dynamic capabilities that yield multiple innovations—creating a "golden goose" rather than a single golden egg.

PUTTING INTELLECTUAL PROPERTY IN PERSPECTIVE[1]

The overemphasis on intellectual property distorts discussion and analysis in three significant aspects. First, it contributes to an impression that patents and other legal protections are the dominant means that innovators employ to capture returns. This is very far from the case. While there are industries in which patents do play a major role, and scattered examples of important patents elsewhere, the general effectiveness of patents varies substantially across industries. Other mechanisms generally play a more important role. The most recent systematic evidence on this point comes from a 1994 survey of R&D managers conducted by researchers at Carnegie Mellon University.[2] For product innovations, the CMU study concludes that "patents are unambiguously the least central of the major appropriabil-ity mechanisms overall, reflecting their subordinate role in most industries."[3] For the most part, other mechanisms provide shorter periods of effective protection than patents *promise*—and thus seem inferior to patents *prima facie*. In many cases, the period of protection provided is not, however, in-ferior to what patents actually deliver.

Second, the overemphasis seems to contribute to a misallocation of attention and resources. Resources and attention are focused on making the most of available legal protection, while energetic and sophisticated use of the alternatives is neglected. Patent litigation can be costly in technical and financial resources, and the effort to obtain legal redress for the in-juries of the last competitive round may amount to a self-inflicted injury in the next round. Survey evidence shows that lead time appears to be much more effective across industries than patents. This suggests it may be more effective to get the researchers off the witness stand and back into the lab.

Third, this perceptual distortion leads to mistakes in selecting projects to pursue. It lends superficial credibility to the idea that "if we can't count on getting intellectual property protection for the results of this project, the project is not worth doing." This assumption is quite wrong. Whether the returns are protected by intellectual property law or some other mecha-nism does not matter. And, in an important sense, it does not really mat-ter how well the returns from the innovation are protected. That is, it does not matter if undeserving imitators leap in and grab 90 percent of the hy-pothetically available returns, if in fact the costs are only 5 percent of the returns: ten over five is still doubling your money. It may be frustrat-ing to speculate that if those imitators could be fended off to the point where they grabbed "only" 85 percent of the returns, you would be tripling rather than doubling your profits. The unavailability of the tripling is not, however, rational grounds for passing up the doubling.

This last point is certainly appreciated by managers who innovate in ways that the intellectual property system does not significantly protect. Or rather, since they have low expectations about what the legal system will do for them in the first place, they are not subject to the pain of contem-plating what a perfect system could do for them. They rightly consider themselves innovative if they succeed in making money by innovating; there is no occasion to focus on the fact that they might be making more money if they had more support from the law. Consider, for example, the comment of Howard Schultz, legendary CEO of the Starbucks coffee chain:

We had no lock on the world's supply of fine coffee, no patent on the dark roast, no claim to the words *caffé latté* apart from the fact that we popularized the drink in America. You could start up a neighborhood espresso bar and compete against us tomorrow, if you haven't done so already. . . .

What we proposed to do . . . was to reinvent a commodity. We would rediscover the mystique and charm that had swirled around coffee throughout the centuries. We would enchant customers with an atmosphere of sophistication and style and knowledge. . . .

The best ideas are those that create a new mind-set or sense a need before others do, and it takes an astute investor to recognize an idea that not only is ahead of its time but also has long-term prospects.[4]

Starbucks relied not upon patents but on lead time leveraged by complementary assets to appropriate gains from its innovation.

Part of the problem of flawed perspectives is that strong patents, in the relatively rare cases where they can be secured, define a high standard of protection that is (wrongly) sought in other contexts. The absence of such protection leads to the speculation that, if the higher standard of protection were achievable, the prospects for profitable innovation would be much better. This argument is misleading.[5] The fact that a company would profit more from innovation X if it (alone) had stronger protection does not establish that it would profit more from X if protection were *generally* stronger in its industry. That might mean, on the contrary, that a rival firm (responding to the strengthened incentives) comes up with an innovation Y that makes X obsolete. Improving appropriability in a particular activity or industry tends to raise the standard of innovative competition—which may or may not be socially desirable, but cannot be presumed to be more favorable for profits for a particular company than any other rise in competitive intensity.

FOUR MECHANISMS OF APPROPRIABILITY

To create a broader strategy for appropriating gains, managers need to see intellectual property protection as just one of the mechanisms for protecting innovation gains, which are described here under four headings:[6]

- Patents and related legal protection.
- Secrecy.
- Control of complementary assets.
- Lead time.

The discussion here concerns the capture of gains from a potential innovation that is already "in hand." It may be an invention or a cluster of related inventions; it may be a new way of organizing internal functions or relations with customers or suppliers. Whatever else it is, it is not yet an "innovation" because the "putting into practice" that traditionally distinguishes innovation from invention has not yet happened, and in fact the details of putting into practice will be determined in part by the capture strategy. For simplicity, the term "invention" is used here for the generic potential innovation, and alternative possibilities are explicitly referenced when needed. It should be emphasized, however, that innovation very often takes the form of procedural and organizational changes that are not "inventions" as commonly understood, and certainly not patentable inventions.

Patents and Related Legal Protection

History demonstrates the power of patent protection, particularly for inventions related to the early development of new technology-based industries. For example, Alexander Graham Bell's basic patents on telephony permitted the Bell Telephone Company to achieve profitable dominance of the new U.S. telephone industry until those patents expired in 1894—after which the number of independent rivals grew over one hundred fold in less than 10 years.[7] More recently, the Cohen-Boyer patent covering gene-splicing techniques, a basic element of biotechnology, is estimated to have earned over $220 million for its owner, Stanford University. There are also examples of strong and very valuable patents on discrete inventions in established industries.

Patent protection, in the typical form of "utility" patents, is available for four types of subject matter—(1) machines, (2) products of manufacture, (3) compositions of matter, and (4) processes—plus improvements thereon. To be patentable, an invention must meet standards of usefulness (utility), novelty and non-obviousness. In practice, little bureaucratic effort is expended on critical evaluation of the inventor's contention that the invention is useful,[8] but a lot of systematic effort is expended on the "novelty" test, and especially on the narrow question of whether the applicant's invention is familiar or has already been patented.

Since 1995, the duration of a utility patent has been 20 years from the date of application.[9] During that period, the inventor has the legal right to prevent anyone else from using the invention—regardless of how it was obtained. That is, others are precluded from using the invention even if they created it themselves with no reference whatsoever to the patented version, and even if they were entirely unaware of the patent until notified that they should desist from infringing it. It is important also to emphasize the negative character of the right. If two or more legally distinct parties own patents on complementary inventions required in a single innovation, no one of them has the right to carry out the innovation unilaterally. A licensing deal must ordinarily be made if the innovation is to be brought about.

Limitations of Patent Protection.   The effectiveness of patents is an issue that is considerably murkier and more ambiguous than the high-profile success stories of Bell Telephone or Cohen-Boyer suggest. Among the limitations are the high legal costs of defending patents and opportunities for rivals to "invent around" the patent. Patents vary greatly in both strength and value; some are valuable only when arrayed in mass formation with other related patents, some are relevant for a period much shorter than that of the patent grant, and some are little more than scarecrows to frighten off the most timid of the poaching imitators. Among the significant limitations of patents are:

- *Legal costs:* Indeed, the Bell patents themselves illustrate one dimension of this complexity. The patent grant per se did not confer dominance on Bell Telephone; that result was accomplished through the defense of that right by very protracted litigation involving more than 600 patent suits.[10] When the second component of a patents-and-litigation strategy turns out to be costly, a significant portion of the rents may be diverted to lawyers' fees and other legal expenses. Another example is Eli Whitney, inventor of the cotton gin, who spent much time and money defending his patent in court while his invention stimulated a boom in the value of cotton land, from which others made great fortunes.[11]
- *Limited effectiveness in some industries:* Economists have long been aware that the effectiveness of patents tends to vary systematically

across industries, as shown in Table 11.1. The table shows the results of two studies that support this picture of inter-industry variation—the "Yale survey" in the early 1980s and the "CMU survey," referenced earlier, conducted in 1994.[12] Table 11.1 illustrates the consistent (and unsurprising) finding that patents are quite effective in the pharmaceutical industry—and also that they are only of middling effectiveness in some industries usually considered highly progressive, such as computers and semiconductors.

- *Inventing around:* When R&D managers in these surveys were asked why they thought patents were of limited effectiveness, managers in both surveys cited the fact that competitors can legally "invent around" a patent as one of their greatest concerns.[13] That is, competitors can come up with their own inventions that do not infringe the patent but do succeed in appropriating some of the potential returns of the focal invention.[14]

The notion of "inventing around" offers insights into the permeability of patent protection and the broader challenges of establishing effective property rights in knowledge. To be invulnerable to inventing around, a patent must protect the key idea of the invention. Ideas, however, are considered too

Table 11.1
Patent Effectiveness Scores for Product
Innovations, Selected Industries

Industry	Yale[a]	CMU[b]
Drugs	6.53	50.2
Plastic products	4.93	32.7
Medical instruments	4.73	54.7
Semiconductors	4.50	26.7
Petroleum refining	4.33	33.3
Aircraft and parts	3.79	32.9
Computers	3.43	41.7

[a.] Effectiveness of patents to prevent duplication, mean response on a 1 to 7 scale.

[b.] Percentage of product innovations for which patents considered effective in protecting competitive advantage, mean of responses.

vague to be patentable per se. An idea, unrelated to a specific physical means for carrying it out, cannot be patented. On the other hand, a patent may readily be obtained if such means are detailed very precisely, assuming other criteria of patentability are met. But such precision often leaves abundant room for an imitator to achieve a similar result by somewhat different means, thus, inventing around the patent. In effect, an overly-narrow patent of this sort achieves a result analogous to what is achieved by copyright, which does not protect the substantive content of a work, but only the "form of expression." Borrowing ideas from a copyrighted work without violating the copyright is typically (and intentionally, from a policy view point) easy because there are many ways to express the same idea.

Patents can come closer than copyright to protecting ideas because an imitator's trivial "cosmetic" changes in the specifics of an invention are not sufficient to avoid infringement. This is the import of the "doctrine of equivalents" in patent law. As the Supreme Court said in 1950:

> [I]f two devices do the same work in substantially the same way, and accomplish the same result, they are the same, even though they differ in name, form, or shape.

The ability to recognize that two devices "do the same work in substantially the same way" closely approximates the ability to recognize "the same idea."

In practice, courts are more likely to recognize an "equivalent" when there is greater similarity in the physical expression of an idea. For this reason, patents tend to be relatively invulnerable to inventing around in areas of technology where the physical expression of ideas is relatively inflexible.

Drug discoveries provide important illustrations of this logic. Until the 1970s, pharmaceutical companies pursued new drugs by a "random screening" approach. They explored the medical efficacy of molecules by the thousands, looking for ones that would have therapeutic effect against some condition or other, and with little or no guiding idea for the search. The efficacy of a drug discovered by such a process is virtually an isolated fact; an imitator can attempt to make "cosmetic" changes in the molecule while retaining efficacy, but such attempts tend to be successful only when they are so trivial as to provide no defense against a charge of patent infringement. By contrast, the process of "guided" drug discovery that has become prevalent since the 1970s is guided by scientific ideas about mechanisms of action, such as looking for a molecule that blocks the action of a particular enzyme.[16] If the validity of the idea of treating a particular condition by blocking a particular receptor is demonstrated by a molecule identified by inventor A, inventor B may borrow the idea and develop another molecule that does "the same work" but in a quite different way—thus inventing around A's patent. Thus, "guided" drug search, or "rational drug design" is likely to produce a regime where pharmaceutical patents are more susceptible to inventing around than they were under the "random screening" regime.[17]

The knowledge context and the physical constraints of the subject matter—along with the evolution of the law—determine the ease of "inventing around," with different results in different areas and great variability at the level of the individual invention. While competent patent attorneys can shape the scope of the claims in an individual patent application, they cannot reshape the underlying knowledge conditions and physical constraints. Still less can they stop the leakage of other key information about the innovation that strongly affects the level and quality of rival efforts to encroach upon the gains, such as the demonstration of the existence of a previously unmet need and the identification of where that need resides. Some of this information may leak out in the patent application itself.

Some patents manage to resist this process of inventing around, continuing to be highly valuable for their full legal duration. The end of legal protection then brings a precipitous decline in the rent stream. For example, Glaxo Wellcome's anti-ulcer medication, Zantac, became the world's largest selling prescription medicine, with sales peaking at $3.8 billion in 1994.[18] When the U.S. patents on the drug expired at the end of July 1997, generic competitors captured over half the prescriptions within one quarter. From 1996 to 1998, Glaxo Wellcome's sales of the drug declined from $3.01 billion to $1.26 billion.

In many cases, however, science and technology catch up with the patent in a period substantially shorter than the patent life. Zantac itself was the villian of such a tale for Tagamet, the anti-ulcer medication that Zantac surpassed as the leading prescription drug in 1986, eight years before Tagamet went off patent. Outside the pharmaceutical industry, examples of rent streams shorter than the patent life seem to represent the typical case. Both the Yale and CMU surveys inquired into the lengths of "imitation lags" for both patented and unpatented innovations, and both surveys yielded evidence that these time periods are remarkably short—with "imitation lags" of less than five years, and often much shorter.[19] (It should be noted that the term "imitation lag" is used rather loosely here: The survey did not inquire directly into imitation, but asked how long it was before another firm introduced a "competing alternative" for a recent major innovation of the respondent's firm.) What patents accomplish, most typically, is a marginal improvement in lead time.

Frustration with the imperfections of the patent system, stimulated perhaps by persistent visions of the rents that got away, has led to calls for improvements. A certain utopianism often appears in this commentary. Such radical reforms appear unlikely, given that patent laws are implemented by government bureaucracies like the U.S. Patent and Trademark Office (PTO), and this is a time of budgetary stringency, diminished respect and non-competitive pay for government employees, rapid technological change and burgeoning growth in patent applications. Rather than railing against the weaknesses of the patent system, managers may more profitably put their energies into other mechanisms for appropriating innovation gains.

Secrecy

Secrecy is the analogue in the knowledge realm of a fence around real property—a straightforward but limited type of protection that the owner can provide without aid from a government.[20] As such, it plays a gap-filling role in the strategic repertoire of a private entity: It is the available private means to provide protection that public institutions have failed to afford. Its importance tends to be greatest when the performance of public institutions is the weakest (e.g., in less-developed countries).

In the knowledge strategy of an innovative company, there is virtually always some role for secrecy. Whether it can or should play the central role in appropriability strategy for a particular innovation is generally a harder call. Sometimes it cannot, and sometimes it should not because the sacrifice of other objectives is too great. The potential protection offered by secrecy varies dramatically, depending on attributes of the discovery and the circumstances of its use.

Protecting innovative products through secrecy is difficult. Aspiring imitators often can acquire the product—either by posing as legitimate customers or by inducing legitimate customers to pass along the product or information about it—and then reverse-engineer it. This vulnerability typically emerges only after the product is available in the market, which gives the innovator the advantage of lead time, particularly if secrecy is maintained during product development. Manufacturers also may make their products resistant to reverse-engineering. For example, semi-conductor manufacturers have sought to protect their devices from reverse-engineering of circuit design by enclosing them in epoxy resins, so removing the coating destroys the circuit.[21] Other companies retain physical control of the innovative product and sell its services, as may be possible for the equipment in manufacturer-operated repair facilities or for equipment supporting business services such as data processing.

In general, it is easier to keep secrets about processes than about products. Production processes can usually be conducted behind the company's walls, out of sight of potential imitators. Companies also can discourage loose talk or outright defection by the personnel involved. Some business processes are inherently hard to protect, however. Some are "observable in use," for example, the manner in which service personnel relate to customers. Others rely on secrets that can be conveyed very simply, however difficult they may have been to discover in the first place. For example, it has been reported that one of the "secrets" of McDonald's famous french fries is to cook them "until the oil temperature has climbed three degrees above the low temperature reached when the potatoes are put in the oil."[22] Such a simple rule is obviously a hard secret to keep. The situation is only slightly different when the secret information cannot easily be expressed in ordinary language, but can be rendered precisely in an appropriate "code" that numerous experts understand—for example, the reaction path of a chemical process.

At the other extreme, the risks of leakage may be minimal because the secret is too large, too complex or too resistant to articulation to transfer easily. The detailed operating techniques that account for high yields in a semiconductor fab are one example, the combined wisdom and experience of a team of expert consultants are another.

Efforts to facilitate internal knowledge transfer by codifying the key information may have the unintended consequence of facilitating the escape of the secrets to imitators.[23] The severity of this effect depends on the extent to which the knowledge and capabilities required to interpret and exploit the codified account are widely available among potential imitators.[24] Rarely, however, can a discovery be effectively exploited while its principal secrets remain locked up in a big safe or a secure R&D lab. Large scale exploitation is likely to require that many people know something about those secrets, and/or that a large number of documents record them at least in part. It would be self-defeating to make major sacrifices in the breadth of the rent stream in the attempt to extend its length through secrecy. But managing such strategic trade-offs can be painful, especially for a small company aware that large rivals can bring to bear many strengths in exploitation if they once get hold of the discovery itself.

It is not just the current innovation but also the future productivity of the company's R&D activity that can be hampered by too much secrecy. Within the company, attempts to enhance secrecy through compartmentalization of information may reduce R&D efficiency in a variety of ways. Heavy-handed enforcement of secrecy rules may produce disaffection in key employees.[25] Either success or failure in one project may provide useful input to another, but strict isolation of projects will postpone the sharing of success and perhaps sacrifice the learning from failure. Numerous small "wheels" may have to be reinvented time and again. The byproducts that are useless in their project of origin may provide the keys to success in another project—as in the famous case of the failed adhesive that was the technical key to 3M's Post-It Notes.[26] Such benefits will not be exploited if a penchant for secrecy consigns the information about them to the burn bag.

Similar issues arise with respect to communication across company boundaries. Secrecy aims to prevent valuable information from leaking *out,* but the sealants used may be equally effective in preventing valuable information from leaking *in.* There is good reason to believe that informal professional networks contribute importantly to the productivity of scientists and engineers. (See, for example, Lori Rosenkopf's discussion of knowledge networks in Chapter 15.) Informal norms of trust and reciprocity guide interactions in such networks—A's willingness to share information with B depends on A's perception that B won't make him regret it in the short term and might well reciprocate in the long term.[27] The more significant the information, the more prominent these norms are likely to be. As a result, company policies that impede intellectual trade may be beneficial in the short term—because leakage is stopped while inflows temporarily continue—but counter-productive in the long as new ideas cease to flow into the organization.

There appears to be increasing awareness among managers of the importance of secrecy in appropriating innovation gains, as seen in a striking contrast between the results of the Yale survey and those of the CMU survey eleven years later. Secrecy ranked last overall in the Yale survey among the major means for capturing returns from product innovations, but in the CMU survey secrecy ranked at the top, in a virtual tie with lead time. Adjusting for difference in industry definitions, the CMU survey shows secrecy ranked first or second in 24 out of 33 comparison industries, as against none in the Yale survey.[28] This change may reflect a general intensification of competitive pace and pressure, including a heightened awareness of the hazards of industrial espionage. (It is also possible that this seemingly strong result is partly an artifact of differences in the details of the two surveys—for example, differences in the specific question asked—but the survey results do indicate the importance of secrecy in protecting innovations, particularly in the early stages.)

A company's efforts to protect its secrets are "leveraged" by the law of trade secrets. The two key conditions for information to be entitled to protection as a trade secret are that it affords some competitive advantage (if not, who cares?) and that it be treated as a secret by its owner. The latter condition makes improper acquisition of trade secrets an offense akin to the crime of breaking and entering: there is, by definition, no such offense if the door was left unlocked. Thus, by such measures as maintaining physical security and limiting access to sensitive premises, avoiding inadvertent disclosure, insisting on confidentiality, nondisclosure and noncompete agreements with employees and others exposed to the information, a company not only raises a direct barrier to the appropriation of its secrets but also satisfies a crucial precondition for legal action against anyone who might surmount the barrier.

Trade secret law affords no protection, however, if a rival independently conceives of the same information or discovers it by parallel research—legitimate activities that do not require the transgressing of any barrier set by another "owner." This represents an important distinction between secrecy and patent protection, since in principle patent protection does bar the use of the invention by an independent inventor.

Complementary Assets

Controlling complementary assets can help to capture the gains from an innovation. As discussed by Mary Tripsas in Chapter 8, complementary assets include resources such as access to distribution, service capability, customer relationships, supplier relationships, and complementary products. The firm that controls these assets may be more likely to benefit commercially from the innovation.

Perhaps the simplest example of the effect of complementary assets arises when a new product lacking patentable features requires specialized facilities and capabilities for its manufacture. If the innovator is the sole owner of such complementary assets, the lack of patent protection may matter very little. (As this example suggests, the term "assets" is to be understood expansively in this discussion, and includes asset services and organizational capabilities as well as assets in the narrow sense.) Unlike patent protection, which the law offers only to the "original and true inventor" or his assignee, the opportunity to capture returns through complementary assets is open to anyone who holds such assets—whether the innovator or someone else.

An innovator who lacks some of the critical complementary assets may have great difficulty capturing the gains.[29] Suppose, for example, that the innovation is "co-specialized" to assets held by one or a few other companies—not useful at all except in conjunction with those assets. In that event, even if the legal protection for the innovation is secure, those other actors are in a position to insist on a slice of the pie and are not likely to pass it up. On the other hand, when intellectual property protection is strong, complementary assets are of a generic character, or services are available in a reasonably competitive marketplace, innovators who lack complementary assets may not pay too high a price to obtain them.

It would be hard to imagine a more emphatic demonstration of the relevance of complementary assets than is provided by the biotechnology industry, and its pharmaceutical branch in particular.[30] In the earliest days of the industry, new technology emerged from university laboratories by way of a large number of entrepreneurial start-up companies, usually with university links. The first, and one of the most successful, of these new firms was Genentech, co-founded by Herbert Boyer, the Stanford University scientist who was also co-inventor of the recombinant DNA technology basic to the biotechnology industry. These small companies had R&D capabilities and programs, or dreams, and very little else. Some developed new drugs for which they obtained patent protection, but the complementary assets and capabilities required for actually getting new drugs to market lay in the hands of the established pharmaceutical companies. Those companies had the ability to arrange clinical trials, obtain FDA approvals, establish manufacturing facilities and get regulatory approval for processes, and ultimately market and distribute the drugs.

The ability of the established pharmaceutical companies to come into the biotech game relatively late, and profit from it, is a strong confirmation of the role of their complementary assets in appropriating the gains of biotech innovations. At the same time, however, patent protection of the start-ups (and their control over research capabilities that could be valuable in the future) helped them retain the gains that otherwise might easily have been seized by their giant partners.

The power of complementary assets also can be seen in the strength of post-war IBM.[31] IBM developed a strong sales force and service/support organization, and innovation and manufacturing capabilities. By leveraging these pre-existing capabilities and building new ones, IBM arrived as of the mid-1950s at a domestic market share of about 85 percent—which it then maintained for three decades, despite waves of technological change. As product and process innovations flowed through the industry, IBM's position gave it the opportunity to make about five times as much use of any given innovation as the rest of the domestic industry combined. The fact that innovation overall was so rapid meant that an occasional innovative success outside the company could not (for a long time) imperil IBM's position. Other companies and independent inventors found they needed IBM for manufacturing and market access. As a result, IBM was in a position to pick and choose from the streams of innovations, and it captured a substantial portion of the gains from those innovations. Technical and financial success naturally fed back on itself, leading to the accumulation of even more complementary assets.

In this multidecade saga of industrial leadership, patents apparently played no significant subsidiary role. Most of the scholarly accounts of IBM's dominance make no mention at all of patents or other intellectual property considerations, or mention them only for the purpose of discounting their importance.[32]

As the example of IBM illustrates, the most powerful role of complementary assets is not in appropriating gains from a single innovation but across a wide range of innovations. IBM created a strong *dynamic capability*—the ability to make continuing improvements in a related set of products and processes by re-deploying related assets and capabilities in R&D, manufacturing, marketing, and other functional areas.[33] In the context of such strength, successful appropriation of gains from individual innovations is largely pre-ordained—and when it is not, it doesn't matter much because strong dynamic capabilities are likely to prevail again in the next round. In the large firm, dynamic capability is the proverbial goose that lays the golden eggs of innovation. As long as the goose is secure, it doesn't matter much if somebody makes off with a few eggs occasionally. Of course, that's not to say that we *like* having somebody steal our eggs.

Lead Time

The CMU survey ranked "lead time" as the most effective appropriability mechanism for product innovations, slightly higher than secrecy (based on the percentage of innovations for which the mechanism is considered effective).[34] As in the case of other mechanisms, there is substantial variation across industries in the effectiveness of lead time. For example, respondents rated it as effective for almost two-thirds of product innovations in communications equipment, auto parts and cars and trucks, but for only about one-third of innovations in electrical equipment.[35]

The time scale of lead time advantages also varies across contexts. Achievable leads tend to be quite short when the opportunity for a particular type of innovation is, so to speak, on a public agenda that is visible to numerous competent players. When novel resources offer innovative opportunity to all comers on reasonably similar terms—such as a latest-generation microprocessor for PCs—the company that moves fastest in developing the innovative application will be able to gather rents until followers appear to chisel the advantage away. The length of the lead time is determined by the good luck, flexibility and skills of the leader as well as the bad luck, inertia and incompetence of the others.

An innovative leader can obtain longer lead times if the innovator is not merely the inventor of the winning entry in a widely recognized contest, but the creator of the contest itself. Few if any rival efforts get under way until the innovator's experience provides convincing demonstration of the opportunity.[36] Lead times then tend to be longer, to appear in combination with other innovative advantages of the leader, and to provide opportunity for strengthening appropriability further through complementary assets or other means. The extent and durability of such competitive advantages has been the subject of extensive discussion about the strength of "first mover advantages" and the impact of "hypercompetition."[37] Similarly, the question of what being first is worth per se has been difficult to answer.

It is clear, however, that the dramatic successes achieved by many "first movers" do not constitute a convincing case that all aspiring innovators should try harder to be first. On the one hand, leaders often have other advantages that contribute to their success. On the other, extreme efforts to rush a product to market may lead to an excessive sacrifice of quality, reliability, and readiness to provide service support.

Lead time can be coupled with learning capabilities to appropriate gains from innovations for longer periods. For example, when Nucor introduced

compact strip production (CSP) process in steel making, its German equipment supplier, SMS Schloemann-Siemag (SMS), controlled the intellectual property for the technology.[38] This created the significant risk that
Nucor's learning from the initiative might well be captured by SMS and
then transferred along with new equipment installations to Nucor's steel industry rivals. While the new technology would enable Nucor to produce
higher quality flat steel and move into higher markup markets, this also
had the potential to incite the wrath of large integrated rivals such as USX
who dominated those markets. Nucor went ahead nevertheless. Learning
proved difficult and expensive, but a few years later Nucor had mastered
the technology while the rest of the industry stood aside and watched. CSP
and other innovations helped Nucor generate an annual total return to investors in excess of 27 percent from 1984 to 1994, a far better performance
than its principal rivals.[39]

Although it initially seemed competitors could easily follow Nucor's
highly visible lead, this example shows that the protections of lead time are
stronger than an abstract view might suggest. Nucor's ability to profit
from CSP rested on its substantial dynamic capabilities as a process innovator in steel production, which endowed it with a potential lead time advantage more fundamental than a mere decision to go first. And in braving
the hazards of retaliation from large integrated firms, or of finding the
fruits of its costly learning dispensed free to other rivals by SMS, Nucor
drew on well-founded confidence in its learning abilities and problem-
solving culture.

The impact of lead time also is affected by product characteristics, particularly durability. The more durable the product, the more valuable a
given lead tends to be. This is because the appearance of an innovative version of a product stimulates a burst of purchasing, after which demand is
largely driven by replacement. Airlines, for example, typically only buy
new aircraft to replace planes in their existing fleets, or to support an expanded schedule of flights. When innovation improves the price/perfor-
mance significantly, the result is a wave of obsolescence of the existing
stock, and increase in purchases. Once this bulge in demand works through,
the manufacturers confront only replacement demand again. Thus, the
company that can be first in bringing to market a major improvement in
the product may be able, if its lead time and manufacturing capability allow,
to be the supplier for a major fraction of the stock adjustment. Given some
willingness of buyers to accept delivery lags from the innovator, the lead
time required to capture most of the gain may be only on the order of a

year or two, or even less. A series of relatively minor wins of this sort can lead over time to the dominance of an industry, as the history of the commercial aircraft industry illustrates.[40]

Lead time advantages also are increased when the innovator can establish a strong reputation or customer switching costs are high. Particularly in consumer markets, there is enough inertia in demand to confer surprisingly large benefits on the producer who meets the demand first. Aggressive competition from otherwise qualified followers is largely forestalled simply because the price or equivalent concessions that they must make to attract the customers are too large to make it worthwhile.[41] The mutual fund industry provides a striking illustration of this point. No innovative product is more readily "reverse engineered" than a novel type of mutual fund. Yet it appears that the "first mover" advantage in the industry is quite significant,[42] though it seems to rest on nothing more substantial than the inertia and friction that retard the flow of customers toward the best offer.

Finally, lead time may offer not only the opportunity to do some profitable business before rivals appear on the scene, but also to lock up complementary assets—with the result that the rivals find life quite frustrating, even when they do appear on the scene. As mentioned above, Starbucks' ability to make high quality coffee beverages is easily imitated but the company developed a capability to open new stores at a remarkable pace, now on the order of a store a day, while maintaining its high quality standards in all aspects of its operations. That capability, not the ability to make good coffee drinks, was the "inimitable" aspect of its innovative success. By identifying the best store locations, acquiring them and putting attractive and well-functioning stores in them, Starbucks altered innumerable local environments in ways unfavorable to late-coming competition.[43]

APPROPRIABILITY ISSUES IN EMERGING TECHNOLOGIES

How can companies use these four mechanisms to appropriate the gains from emerging technologies? There is no magic formula, no one-size-fits-all answer to the question of how to capture gains, either in general or—even more emphatically—for the specific context of new industries arising from emerging technologies. Much of the art of successful strategy, in this area as elsewhere, lies in the accurate matching of the specifics of the solutions to the specifics of the problems. Because the diversity of the problems arising from the variety of company situations and contexts is enormous, a useful handbook can hardly be provided in a few pages. The exhortations and observations in this section are put forward in the hope of providing useful guidance for this matching process—and in the spirit of this chapter's general warning against locating capture strategy in the cramped quarters of the intellectual property rubric.

Identify the Uncertainties

Uncertainty tends to be pervasive in industries based on emerging technologies and appropriability conditions share that tendency. As one example, intellectual property law evolves with court decisions and legislative actions, which can increase or erode the protection given to technology development already in progress. For example, the Supreme Court decision in *Diamond v. Chakrabarty* was an important legal milestone for the biotechnology industry because it opened the way to patenting of things that, though not naturally occurring, were derived from living things. This decision came in 1980, eight years after the patent application was filed. It may have helped to trigger the boom in biotechnology start-ups that occurred in that year, but a substantial number of such firms were already pursuing research programs whose value was likely to be affected by the ruling.[44]

There is also considerable uncertainty about the effectiveness of secrecy in protecting competitive advantages, particularly in the early stages of emerging technology industries when markets for personnel holding key (and arguably proprietary) know-how are unusually fluid. For example, one leader of a very successful information services start-up bemoaned the loss of a talented young programmer. The programmer was lured away with an annual compensation offer that leaped from $75,000 to $350,000 in a single bound—a jump that may have reflected his possession of key knowledge of the practices of his former employer. Trade secret law may provide some protection against losses by this mechanism, but in the tumultuous context of a young industry, the protection is often weak.

Similarly, substantial uncertainty surrounds the claims that holders of complementary assets may have against innovation rent streams. It is generally fairly clear what sorts of assets are relevant to a particular line of innovative activity—but it is not at all clear how supply and demand for those assets will develop in a highly dynamic context. That is what determines the rents that such assets can command. For example, it is clear that innovative online retailers like Amazon.com need the support of the more conventional assets and capabilities required in the physical warehousing and distribution of the goods. If competition for these assets intensifies, it is quite conceivable that much of the gains from online retailing will be transferred to firms like Ingram's and UPS that are well positioned in the complementary assets—or the value could ultimately go to consumers.

Companies pursuing lead-time advantages in an emerging technology-based industry have even less certainty. A dense fog of uncertainty obscures the positions of the contestants on the various technological racecourses; indeed, some of the racecourses themselves may be invisible. A company trying to establish its position in assets complementary to its own technological approach, in the hope of assuring long-term dominance, may find its approach rendered obsolete and its investments devalued by some entirely different approach. Even the brief lead-time advantage that its approach might have enjoyed may be sacrificed in the quest for long-term advantages. Such scenarios seem to play out recurrently in the telecommunications industry, where the quest continues for more and cheaper bandwidth to support the needs of the Information Age and both technical and regulatory uncertainties abound. Worst of all, the firm that picks what turns out to be the right technological goal and succeeds in getting there first may find that it has arrived too early for the party—when the technology is immature, Wall Street is not yet impressed, and the customers have not caught the fever yet. Its followers will learn from its mistakes.

Although uncertainty is largely inescapable where emerging technologies are concerned, the effort to identify the key uncertainties affecting the gains from a specific innovation is still worthwhile. It is often possible to develop strategies to address at least some of those uncertainties—by buying real options or making hedging investments where they are not too expensive, by devising probes for answers to the most critical questions, or by refining monitoring mechanisms to trace developments more effectively. Beyond that, stay alert and wait for the dust to settle.

Assess and Reassess the Knowledge Environment

The effectiveness of the various appropriability mechanisms depends on the character of the knowledge environment. Managers can ask the following questions to better understand their environment and its impact on appropriating innovation gains:

- *Who are your competitors?* Companies may be the lone runner in a particular new field or face a close race among many rivals seeking to reach the market with a new generation of an established product.

The company's position on this continuum affects its appropriation strategy. Crowded fields tend to generate close races, with obvious implications for lead time and patent decisions. If there are no products out on the market, how can these rivals be identified? Although these competitors are often invisible in the early stages, they can be imagined by careful consideration of the background features that led to the pursuit of the innovation. Specifically, managers can answer the question: "How did it happen that we got involved in this particular quest?" If enough humility can be mustered so that the answer "brilliant creative insight" can be set aside, some more specific explanations may surface. How many other people might have read both of those journal articles that we insightfully juxtaposed? How many others might have suffered a particular frustration or problem in research or in everyday life? How many saw the other company's machine at the trade show and wondered about their design choices? Who also trained with the Nobel Prize winner who dominates the relevant branch of science? By this path, it may be possible to gain a sense of the likely number of invisible rivals in the innovative race—rivals who are out there but perhaps taking care not to make themselves visible.

To take a less defensive approach, managers might ask: "Where should we look for the innovative ideas that are *least* likely to be under active exploration by someone else, and thus most likely to bring us lead time advantages or a successful patent application?" A possible answer is to identify the innovative ideas already known to us but not known to the world. This points to the importance of stepping quickly on the stepping stones of our own achievements, searching carefully for the new possibilities that such achievements open up—before the rest of the world gets a look at them.

- *Where are the headwaters of the new knowledge that flows in this field?* This question is an extension of the sort of analysis just mentioned. Biotechnology is the leading contemporary example of an emerging technology with deep and recent roots in government-funded academic research. Companies hire recently trained PhDs, who then pursue commercially valuable inventions and innovations in areas closely related to—or even identical with—their dissertation research. The aspiring innovator in such a field, where the headwaters are well defined and knowledge at the headwaters is highly relevant to innovation, should be in touch with what is happening at the headwaters. Otherwise, there is the risk that the apparent opportunity spotted floating by a downstream location has already been seen upstream by other more qualified participants, who are either actively working on it, hence ahead in the race, or have convinced themselves that it won't work. Access to the knowledge flows from the headwaters often provides general guidance to a company's innovative efforts that goes beyond the specific information acquired. If an assessment made with good upstream knowledge suggests that the concept under consideration is highly original, a major break with standard thinking, it is unlikely that the field is very crowded.

In contrast, the headwaters of information technology are harder to identify because there are probably more college drop-outs than PhDs active in innovation. The headwaters are everywhere and nowhere (although Silicon Valley might be a good place to start looking). There is little that can be done to generate reasonable confidence that an apparently innovative concept has not occurred to someone else—and such confidence is rarely warranted. The result is that speed and quality of execution count for virtually everything, and rarely are there more than short leads available for the winner to exploit.

- *How fast are the rapids?* When knowledge is advancing rapidly, the rent streams that can be captured from individual innovations tend to be short regardless of the means adopted to try to protect them. Even a position well fortified by strong patents can simply be bypassed by the technological wave front. For example, Kodak and Polaroid were slugging it out in chemical photography while the action moved to digital. Under such conditions, emphasis should be given to effective exploitation and leveraging of whatever lead time can be gained, and to complementary assets and dynamic capability advantages that are more durable—as the IBM example vividly illustrates.

- *What are the prospects for strong patents?* As discussed earlier, patents are not the sine qua non of successful rent capture that they are often made out to be, but strong patents are certainly useful when they are attainable. Strong patents are critically important for a small, highly innovative firm that comes up with a new product so advanced that it may even be ahead of its time. It needs long-lasting protection if it is to capture a reasonable share of the rents, lead-time advantages are likely to be slim because the product is too early relative to the demand, and in any case the potential rent stream is a long one. Secrecy is of little avail; imitators will have time to reverse-engineer the product and still serve the first wave of demand. Being small, the company is likely to have to make deals to access the assets and capabilities complementary to its own innovations. It also is unlikely to have resources to risk on extended patent litigation or to have the time and resources to accumulate a large portfolio of supporting patents. In this situation, strong patent protection is about the only resource that may be powerful enough to keep the larger, stronger late coming rivals at bay. The unambiguous "need" for such strong patents, however, by no means guarantees that they are attainable. When the outlook is bleak, the realistic prescription is to prepare to be content with less than a big win.

There are two considerations that may be indicative of the outlook for obtaining strong patents. They both bear, in different ways, on whether the patent is relatively invulnerable to "inventing around." First, if the courts are likely to consider the patent to relate to a "pioneer invention"—defined by the Supreme Court as "a patent covering a function never before performed, a wholly novel device, or one of such novelty and important as to mark a distinct step in the progress of the art"[45]—they will apply the doctrine of equivalents in a manner generous to the patent holder and "stretch to find infringement even by a product whose characteristics lie considerably outside the boundaries of the literal claims."[46] The second consideration is whether, apart from its possibly pioneering character, the invention embodies ideas that are relatively inflexible in their technological expression. The less flexible they are, the harder it is for the imitator or independent inventor to accomplish the result in a manner that does not bear a strong resemblance to the protected manner. This may well be the underlying consideration that accounts for the fact that patents tend to be strong in the chemical industries; in particular, molecules are relatively inflexible ways of expressing the "ideas" associated with their functionality.

Guard the Golden Goose

There is no better hedge against failures of rent capture for a particular innovation than the ability to come up with the next innovation at modest incremental cost. When the fund of technological opportunities is rich and the pace is rapid, it is foolish to focus overmuch on the individual battles and neglect the prosecution of the war. It is wiser to turn attention to building and sustaining the dynamic capabilities that permit the company to track and sometimes lead the portions of the technological frontier that are most relevant to its business. Long-term survival depends on figuring out what to do for an encore, and then for an encore to the encore, and ultimately on building the dynamic capability to support sustained innovative accomplishment. Companies that succeed in doing this create a moving target for the competition that is much harder to hit than the stationary targets represented by the individual innovative achievements. Also, the subtle internal dynamics of a successful R&D organization are much harder to reverse-engineer and imitate than the specific products the lab turns out.

A key strategic question here is the scope of the dynamic capability, the section of the technological frontier in which the company seeks to maintain a significant presence. If the scope is too narrow, there is the risk that a relatively minor shift in the conditions of technological opportunity will make the company's capabilities obsolete. Also, opportunities for fruitful cross-fertilization and recombination among the company's different capabilities are limited. If the scope is too broad, there is the danger of losing focus and flexibility; the managers who make top-level resource allocation decisions become too remote from the action to make informed judgments—or are close to only part of the action and hence find it difficult to make impartial judgments. Large companies tend to fall into this trap, while smaller companies tend to suffer from too narrow a focus on a single innovative success that cannot be sustained forever.

Emerging technologies face an even more basic question: Which goose will produce the golden eggs? Even a very broad and strong dynamic capability may not be worth further investment if the very foundations of competitive advantage are being undercut by "competence-destroying" change.[47] And since it is likely to be impossible to master a new field comprehensively and simultaneously, the beginnings of dynamic capability in a new technology are almost necessarily of narrow scope.

Managing the dynamic capabilities needed for success in emerging technologies raises increasingly complex human resource challenges, as discussed by John Kimberly and Hamid Bouchikhi in Chapter 18. The mobility of workers and growing need for organizational flexibility not only are redefining working relationships, but also mean there is a general tendency for the rents to shift to the technical talent as the whole compensation profile floats upward with the salaries of the stars. Successful response to these contemporary threats to the foundations of sustained innovative achievement may do more for long-term profitability than the effort to surround the quicksilver of intellectual property with legal fences originally designed for real estate.

PART IV

INVESTING FOR
THE FUTURE

By 1999, pharmaceutical giant Merck & Co. was pouring more than $2 billion per year into its research labs in pursuit of breakthroughs that it could take to market. The company continued to broaden the scope of this research into areas such as combinatorial chemistry, polynucleotide vaccinations and genetic research. It also continued to expand its applications, from 11 therapeutic categories in 1992 to a projected 24 categories by 2002. These investments would not produce payoffs for years—if ever.[1]

"Will we be successful in developing a new drug? Will it be successful in the market?" asked Tom Woodward, Executive Director of Financial Evaluation & Analysis at Merck during a conference at the Wharton School. "The whole process is inherently fraught with risk. You don't really know when you begin if there will be a payoff in the end."

Traditional financial analysis would frequently reject such risky investments out of hand, but company leaders know this research is crucial to the firm's long-term success. Woodward described how Merck used a combination of financial and analytical tools, including real options analysis, scenario planning, Monte Carlo simulations and sensitivity analysis to provide multiple estimates of the value of these highly uncertain technology initiatives. "Traditional valuation methods may lead to a decision not to pursue a project when management knows the project has strategic value," he said.[2]

DEFYING LOGIC

Emerging technologies, like Internet IPOs, often seem to defy standard evaluation logic. Given the tremendous investments needed to succeed and the uncertainty of the future value that might be produced, it usually is difficult to put a realistic figure on the value of an emerging technology investment. The technology isn't proven. The markets are embryonic. Cash flows are speculative. There is no net present value to forecast. Yet seemingly on a wing and a prayer, managers are expected to commit millions or billions of hard dollars to these "soft" projects that offer so much potential for both profit and risk.

What is a given investment in an emerging technology really worth? How can managers effectively analyze and allocate resources to competing technology projects? How are these investments best structured through various forms of financing to ensure the capital to stay the course and the flexibility to change it? This section offers insights into using an options framework and developing strategies for financing emerging technologies.

There is increasing recognition of the weakness of NPV and other financial approaches in dealing with the uncertainty involved in emerging technologies, and growing interest in the use of real options. Where NPV discounts for uncertainty, real options provide a way to recognize the increased flexibility created by options. But, as William Hamilton points out in Chapter 12, options valuation based on a strictly financial model usually leaves out important pieces of the picture. To fully develop the real options perspective, managers have to recognize the existing options in the first place, find ways to create new options by structuring the decisions properly and have to be skilled at implementing the ensuing options. Managing this dynamic process of identifying, building, and exercising options is where the true strategic value can be created.

Finding money to finance emerging technology businesses is an on-going challenge. Within large firms, traditional resource allocation models that set hurdle rates for existing and new businesses often are comparing apples and oranges. Companies need to develop new resources allocation models, as Franklin Allen and John Percival point out in Chapter 13. They emphasize that the seeming advantage of being cash-rich may in fact lead established firms to be cavalier with their funds rather than cagey and disciplined. Allen and Percival also discuss the external venture capital markets, IPOs, and other financing strategies that are vital to the growth of start-ups.

Nowhere has financial strategy been pushed more to its limits than in biotechnology, with its massive appetite for cash, long payback periods and high uncertainty. The industry has spawned financial innovations such as off-balance sheet financing strategies, which allow investors to target their investment to a specific research initiative rather than investing in the entire company. In Chapter 14, Paul Schoemaker and Alan Shapiro explore the strengths and weakness of these innovative approaches to financing technology. They examine how Centocor used this investment vehicle to fund its research. Schoemaker and Shapiro especially explore the impact of information asymmetry (managers generally know more about the technology and company than investors) and principal-agent conflicts (incentives of managers and investors not being aligned well) on the various financing options available to new ventures.

Evaluating and valuing emerging technologies will probably never be a precise science. There will always be a large measure of art and intuition involved. In the face of so much complexity, the pursuit of very neat numerical answers is probably foolish. There are tools and frameworks, however, that can help managers develop assessments that, while not precise, are at least more rigorous. The key is to find a route that lies between going on blind faith or gut feel and the misplaced precision that stems from myopic or static financial models. It is about being roughly right, rather than being precisely wrong. It is also about being innovative in creating strategic options and being smart about designing a suitable financial structure for the opportunities at hand.

CHAPTER 12

MANAGING REAL OPTIONS

WILLIAM F. HAMILTON
The Wharton School

The primary value of investments in emerging technologies is in the options created through opportunities for future development and profitable commercialization. Despite increasing interest from both academe and industry in the potential offered by "real options," the dynamic nature and considerable power of this approach have been obscured by analytical complexity and a primary emphasis on valuation. In this chapter, William Hamilton discusses the process by which real options are developed and exercised: recognizing opportunities through an "options perspective," structuring decisions formally to create future managerial flexibility, valuing the options and then actually realizing that value through systematic implementation. To manage emerging technologies and capture their full value successfully, managers need to move beyond traditional methods to this dynamic process of creating and exercising real options.

The new CEO of a high-technology company faced a difficult decision. His company had an opportunity to enter a joint technology development partnership with a small high-technology company with considerable expertise in a potentially important emerging technology. The technology had clear applications in the large firm's current markets and offered the potential to move into attractive new markets. However, the project would require a $2 million investment up front for R&D and a significantly greater investment for later scale-up and commercialization. The subsequent commercial venture would require payment of licensing fees or a share of revenues, depending on the respective roles of the partners.

The CFO recommended against funding the project. His thorough financial analysis, taking into account all anticipated funding requirements and payoffs, indicated that projected discounted cash flows were inadequate

271

to justify the investment—by a considerable margin. The financial evaluation should have made the investment decision easy, but the CEO had nagging doubts. He was uncomfortable with the financial analysis but reluctant to override its conclusions. At the next board meeting, the CEO presented the investment opportunity for discussion, calling attention to the "intangibles" of the project and possible benefits which seemed somehow beyond the CFO's analysis of discounted cash flows. Following extensive and wide-ranging discussion, the board voted unanimously to approve the investment—contrary to the conclusions of the formal financial analysis.

How did they reach this decision? The CEO and board knew intuitively that this opportunity offered greater value than was reflected in the traditional assessment. The CFO's financial analysis did not appreciate that the most important value of the investment was in the options it created for the future. Some of this value was in the tremendous *upside potential* for expansion into new markets and the *flexibility* to cut losses if appropriate, which appeared to more than offset the shortfall of the strictly quantitative projections. Indeed, in the minds of some board members, the new technology offered a significant, but hazy, prospect for radically transforming the industry in the next decade. Just how and when (and if) this might happen would depend significantly on the outcome of R&D efforts over the next two or three years, and no one could hope to make credible estimates of possible payoffs until then.

Further, several board members noted that they placed considerable value on the new knowledge that would be created through this research as well as the yet unrecognized potential for other new products even if identified applications failed to materialize. This initial investment could also lead to broader collaboration with the small company or even the opportunity to purchase it. A board member characterized this set of opportunities as "the lifeblood of our long-term competitive advantage." On balance, following extensive discussion, the board concluded that the future potential of research collaboration with the high-tech partner outweighed the apparent costs and risks which had driven the formal financial analysis.

Without fully understanding it, the CEO and board correctly recognized that the proposed investment offered valuable real options for future flexibility and growth. The opportunities to expand existing businesses and diversify into new ones based on future developments in emerging technology were only part of the story. The investment also carried with it the prospect of additional collaborations with the high-tech company and, possibly, the option to purchase the company in the future. These and other real options embedded in the investment decision represented substantial value to the company that was not reflected in the financial analysis.

But how realistic were these assessments? In this instance, the investment did ultimately result in substantial strategic and financial benefits to the firm. But in many cases in which executives and their boards leave the well-traveled path of traditional financial analysis, the outcomes are not so positive. The expected strategic benefits of options are never realized or turn out to be worth far less than anticipated. Moreover, many managers do not even get to this point in the discussion. The options embedded in their decisions are not recognized at all. Many investments or acquisitions that offer significant upside potential because of the options they create are not even actively considered. Finally, because the benefits actually derived from exercising options depend on future decisions, realization of their value depends heavily on how and when these decisions are evaluated and implemented.

How can managers do a better job of developing, assessing, and managing real options? While there are no simple solutions, this paper offers a framework to guide managers to a deeper understanding of the real options approach and what is required to employ it in technology investment decision making. It addresses questions often asked by managers attempting to understand the real value of real options: How do I know when I am dealing with an option? Why should I consider using an options approach to technology investing? What is required to implement the real options approach in practice?

REAL OPTIONS AND EMERGING TECHNOLOGIES

The potential returns from investments in emerging technologies are inherently uncertain because of the evolving nature of both the technologies and the markets they address. And, of course, the greater the uncertainties, the more challenging are the decisions to invest. Much, if not most, of the value offered by emerging technology investments is in the real options created. At the heart of this option value is the managerial discretion to take full advantage of favorable developments (such as unexpectedly positive research results) while limiting the negative effects of unfavorable developments (such as slow market acceptance). The greater the uncertainty, the greater the value of such managerial flexibility and, hence, of the associated real options.

Understanding Real Options

The evolving theory and practice of both financial options and real options have attracted increasing attention in recent years.[1] In the financial
realm, a typical call option creates the opportunity, but not the commitment, to make a future financial investment. The purchase of an option
on 100 shares of common stock, for example, gives the purchaser the right
to purchase the underlying stock at a defined "strike price" within a specified period of time. Should the market price of the stock rise above the
strike price, further investment in the stock (exercise of the option) would
result in a profit; on the other hand, should the market price remain below
the strike price, the purchaser can lose no more than the original investment
by choosing not to exercise. An essential characteristic of all options is
asymmetry in the distribution of returns—greater upside potential than
downside exposure. This results from opportunities to terminate the investment or otherwise limit negative outcomes while taking future actions
to exploit positive outcomes fully.

The term, *real option,* was first used to refer to options whose underlying assets are non-financial in nature.[2] In concept, real options are quite
similar to financial options because they both create the flexibility, but
not the requirement, for further action as the future unfolds. For example,
in making technology investments, a company might make a relatively small
bet to develop a new technology that then creates the opportunity for a
larger investment and return through commercialization if the development effort is successful. The *price* of the option is typically the cost of developing or acquiring the technology. Exercising the option is usually the
decision to commit to commercialization and the *exercise price* is the cost
of commercialization.

In practice, real options are considerably more complex and challenging
to implement. Real options take many forms and arise in most technology
investments, but recognizing them requires clear understanding and careful identification. Some real options might not arise naturally, but can be
created through systematic restructuring of decisions. Valuation of real options can also present difficulties because of the uncertainties and unique
characteristics of the underlying real assets. Finally, realizing the full value
of real options typically requires careful management of information flows
and subsequent decisions.

In emerging technology investments, real options include the flexibility
to defer, expand or contract, terminate or otherwise modify projects. The

benefits of such options are operational in nature, relying on active, on-going project management. Another important class of real options are more strategic, reflecting future growth opportunities in both existing and new products, markets and businesses. Such growth options are created, for example, through early stage investments in R&D which can lead to proprietary knowledge, new products, and new business opportunities.[3] Similarly, acquisitions of new technologies can offer future commercialization options in both existing and new markets. Investments in information technology platforms typically open up wide-ranging opportunities for future applications. In each case, the value of the investments depends substantially on the creation and implementation of real options.

Why Use a Real Options Approach?

The limitations of traditional financial analyses in evaluating investments in emerging technologies have been widely recognized.[4] Net present value (NPV) and other discounted cash flow (DCF) approaches largely fail to recognize these value-creating characteristics of emerging technology investments. And, in some cases, DCF methods actually treat them as negatives, not positives. In relatively stable markets with well established technologies and well understood applications, confident projections are possible and discounted cash flow approaches offer a conceptually sound and analytically elegant evaluation of future potential. Not surprisingly, DCF has gained widespread support as a simple and powerful investment evaluation tool in such settings. But the limitations become more problematic in the highly uncertain, rapidly changing environments associated with emerging technologies.

In essence, DCF approaches assume a static investment scenario, with a clearly defined decision path and associated outcomes. The inherently dynamic nature of most decisions and the value of managerial opportunities to adjust future decisions and alter outcomes are largely assumed away. Flexibility is ignored or substantially undervalued.[5]

The options approach recognizes not only that flexibility has value, but that its value increases with increased uncertainty. Indeed, in the absence of uncertainty, options offer no value at all because there is no opportunity for management discretion over time. Most DCF applications, on the other hand, treat increased uncertainty as a negative factor leading to a lower valuation. In practice, managers often purposefully use much higher discount rates when evaluating highly uncertain technology investments than when evaluating more certain, shorter-term investments such as plant expansions. In such instances, DCF application is not merely myopic, but is actually prejudiced against longer-term emerging technology investments and their associated uncertainties.

The real options approach is particularly appropriate to emerging technology investments because they exhibit characteristics typically associated with options value:

- Payoffs are highly asymmetric—the greater the disparity between upside and downside outcomes, the greater the option value.
- Future revenues and costs are highly uncertain—in general, the greater the uncertainties, the greater the value of managerial discretion.
- Initial investments (technology development or acquisition) are relatively small in comparison with future investments (scale-up or full commercialization), increasing the benefits of flexibility.
- Most technology investment decisions proceed naturally through multiple stages, or a sequence of decisions, creating multiple options and increased value.
- Time horizons are often long, allowing increased opportunity for updated information on critical uncertainties and subsequent decisions, increasing options value; but preemptive competitor moves in the technology and/or markets can have the opposite effect.

DEVELOPING AND MANAGING REAL OPTIONS

Although the potential of real options has been recognized for more than two decades, only recently have companies and consultants begun to work through the challenges of applying the real options concept to real world problems.[6] This shift has been driven by rapidly changing environments and increased awareness of the limitations of traditional evaluation approaches. It also has been due in part to a growing understanding that options involve far more than a new analytical model that can simply be substituted for DCF analysis.

Most discussions of real options have focused on *valuing* options. Valuation is fixing a specific value on a particular option, pinning it down. But managers who look for a real options formula to substitute for standard DCF valuation have missed the point. Options don't just represent a new framework for valuing decisions. They represent a different *process* for structuring and managing these decisions. A different approach to valuation is important, but is only part of the picture.

Instead, the options approach might be thought of as a cycle, as illustrated in Figure 12.1. This process includes:

- *Adopting an options perspective.* A necessary first step in developing and managing options is to see them. Most business decisions present options, but these are frequently ignored or undervalued when decisions are framed from a traditional financial analysis perspective. Without a fundamental shift in managerial mindset, future opportunities may not even be recognized as real options. And options that are invisible cannot be managed or valued.
- *Creating and structuring options.* Real options are not just a given. Some are inherent in the investment, but others can be created by building additional flexibility into the decision process. By structuring decisions to increase future discretion, managers can generate new options and increased value.
- *Valuing options.* Once options are recognized or created, they can be valued. Even here, the valuation is not a one-shot deal. As each

Figure 12.1
Dynamic Real Options Framework

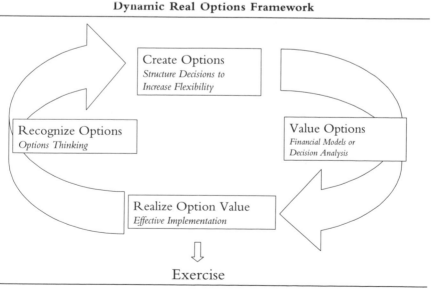

Create Options
Structure Decisions to Increase Flexibility

Recognize Options
Options Thinking

Value Options
Financial Models or Decision Analysis

Realize Option Value
Effective Implementation

Exercise

decision is made and intermediate outcomes are known, the value of remaining options will change, so valuation is an on-going process.

- *Implementing the real options approach.* What looks good on paper can sometimes be quite disappointing when implemented. Real options are focused on future value which, by definition, does not exist at the time the options are assessed. It only is realized through careful management and exercise of real options over time. Unlike financial options, where information requirements are minimal and exercise is generally quite straightforward, exercising real options requires constant monitoring, information updating and timely decisions. The devil is in the details.

ADOPTING AN OPTIONS PERSPECTIVE

Virtually all investment decisions involving emerging technologies—and most decisions involving real assets—present opportunities for future managerial flexibility. In fact, the real options framework reflects the general case that *all* decisions are options decisions; it is highly unusual that a nonfinancial investment offers no possibilities for management discretion to affect future developments. This recognition is the essence of the real options perspective.

Traditional DCF-based investment analysis has been a foundation of management thinking and practice for decades. But a shift is clearly in order, particularly for emerging technology investments, as summarized in Table 12.1.[7]

Table 12.1
Traditional Financial versus Options Perspectives

Traditional DCF Perspective	*Real Options Perspective*
Views uncertainty as a risk that reduces investment value	Views uncertainty as an opportunity that increases value
Assigns limited value to future information	Values future information highly
Recognizes only tangible revenues and costs	Recognizes value of flexibility and other intangibles
Assumes clearly defined decision path	Recognizes path determined by future information and managerial discretion

Managers who view their decisions through a traditional DCF lens will be blinded to real options. Many more options—and increased value—would be apparent if only managers knew where and how to look. Managers need to work actively to counter these options blindspots and to actively identify the options that are inherent in virtually every decision. Most decisions, for example, offer opportunities to delay, accelerate or abandon proposed investments depending on future circumstances. These need to be understood and treated as real options.

CREATING AND STRUCTURING REAL OPTIONS

Once managers recognize that real options are an integral part of investment decision making, they can focus attention on creating new ones. Beyond the options that arise naturally, managers should consider how to structure their investments purposefully to provide additional opportunities for future managerial discretion and increased value.

Some options may be designed to create future operating flexibilities. For example, developing an option for flexible capacity expansion may require design modifications to a planned manufacturing facility. Other options, often characterized as "growth options," can be created to allow for future expansion into new products or markets.[8] In the example cited earlier, investment in the joint development project, if appropriately structured, could offer the option for future diversification into important new markets based on technology advances. Expansion into new but uncertain markets might also require consideration of possible additional distribution channels or marketing partnerships depending on future developments. Similarly, decisions about internal R&D investments in the pursuit of new technologies or products might be structured to include possible joint development ventures or access to complementary technologies in the event of highly successful outcomes. At the same time, options to purchase or license competitive technologies in the event of disappointing R&D progress might also be considered.

Two general approaches can help managers identify ways to create and structure additional options:

1. *Look for opportunities to unbundle decisions.* Most investment projects involve multiple decisions, or sequences of decisions, during the period from initial commitment to ultimate completion. Unbundling

these decisions allows structuring investments as multistage decisions with appropriate milestones. This, in turn, formalizes management's discretion to alter the scale, scope, and direction of projects at different points in time, capturing the dynamic nature of the decision process and its associated option value.

2. *Expand consideration of additional possibilities for future action.* Opportunities for future managerial discretion abound in most investments. Careful consideration of both complementary and competitive possibilities often reveals the potential for wide-ranging additional options. Acquisitions, divestitures, strategic partnerships, technology licenses, and a wide variety of expansion and diversification alternatives are among the most common candidates for options creation.

To get started, managers might ask themselves a series of questions such as: At what points in time (or in the evolution of the project) might we be able to alter the timing of revenues, costs, and other outcomes? Can we defer or accelerate portions of the project? What actions could we take to capitalize on better-than-expected outcomes? How might we mitigate the effects of worse-than-expected outcomes?

Returning to the technical venture example, such questions might lead to structuring the investment into multiple phases—exploratory research, full R&D commitment, pilot plant scale-up, initial product development/market test, full-scale commercialization and so on. At each stage, opportunities for management action-based on updated information could add value beyond what would be reflected in a static DCF analysis. Early in the project, for example, it might become desirable to accelerate the acquisition of knowledge through expanded research efforts because of rumored competitive activity that might shrink the window of opportunity for successful commercialization. Later on, delaying the decision to commit major investment funds to full scale commercialization until significant market uncertainties are resolved may be appropriate. Furthermore, acquisition of the high-tech partner to capitalize fully on successful technology development is a clear future possibility, whether or not a formal option to acquire is negotiated in the partnership agreement.

To the extent that the initial joint R&D results are highly promising, internal R&D or acquisition efforts directed at complementary technologies may deserve consideration. So might a pre-emptive move to acquire exclusive rights to a potentially competitive technology that could offer strategic parity to others. If the development of the initial technological platform is unsuccessful, these alternatives might give the company another technical platform for future growth.

These and other options for managerial discretion can be created and structured through systematic unbundling and expansion of investment decisions. But some real options come at a cost, which must be taken into account in assessing their value. A manufacturing facility designed for flexible production or future expansion may cost more—in capital or operating costs or both—than one without such flexibility. Similarly, accelerating initial research efforts requires additional resources. And securing a formal option to acquire technology rights or other assets has a price.

At the same time that creating additional options can increase value, it also increases complexity. The shift from static assumptions in traditional financial analysis to the dynamic recognition of wide-ranging real options carries with it significant complications in structuring, valuing, and implementing investment decisions. Decision analysis offers some help in meeting these challenges.

A broad range of decision analysis applications to emerging technology investments, including R&D investments and acquisitions of new

Figure 12.2
Hierarchical Tree Structure of a Technology Partnership

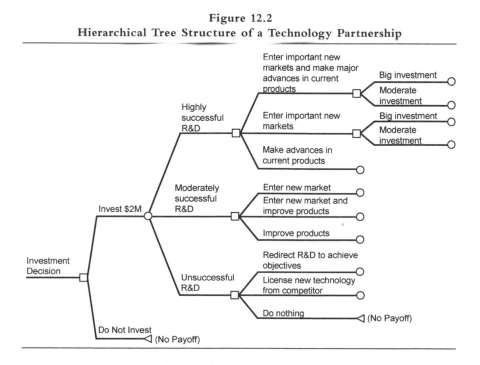

technology, have been discussed and illustrated elsewhere.[9] As illustrated in Figure 12.2, the technology partnership investment discussed earlier can be formalized in a hierarchical tree structure which allows explicit representation of opportunities for managerial discretion. The branches of the decision tree present the decision alternatives and associated outcomes which might arise over time. The first decision is pretty straightforward: invest or not. But future decisions depend upon the success of R&D, whether the technology is applied to new or existing products, the development of new markets and future investments. Only if these subsequent decisions are recognized at the outset can they be identified and valued as options. Creating and structuring real options is an essential step toward realizing the full value of future managerial flexibility in investment decisions.

VALUING REAL OPTIONS

How much are real options worth? Valuation is by far the most explored and discussed aspect of real options, with an extensive literature devoted to alternative valuation models and methods.[10]

Most commitments to new technology offer several potential benefits. The first, and most obvious, is a *financial return* generated by future cash flows from successful commercialization of the technology. A second source of value derives from advantageous *strategic positioning,* providing opportunities for future strategic initiatives or building new distinctive competencies. And the new *knowledge* generated by investments in emerging technology can be of significant value in guiding future investments in related technologies and businesses.

Each of these potential benefits represents an important class of real options, and can in concept be structured as a real option for valuation purposes. But it is important in practice to recognize that quantitative assessment is extremely difficult, if not impossible, for the majority of strategic positioning and knowledge options. For this reason, most options valuation efforts focus on the financial returns which can be directly associated with the investment.

Financial Models

The most extensive literature on real options deals with financial valuation models.[11] For the most part, these reflect extensions and variants of the highly successful Black-Scholes model for valuing financial options. The analogy between financial call options and real options is attractive and powerful. In its simplest form, an investment (the cost of R&D) in a real asset (an emerging technology) offers a future right, but not the obligation, to exercise the option (commit to commercialization) at the exercise price (the commercialization cost). Not surprisingly, the success of the Black-Scholes model in valuing financial options has led many managers to the false hope that a similar mathematical formula can be substituted for traditional financial models to value real options. But investments in real assets, particularly in emerging technologies, are often considerably more complex than financial options in their structure (in the number, variety, and timing of interdependent decisions) and in the availability of data inputs. This makes the application of financial options valuation models difficult at best and, often, even misleading.[12]

In general, financial models approach option pricing from a financial market perspective, or that of shareholders or investors. The value of an investment is considered to be its contribution to the market capitalization of the firm. This is estimated as the value of an identical (or twin) asset, or portfolio of assets, that identically reflects the payoffs and risks of the option. Following this approach, for example, an R&D project to develop a technology would be valued on the basis of an equivalent or twin technology that has already been developed and whose payoffs and risks are already understood. This simple example highlights the apparent weakness of such an approach for truly new, or emerging, technologies and their commercial implications. The newer the technology, the less likely it is that relevant outcomes can be fully identified, let alone well understood based on other technology projects. The search for equivalent assets on which meaningful estimates can be based is exceedingly difficult if not wholly inappropriate.

While such financial models have been used successfully to value selected and often highly simplified real options investments, their practical use is severely limited in situations involving emerging technologies. As an alternative to making heroic estimates and simplifying assumptions to conform to the requirements of analytic financial models, some have suggested the use of approximation techniques or decision heuristics to generate reasonable estimates of option value.[13]

Decision Analysis

Decision analysis, discussed in creating options earlier, can also be used in valuation. It is sometimes preferred by managers because of the power it offers in both structuring and evaluating complex investment decisions.

In contrast to the market perspective reflected in financial valuation models, decision analysis typically reflects the risks and values anticipated by the decision maker. Decision analysis in general, and decision tree analysis in particular, is driven more by the decision maker's understanding of the structure of the investment opportunity than by how it might be expected to affect the market value of the firm.[14]

This has two apparent advantages over financial models for managers dealing with investments in emerging technologies. First, as noted earlier, it is unlikely that the underlying value of an entirely new technology or product can be estimated through the financial markets when no comparable technology or product has ever been developed before. Second, insiders generally have access to information about the technologies, products, and markets associated with an investment that are not reflected in the financial markets.

Once all possible decision paths along with their associated probability estimates and outcomes are laid out in a hierarchical "tree" structure as illustrated in Figure 12.2, the value of an investment—including all identified options—can be computed by "folding back" the branches of the tree to determine its expected present value.[15] Monte Carlo simulations and extensive sensitivity analyses are commonly conducted in practice to help assess the implications of different assumptions and scenarios represented in the decision tree.[16]

Decision tree analysis is not without its difficulties and limitations. First, most complex investments with multiple real options can lead to very bushy and somewhat unwieldy tree structures which present both structuring and computing challenges. But this is the inevitable price of full representation of the multiple opportunities for future managerial discretion and associated option value. Second, discount rates may be required for different project phases or decision tree branches to account for differential risks. And, because the computed investment valuations can be highly sensitive to management estimates of probabilities and outcomes, great care must be taken to develop and test realistic model inputs.

There has been some debate about whether financial modeling or decision tree valuation approaches are superior. In many cases, they can yield somewhat different results because of their fundamental differences in perspective and inputs. The choice depends largely on whether truly equivalent "twin" assets can be identified and valued in the market, and on the confidence with which managers can estimate project payoffs. But when both reflect a similar market perspective and are correctly applied, financial option pricing and decision analysis produce consistent results.[17]

Threshold Assessment

Despite the fondest hopes of academicians and practitioners alike, none of the available quantitative techniques are completely adequate to deal with the complexities which characterize many investment decisions presenting multiple real options. It is exceedingly difficult in practice to compute a precise value for most real options. But quantitative techniques do offer important and valuable support for the managerial judgment on which all significant decisions must ultimately rest. Judgment is always involved in investment decisions whether or not financial models or decision trees are used. From the choice of the particular quantitative model to the selection or estimation of input data and the interpretation of results, some degree of judgment is clearly required.

A number of companies have placed primary, if not complete, reliance on managerial judgment in selected investment situations designated as strategic. This designation is used to ensure that major projects with significant upside potential due to embedded real options are not inappropriately undervalued and, perhaps, rejected because of the inadequacies of financial evaluation methods. Such an approach does call attention to the value offered by future flexibility by excusing the decisions from the normal financial criteria. But the "special" and, often, more favorable treatment accorded strategic projects has led in some companies to what one manager characterized as a "formalized way to get approval for weak proposals."

Threshold Assessment is a related, but more formalized, approach to investment decision making that explicitly combines both managerial judgment and quantitative analysis. Rather than attempt to value future flexibility directly, managers first compute the value of the investment using conventional DCF techniques, recognizing that the value of embedded options may be ignored or substantially understated. To the extent that the result falls short of an acceptable threshold level, the value of both strategic positioning options and new knowledge is considered to answer the question, "Is the value of future flexibility and knowledge sufficient to compensate for the shortfall?"

In essence, this is what happened informally in the board discussion of the joint venture opportunity at the opening of the chapter. When the proposed project failed the standard financial assessment, the CEO and board carefully considered whether the additional value represented by future business opportunities and knowledge was greater than the apparent financial shortfall presented by the CFO. In this case, the value of future discretionary opportunities offered by the investment, or growth options, along with the knowledge to be gained from the joint venture research program, was judged to be more than sufficient to justify supporting the project.

By framing the question in this manner, the focus shifts from attempting to compute an absolute value for the real options embedded in the decision, which is often extremely difficult. Instead, Threshold Analysis directs attention to the real issue: Is that value enough to justify the investment? This requires careful, rigorous examination of the embedded options and a judgment call about their value relative to the decision threshold, or minimum acceptable level of return required to initiate an investment project.

IMPLEMENTING THE REAL OPTIONS APPROACH

Just as managers can increase value by creating and structuring real options, they can also allow this value to erode through inadequate attention to implementation. However thoughtful managers may be about real options, however carefully they structure their investment decisions to reflect embedded options, and however appropriate the valuation, the actual value of real options can only be realized through effective implementation over time.

Real options and their values are not static. Changing markets, competitor actions, unexpected research outcomes, shifting strategic priorities and a host of other external and internal developments may affect subsequent decisions and the value of embedded options over time. Most discussions focus on the nature and value of real options *at a given point in time*. But the development and assessment of real options must be ongoing. Because options can only deliver value if they are exercised in a timely and appropriate manner, the organization needs to have structures and processes in place to continuously evaluate and re-evaluate options. To manage the options process, managers need to pay careful attention to the nature and timing of three key activities:

1. *Monitoring progress.* Regularly updated information on project progress is essential to support decisions to exercise or defer options. It is particularly important in structuring decisions to identify the major uncertainties which should be monitored over time.

2. *Testing and updating assumptions.* Key assumptions and decision points need to be identified at the outset and tested against current conditions, either at regular intervals or when key project milestones or "event triggers" signal the need for re-evaluation. An event trigger might be a competitor action or an unexpected project outcome that would signal the need to review assumptions. Each re-evaluation iteration may require a new or revised decision tree structure to reflect the changes. The longer the development period and greater the uncertainty, the shorter the periods should be between project reviews.

3. *Exercising options.* Continuous updating and review provide the basis for management decisions to exercise, defer, or abandon identified real options. As projects and market conditions evolve, prior assumptions and expectations may need to be revised in response to updated information and these decisions will change. Furthermore, because options often have limited windows of opportunity due to competitive actions or other shifts in the environment, timing is critical. An option exercised too late may have little or no value.

These activities present difficult questions and management challenges. For example: What information is required to properly judge the disposition of real options and how frequently must it be updated? Which options should be exercised and when? Who should be responsible for updating, review, and exercise?

Organizational inertia and other sources of resistance to changes in plans can make implementation of real options difficult in practice. Because projects often develop their own momentum once they have been set in motion, the exercise of some options—especially those requiring shifts in personnel or downsizing—may demand sustained management attention until fully implemented. In many cases, special reporting procedures, incentive programs, and other organizational practices are appropriate to support the full realization of option values.

Because the real options approach is so different from traditional approaches, it represents a challenge to almost every aspect of typical corporate culture and processes. "Playing it safe" has become the norm in many organizations where control systems penalize those whose decisions turn out to be unprofitable, but not those failing to pursue opportunities which might have been great successes. How do organizations maintain accountability or establish performance-based bonuses when projects with inadequate discounted flows are allowed to proceed? New accounting and compensation structures are needed, but the inertia of existing practices often makes such changes quite difficult.

It is hard to underestimate the managerial and organizational complexities of implementing the real options approach. On the other hand, this complexity is precisely what gives an advantage to organizations that have developed the capability to create, evaluate, and implement real options.

CONCLUSIONS

Although the concept of real options is gaining increased currency, real options approaches are far from widely used. The weaknesses of NPV and other discounted cash flow approaches are well documented, but successful application of the options framework has been limited in practice by a predominant focus on valuation techniques and their complexities. Managers may rely on their judgment and intuition to overrule the conventional discounted cash flow-based evaluations and incorporate intangibles or strategic value into their assessments. More often than not, managers are forced to choose between the tightly reasoned analysis of the CFO and a gut sense of the value that can't be captured by traditional financial analysis.

How to cope with uncertainty is at the heart of this issue. The real options approach does not eliminate or even reduce the uncertainties and risks inherent in technology development, market evolution and competitive activity; these remain as real as ever. But the framework does focus explicit attention on such uncertainties and how their implications can best be exploited or limited. Indeed, the greatest benefit is often that managers are forced to confront uncertainty head-on and to formalize the full range of decisions that might be considered over time. This requires attention to far more than options valuation. As argued in this chapter, actually achieving full value also requires systematic creation and structuring of real options along with careful attention to organizational processes supporting timely implementation.

Employing the real options approach requires continuous re-evaluation of alternatives and expectations. And this can only be accomplished with appropriate changes in decision making processes and associated organizational practices. But the companies able to make these shifts can be assured of real returns from real options.

CHAPTER 13

FINANCING STRATEGIES AND VENTURE CAPITAL

FRANKLIN ALLEN
The Wharton School

JOHN PERCIVAL
The Wharton School

With access to greater financial resources, established firms should be at an advantage over start-ups in financing emerging technologies. But the internal resource allocation systems of large companies often put them at a disadvantage over companies that rely on external sources. Corporate allocation is usually based on earning returns greater than the firm's cost of capital. The adoption of emerging technologies and the higher risks and potential returns usually associated with them requires a more sophisticated approach. In this chapter, the authors examine some of the obstacles for established firms, alternative models for resource allocation, and strategies for external financing.

A telecommunications company is considering a shift in strategy. Top managers feel that traditional telecommunications is not a growth business for the future. They believe that the future of the industry lies in information technologies. New ideas, not the old wired networks, will be the source of value in the future. The managers think that it is important to make a move in that direction now, before it is too late, so the company creates a new information technology division and develops a new strategy for the business.

To support the new information technology initiative, managers could provide venture capital to a separate company or create its own "external" investment through a spinoff funded by corporate capital. Alternatively, the company might employ internal financing. The telecommunications business is currently very profitable and generates a large operating cash flow, which has historically been reinvested in the business, used to fund acquisitions to diversify out of telecommunications, or to pay dividends to shareholders. Management decides that the telecommunications business might now be used as a cash cow to fund the long-term shift in strategy to information technology.

Historically, the company has allocated resources based on its cost of capital, which in recent years has been approximately 10 percent. This number has been used as the *hurdle rate,* or minimal acceptable rate of return. Businesses within the old telecommunications business have been expected to earn at least 10 percent rate of return. It is expected that in the future, the information technology division will have many investments that will offer rates of return in excess of 10 percent and that resources will gradually be redeployed into the information technologies division and away from the telecommunications division.

Managers in the information technology business might argue that the new business should not be held to the 10 percent hurdle, particularly in the short run. Strategy should drive resource allocation and not vice versa. If the company's vision is to move into emerging technologies to provide a future for the firm, then resource allocation should simply support the vision.

But the management of the old telecommunications division argues that there is a fundamental flaw in this resource allocation system. The telecommunications division has successfully earned in excess of the company's cost of capital in spite of historical regulatory constraints. The division feels that the business has potential competitive advantages for the future as it is deregulated. The old-business managers argue further that as the company shifts to this emerging technology, the overall risk in the company's operations will increase. This will lead to an even higher cost of capital and necessitate a higher rate of return.

Continuing to allocate capital using the current resource allocation system will, de facto, lead to the telecommunications division becoming a cash cow to feed the growth of the information technology division. That may seem appropriate given the strategic vision of the telecommunications company but it may be inappropriate if the company has no competitive advantage in information technology that will allow it to earn the likely higher cost of capital associated with this new and probably more risky activity. The risk of the emerging technology may not only be higher but also fundamentally different from the risk of the traditional telecommunications business.

Both the managers in the old communications business and the managers of the information technology business have valid points. How can senior managers best build for the future without undermining its past strengths? How can they provide good stewardship of shareholders' funds while launching new strategic initiatives based on uncertain technologies? These are some of the central challenges and trade-offs facing companies as they begin to develop emerging technologies. A legitimate role of the discipline of resource allocation is to try to make sure that capital is allocated to businesses where there is a believable story that risk-adjusted cost of capital will be earned and economic value created. This chapter examines these challenges for internal resource allocation and also explores strategies for external financing of new technologies, using strategies such as venture capital and initial public offerings (IPOs).

FINANCING EMERGING TECHNOLOGIES IN CORPORATIONS

The financial objective of a corporation such as our telecommunications firm should be the attempt to create sustained, and sustainable, economic value. In considering any investment, the company needs to be able to ensure that investor funds are being used responsibly. In other words, the expected returns of the investment need to be greater than some minimal return, usually referred to as the cost of capital. This cost must reflect the risk inherent in the investment and therefore will differ across different divisions. In our example, the telecommunications division should use a cost of capital reflecting risk in the telecommunications industry and the information technology division should use a cost of capital reflecting risk in that industry.

The cost of capital in a particular industry is calculated by considering "pure play" firms that are entirely focused on the industry. It is found by taking a weighted average of such companies' current after-tax cost of borrowing and their cost of equity. In this way the management sets the cost of capital equal to the opportunity costs of investors. In other words, to justify retaining funds, the reinvested earnings must be expected to provide at least the return that the shareholders could have obtained if the earnings had been distributed in dividends and invested in another company at about the same risk.

The key to sustained value creation is earning more than cost of capital in both good years and bad in all divisions. Earning cost of capital in competitive business requires the existence of a competitive advantage that a company can utilize to earn this rate of return. An inevitable process of change takes place in most businesses over the years as the customers, competition, economy, and technology change

From Physical Assets to Intangible Assets

One of the greatest challenges of resource allocation in emerging technologies—and one of the reasons the telecommunications managers had trouble seeing eye to eye with managers in the new information technology business—is that the sources of value for emerging technologies tend to be quite different from the sources of value for traditional businesses. One hundred years ago, successful companies such as U.S. Steel had most of their people involved in managing physical assets. There were a few people who were looking for different ways to fuel the blast furnace, or trying different types of raw materials to see what would happen, but most of the value was derived not from knowledge but from physical assets.

Today's successful companies are firms such as Microsoft that can continuously create new ideas. Almost everyone in this firm is engaged in trying to find new formulas and better ways to do things. There are some people involved in managing physical assets, such as putting disks into boxes, but those people are a very small part of the process of value creation. By far, the greatest value is from new ideas.

Economist Paul Romer points out that the underlying quantity of raw materials used in our economy has not changed that much over time, yet we are much wealthier than we were 100 years ago.[1] Where did the added value come from? In spite of limited quantity, the real prices of these raw materials have declined over the past 100 years. We have taken this raw material that was available to us and rearranged it in ways that made it more valuable. It is not the raw materials but the rearrangement that has created this wealth. Underlying this process of rearrangement are sets of instructions, formulas, recipes, and methods of doing things that are often classified as intangible assets. We need to think beyond the distinction between human capital and physical capital to the importance of these ideas.

Because of these differences, the financial snapshot of knowledge-based industries such as biotechnology and software is quite different from asset-intensive businesses. The knowledge-based firms tend to have high P/E ratios, low dividend payout ratios, low debt/capital ratios and varying levels of current profitability, as illustrated in Table 13.1. More traditional

Table 13.1
Industry Financial Characteristics

	P/E	Payout	Debt/Capital	ROE	Beta
Biotechnology					
Amgen	23	0	5.3	30	1.02
Chiron	51	0	32.7	8	1.02
Genentech	67	0	7.0	6	0.27
Drugs					
American Home	39	55	43.3	25	0.55
Merck	33	47	9.8	37	1.17
Warner Lambert	50	60	40.2	31	0.88
Software					
Computer Associates	26	4	31.9	24	1.55
Microsoft	31	0	0	32	1.17
Oracle	37	0	10.9	35	1.34
Computer Systems					
AOL	NM	0	21	NM	2.07
Cognizant	27	6	0	39	N/A
Oracle	37	0	10.9	35	-0.22
Peripherals					
EMC	35	0	19.9	23	1.44
Lexmark	19	0	2.3	30	0.82
Seagate	NM	0	15.8	NM	1.69
Computers					
Compaq	21	5	0.9	26	1.06
Hewlett-Packard	20	18	16	17	1.33
IBM	16	13	39.2	31	1.19

(*continued*)

tangible asset-based industries such as chemicals and telecommunications have low P/E ratios, high dividend payout ratios, high debt/capital ratios and more consistent levels of current profitability.

Many businesses based on traditional asset-based value (i.e., holding and using tangible physical assets such as plants and equipment) are moving to more knowledge-based value, and need to adjust their evaluation methods accordingly. Chemical companies and chemical-based pharmaceutical companies are moving into biotechnology. Telecommunications companies are moving to information technologies. Financial institutions

Table 13.1 *(Continued)*

	P/E	Payout	Debt/Capital	ROE	Beta
Electrical Equipment					
AMP	20	52	5.5	16	0.53
Emerson Electric	25	46	9.2	21	0.85
Thomas & Betts	20	40	43.0	16	1.14
Networks					
Bay Networks	NM	0	7.5	NM	N/A
Cisco	50	0	0	24	1.19
3 Com	35	0	6.0	25	1.52
Semiconductors					
Applied Materials	18	0	7.8	17	2.41
Intel	17	3	1.8	36	1.24
Texas Instrument	12	7	19.6	31	1.86
Scientific Equipment					
Eastman Kodak	NM	NM	12.6	0	0.26
Honeywell	21	30	33	20	1.15
Minnesota Mining	17	42	9.5	36	0.75
Chemicals					
Dow	11	44	32	23	0.93
DuPont	29	59	24.1	21	0.83
Rohm & Haas	15	30	21	23	1.13
Telecommunications					
Bell South	19	44	29	22	0.56
GTE	19	65	53	36	0.67
U.S.West	22	88	49	28	0.37

Sources: Fortune, April 1998; *Forbes,* April 1998; Bloomberg.

appear to want to make the transition to knowledge- and technology-based financial services companies. Federal Express is intent on transforming itself from an asset-based package delivery company to a knowledge-based logistics company.

The risk/return trade-offs and appropriate financial structures of these companies will need to evolve appropriately if economic value is to result from the transitions. They will need to earn higher rates of return simply to maintain value since their evolving capital structures and increased business risk will necessitate higher payoffs to obtain or retain capital to fuel growth.

Sometimes, as companies move away from asset-based value, they may justify shaky investments because they are strategic or represent high-growth future markets. Strategic value and growth potential are important considerations, but for a company to earn more than cost of capital and create sustained economic value in a competitive business, more than growth possibilities are necessary. A company must have a competitive advantage to earn more than cost of capital. If the firm lacks the necessary vision, insights, skills, or core competencies, it may be unwise to invest even in the hottest growth market.

In this context, the large incumbent firm, despite greater capital at its disposal than a new entrant, may be at a disadvantage for two reasons. One problem is that it is unclear that a traditional company such as U.S. Steel will have the necessary skills, people, and knowledge to have any competitive advantage, other than the existence of capital, over start-ups focused on "recipe-based" businesses (which are easily copied). The second problem is that the risks involved in these emerging technology businesses are likely to be considerably greater than the risks in more traditional businesses.

These higher risks can be seen in the higher beta coefficients shown in Table 13.1. The beta coefficient is a measurement of risk from the point of view of an equity investor. It measures the undiversifiable, or systematic risk, that is left after an equity investment in a company that is added to a well-diversified portfolio of other investments. Thus, it does not measure the total risk of investing in the company but only the risk that cannot be diversified away by the equity investor. Thus, as the company redeploys capital away from the telecommunications business and into the emerging technology, there is reason to believe that the company's beta coefficient and therefore, perceived risk, will increase. Since these emerging technology companies also have lower levels of debt, the risk of the business clearly appears to be significantly higher than more traditional businesses.

This higher risk leads to a higher cost of capital and the requirement for a higher return to meet this new hurdle. If there is no competitive advantage that would allow a company to earn the higher required rate of return, the investment could very likely destroy value. It is important to be measuring the return properly. Certainly there is a timing element that must be considered. The return does not have to be earned tomorrow, but if the return is delayed it must be sufficient to compensate for the opportunity cost and the opportunity cost is high due to the risk.

The Red Herrings of Synergy and Diversification

Companies making the transition from stable, low-risk but low-growth businesses, often argue that any short-run inability to earn cost of capital will be compensated for by either synergy or diversification. These assumptions should be critically examined, because returns from both synergies and diversification may be difficult to realize in practice.

Companies in telecommunications, such as AT&T, that assumed there were great potential synergies between telecommunications and other information technologies seem to have been tremendously disappointed later. Often, achieving synergies requires competing with companies that used to be customers of the company and the synergy turns out to be negative. There have been large numbers of divestitures and spin-offs of non-synergistic businesses that have resulted. Managers need to carefully analyze expected synergies and determine whether they can actually be realized. There should be a believable story that the company's competence in the emerging technology will be sufficient to earn cost of capital.

Diversification also may be wishful thinking. Forty years of academic research in finance has focused on the fallacy of reducing risk through corporate diversification. In well functioning capital markets, such as those in the United States and in the United Kingdom, such diversification usually can be achieved more easily, and less expensively, by equity investors in their own portfolios. There does not seem to be any real reason to believe that companies are rewarded with a lower cost of capital for such diversification since it is superfluous to shareholders.

If there is to be a justification for diversifying into an emerging technology, it must come from elsewhere. For example, such diversification may provide a way of participating in fast-growing opportunities, it may allow preemption of new rivals who will use the emerging technology as a beachhead or it may replace core product earnings lost to the emerging technology. These types of benefit can translate into high rates of return on investment.

Table 13.1 indicates that companies in emerging technology businesses also tend to have low dividend payout ratios, while companies in more traditional industries have higher payouts. A shift in strategy by the telecommunications company toward emerging information technologies would probably necessitate a shift in dividend policy. Telecommunications companies are cash cows already for their shareholders, because a large amount of cash is currently paid in dividends. If that cash is now to be used to feed the growth of an emerging technology division, the shareholders will have to be convinced that they are better off with lower dividend payouts. Presumably, the justification that management will provide is that the emerging technology has more growth potential for the long run. However, the current shareholders may not be interested in trading current dividends for future growth and it may be hard to convince new equity investors that the current management has the skills to succeed in the emerging technology business. Also, the current shareholders may prefer to receive dividends and then directly invest them in emerging technology companies that have a competitive advantage.

If there is no synergy between the emerging technology and existing businesses within the mature company and if there is no reduction in cost of capital due to diversification, there will probably be pressure from shareholders to divest either the new division or the core business. The shareholders may even prefer that the company be split into several separate, independent entities. This has happened in companies such as Hewlett-Packard, IBM, and AT&T. In the absence of synergy and with no reduction in capital cost due to diversification, the corporate unit is now reduced to resource allocation as a means to create value. If the resources are not being optimally allocated, from a purely economic point of view, the result would be a situation in which the whole is worth less than the sum of the parts.

Strategies for Balancing the Old and New Businesses

The keys to success in emerging technologies can be very different from those in more traditional businesses. Different metrics for measuring performance may be necessary because of the very different looking income statements and balance sheets of emerging technologies. Wall Street analysts find that traditional metrics such as net income and P/E ratios need to be abandoned or interpreted very differently. Thus, it often makes sense for the emerging technology unit to be separated from the more traditional business. Among the approaches companies can use to bypass or modify their resource allocation processes are:

- *Real options.* The use of real options, as discussed by William Hamilton in Chapter 12, can help quantify some of the higher risks of the emerging technology. It adds some rigor to the general arguments of "strategic" value of the new business. It can also help to translate the

potential value of the emerging business, which may be understood intuitively by the managers there, into terms that old-line managers and
investors can understand.

- *Equity carve-out.* Another way to separate the new business from
the demands of the old is through an equity carve-out, essentially a
subsidiary IPO. Thermoelectron and other companies have used equity carve-outs to create public subsidiaries in which the parent company typically retains majority ownership. Subsidiary IPOs do tend
to result in significant increases in managerial autonomy since the
public subsidiaries are generally governed by separate management
teams and boards of directors. Equity carve-outs may improve the
visibility of an emerging technology unit of a mature company, aid-
ing in the ability to use equity financing. The equity carve-out
itself raises new capital. If structured properly the equity carve-
out may preserve the benefits enjoyed by small entrepreneurial organizations without sacrificing many of the advantages enjoyed by
larger firms.

  For the mature company to maximize the benefit of partial ownership of the emerging technology, the management system should
  include some key elements. The first is an incentive structure that is
  tied directly to the equity performance of both the public unit and
  the parent. The second is autonomy in strategic decisions and capital
  acquisition. The third is capital spending aimed at providing the public
  subsidiary with the flexibility to respond to changing market
  conditions.

- *Alliances.* As discussed by Jeff Dyer and Harbir Singh in Chapter 16,
alliances can offer a powerful way to leverage resources of the firm by
combining them with those of partners. If they are structured in a
way in which synergies among the companies can be realized and
shared, they offer a way to reduce the cost and risk of major investments in uncertain technologies.

- *Internal management structures.* These need to be flexible enough to
not impose inappropriate policies, structures, or compensation from
the parent organization, yet hands-on enough to achieve potential synergies. A few companies such as Johnson & Johnson and Merck appear
to have achieved this balance.

Traditional analysis, in which similar kinds of investment are repeatedly
considered, needs to be replaced by new analytical frameworks. Merck, for

example, has successfully developed new systems based upon continually developing new products and dropping old ones. It does this by better understanding risk through real options analysis but requiring higher rates of return where necessitated by increased risk.

EXTERNAL FINANCING OF EMERGING TECHNOLOGIES

Many successful emerging technology businesses do not have the benefit of internal resources. They begin their life in small companies that have no choice but to turn to external markets. Companies such as Microsoft, Apple Computer, and Amgen have shown how this funding can help entrepreneurial firms grow into giants using a combination of venture capital and IPOs. Even large firms can take advantage of external sources of capital through mechanisms such as the equity carve-out discussed earlier. Managers at established firms also need to understand the processes and logic of these external markets that shape the thinking of most of the small firms they might deal with in developing or acquiring emerging technologies.

Asymmetric Information

External financing suffers from the problem of asymmetric information. Lenders are less informed than borrowers about what is going on and, as a result, are unable to ensure that actions are taken in their interest. Borrowers have poor incentives when equity is used because they receive only a portion of the benefits arising from supplying effort whereas they bear the entire costs. They also may take risks that are undesirable from the lenders' point of view because they receive the upside potential but do not bear the downside risk.[2]

Equity will be used when insiders view future prospects as poor, because losses are shared between the new and old equity owners. In anticipation of this, lenders will discount equity when new issues are made, so this type of financing is particularly costly.[3] Raising interest rates on loans may lead to a reduction in the quality of borrowers because the ones that will not care about the high rates are those that anticipate a high probability of default.[4] Such asymmetric information problems provide the motivation for many contractual features that are observed in the financing of small firms.

The Challenge of Emerging Technologies

These arguments suggest that the firms that are easiest to fund externally are those with safe, predictable cash flows so there is symmetric information or assets that have multiple uses so that they provide good collateral. Emerging technology firms have neither predictable cash flows nor good collateral. There is typically considerable uncertainty about the costs associated with new technologies. There is often even more uncertainty about the revenues since the precise uses the new technology may be put to can be unclear. Borrowers are typically better informed about these costs and revenues than lenders such as banks. In addition, new technologies sometimes have minimal tangible assets that can be used as collateral. A substantial part of the value of emerging technologies, particularly initially, comes from the option value to continue if the initial development stages are successful.

External financing can take the form of debt or equity. Which one is preferable is truly in the eye of the beholder. Consider a scientist or engineer who has just discovered a new technology and wishes to set up a company to develop its commercial potential. What kind of financial structure is best for her? She will want to reap the maximum possible benefit from the discovery, so she will prefer the maximum possible amount of debt. If it is successful, she will receive the entire surplus; if it is unsuccessful, she will be able to default and walk away. The other great advantage of debt financing is that it allows her to hold all the equity and the attached voting rights. She will be able to maintain control and develop her ideas in the way she perceives to be best.

How attractive is debt finance of an emerging technology from the point of view of lenders? If collateral can be provided, then this may be an appealing alternative since lenders will be assured of repayment even if there is significant asymmetric information. Collateral is rarely available with early-state emerging technologies, so debt is not an attractive financing instrument for the lender. To compensate for the additional risk, a high interest rate must be charged. This has the disadvantage that it will trigger default in many states and the firm will incur the associated bankruptcy costs in a wide range of circumstances.

From the lender's view, equity is a more attractive financing instrument. It allows the upside potential to be captured when the firm is very successful without triggering costly bankruptcy. It also gives the supplier of capital some degree of control because of the voting rights that are attached to this type of instrument.

Convertible Preferred Stock

There is thus a conflict between the founders of the firm who want to use debt and the suppliers of capital who want equity. In practice, neither pure debt nor pure equity is used but instead *convertible preferred stock* is the standard financing instrument in venture capital deals. Preferred stock is like debt in that it involves a fixed payment. However, if the payment is not made, this does not trigger bankruptcy. Should bankruptcy occur, preferred equity holders have a higher priority than equity holders but a lower priority than debt holders. The convertibility feature allows the venture capitalist to turn the security into equity at a predetermined ratio and capture the upside potential should the firm be successful.

Convertible preferred also means the founder of the firm can formally maintain control. Lenders maintain some control by staging the financing, making it contingent on continued progress that is consistent with the lenders' goals. There may also be complex covenants attached to the convertible preferred which give significant control rights to the lenders.

Venture capitalists typically provide financing for a limited period of time. If a firm is successful, its needs for capital rapidly outstrip the capacity of limited partnerships that are the usual providers of venture capital. An important exit mechanism for venture capitalists is an IPO. The IPO provides the liquidity to allow venture capital to obtain a return on the investment and, because the return comes as a capital gain, it is taxed at a low rate. Even though IPOs are costly, they often represent the best means for initial investors to obtain a return. Another common exit mechanism is outright sale of the start-up to a large firm.

Venture Capital

Emerging technology firms have been supported by venture capital, which accounts for about two-thirds of the private-sector external equity financing of high-technology firms.[5] Venture capital differs from standard forms of financing in that there is much more involvement of investors in an attempt to avoid the problems arising from asymmetric information. Lenders are also concerned about resolving the uncertainty of cash flows. The absence of collateral means they cannot simply leave the entrepreneurs to their own devices. They provide financing in stages to ensure that option value is maximized. These characteristics of venture capital mean that the contractual arrangements for venture capital are much more complex than is usually the case. Typically, they have equity-type characteristics, with both sides receiving part of the upside potential of the project.

Many high-technology companies in the United States have initially been funded with venture capital.[6] Although venture capital has been used for over 50 years it is only in the past 20 years or so that it has become a significant source of funds for new companies. Early venture capital funds had limited success and it was not until regulatory changes in the late 1970s that the venture capital industry started to grow dramatically. In 1979, the Labor Department re-interpreted the "prudent man" provision of ERISA to allow greater investments in new companies or venture capital funds. Legislators also reduced maximum capital gains tax rates from 49.5 percent to 28 percent in 1978 and to 20 percent in 1981. Finally, the widespread use of limited partnerships, offering tax advantages to investors, encouraged the growth of venture capital firms.

The typical stages of venture capital investing are shown in Table 13.2. Venture capitalists may provide funds for all or some of these stages. Usually, the amount invested grows through time. At each stage, the amount invested is expected to carry the firm through until the next stage. By staging the financing in this way the venture capitalists can maximize the option value of the investment by making sure the correct continuation decision is made. The form of security that is usually used in venture capital investments is convertible preferred stock, as discussed earlier.

Although venture capital is a commonly used strategy for start-up finance, it is by no means the only route that can be taken. In fact, venture capital tends to be rather concentrated both geographically and by industry. In 1996, 49 percent of venture funding went to firms in California or Massachusetts while 82 percent went to firms specializing in information technology or the life sciences.[7]

"Angel" investors are one of the most important alternatives to investments by venture capital funds.[8] These are wealthy individuals who invest directly in firms rather than through the limited partnerships used in venture capital. Some of them are very sophisticated entrepreneurs, with considerable experience in the industry, who provide extensive advice. Others have little experience and may be rather naive about what is involved in a start-up. The primary criterion used by angel investors is whether the entrepreneur is known to them or to an associate whom they trust. The angel market in the United States is estimated at between $10 billion and $20 billion annually.[9] This is a substantial market when compared to $6.6 billion for venture capital in 1995 and $20 billion in IPOs in 1995.

Table 13.2
Industry Financial Characteristics

The Stages of Venture Capital Investing

1. *Seed investments.* A small amount of capital provided to an inventor or entrepreneur to determine whether an idea deserves further consideration and further investment. This stage may involve building a small prototype but does not involve production for sale.

2. *Start-up.* Start-up investments usually go to companies that are less than one year old. The company uses the money for product development, prototype testing, and test marketing (in experimental quantities to selected customers). This stage involves further study of market-penetration potential, bringing together a management team, and refining the business plan.

3. *First stage—early development.* Investment proceeds through the first stage only if the proceeds look good enough that further technical risk is considered minimal. Likewise, the market studies must look good enough so that management is comfortable setting up modest production and shipping facilities. First stage companies are unlikely to be profitable.

4. *Second stage—expansion.* A company in the second stage has shipped enough of the product to enough customers so that it has real feedback from the market. It may not know quantitatively what speed of market penetration will occur later, or what the ultimate penetration will be, but it may know the qualitative factors that will determine the speed and limits of penetration. The company is probably still unprofitable, or only marginally profitable. It probably needs more capital for equipment purchases, inventory, and receivable financing.

5. *Third stage—profitable but cash poor.* For third-stage companies, sales growth is probably fast, and positive profit margins have taken away most of the downside investment risk. But, the rapid expansion requires more working capital than can be generated from internal cash flow. New venture capitalist funds may be used for further expansion. At this stage, banks may be willing to supply some credit if it can be secured by fixed assets or receivables.

6. *Fourth stage—rapid growth toward liquidity point.* Companies at the fourth stage of development may still need outside cash to sustain growth, but they are successful and stable enough that the risk to outside investors is much reduced. The company may prefer to use more debt financing to limit equity dilution. Commercial bank credit can play a more important role.

7. *Bridge stage—mezzanine investment.* In bridge or mezzanine investment situations the company may have some idea of the timing of exit and still needs more capital to sustain rapid growth in the interim.

8. *Liquidity stage—cashout or exit.* This is the point at which the venture capitalists can gain liquidity for a substantial portion of their holdings in the company. The liquidity may come in the form of an IPO, an acquisition, or a leveraged buyout.

Sources: Adapted from Table 2, Sahlman (1990), p. 479.

IPO Market.   As shown in Table 13.2, every venture capital investment is made with the expectation of a cashout or exit. Often this exit is facilitated by an initial public offering. It can also be accomplished through a private sale or management buyout.

The IPO market plays a critical role in encouraging venture capital investments.[10] The primary reason venture capital is relatively successful in the United States is the active IPO market that exists there.[11] A comparison of 21 countries found that the existence of an active IPO market is the most important determinant of the importance of venture capital in a country.[12] The United States and United Kingdom have very different systems for venture capital than other countries, which may account for the dominance of these two nations in intellectual property industries and many emerging technologies, such as computers (both hardware and software), biotechnology, and the Internet. Many countries have attempted to spur the creation of new firms by encouraging the establishment of stock markets where IPOs of relatively small firms are possible.

IPOs involve substantial direct and indirect costs. Legal, auditing, and underwriting fees constitute the direct costs. For small issues, these can be high. The gross underwriting spread and out-of-pocket expenses as a percentage of offering price for U.S. registered public offerings during the period 1975–1995 were around 16 percent for issues under $10 million and even for larger issues they do not fall below 5 percent.[13] This does not include indirect costs such as management time and effort necessary to undertake the offering.

IPOs initially tend to be underpriced, in the sense that the offering price is usually below the market price shortly after the IPO. This underpricing can be substantial. In the United States, United Kingdom, Germany, and France it is 15.3 percent, 12.0 percent, 11.1 percent, and 4.2 percent, respectively. In Japan it is significantly higher at 32.5 percent. There have been a large number of theories to explain this underpricing, from there being a "winner's curse" to there being informational "cascades." The underpricing phenomenon is a complex one and many factors are probably at work.

Although they do well in the short run, IPOs tend to underperform in the long run. In the United States during the first three years, new issues underperformed similar stocks by a total of about 15 percent when measured from the offering price.[14] This phenomenon is less well understood than underpricing and most theories that have been put forward to explain it essentially rely on some form of irrationality.

In addition to going to the public markets, investors may exit through an outright sale to a large company, a merger, or a management buyout. One study of exits by U.S. venture capital firms found that 30 percent exited through IPOs, 23 percent through private sales, 6 percent through buyouts, 9 percent through secondary sales, 6 percent through complete liquidations, and 26 percent were complete write-offs.[15] In contrast, a European study found that only 10 percent of exits were through IPOs, while 41 percent were through the sale of the company.[16]

A Symbiotic Relationship

While information technology start-ups may require a few million dollars and a matter of months to produce viable products and find investors for an IPO, biotechnology typically takes much longer. These businesses can require hundreds of millions of dollars of cash "burn" and many years before a product has been developed and approved for sale. This could necessitate going to the equity market many times. It may be hard to take such a company public and may be preferable for a large pharmaceutical company to buy the biotechnology start-up.

Even for information technologies, large companies that are concerned about competition from emerging technologies often buy these potential future competitors. Thus, an ideal environment for emerging technologies to find financing may be a period of venture capital support and IPOs, followed by a combination of investments or acquisitions by large companies.

Given the challenges of allocating resources to emerging technology businesses within large corporations, as discussed at the opening of the chapter, large firms often choose to invest in or purchase start-ups rather than develop an emerging technology business internally. At times, they may take a minority equity position in a start-up to gain access to the new technology it creates. This is a popular strategy for big pharmaceutical firms who lack a biotechnology development capability, but have the necessary marketing, manufacturing, and financing skills.

Another possibility is to wait and acquire a successful start-up. As venture capitalists seek to cash out their investments, the large firm can buy its way into the emerging technology. At this point some of the greatest risks may be behind the start-up and at least some of its potential may be recognized.

There is thus a symbiotic relationship between venture capital markets and large corporations. The large firms give the venture capitalists the funds to exit, while the venture capitalists provide access to the emerging technology.

There are still significant challenges involved. From the viewpoint of the mature firm, one problem is valuing the emerging technology properly and not overpaying. This would require objective unbiased assessments of enhanced revenue from emerging technologies and the higher level of risk involved to allow reasonable prices to be paid. It is too early to tell for sure but preliminary analysis would indicate that AOL's acquisition of Netscape and Johnson & Johnson's investment in Centocor resulted in reasonable prices for the potentially enhanced revenue. If a premium is paid and if there is no synergy, then the acquisition ends up earning less than cost of capital. Another risk of delaying these investments is the risk that the small start-up may have no interest in selling. Instead, it might go to an IPO and then become a formidable competitor against the large company. Or another established competitor may be interested in the same firm and either purchase it first or bid up the price. This strategic risk has to be factored into strictly financial calculations of whether to build the technological capability internally, make small investments in outside efforts, or wait and see and purchase a more mature emerging technology firm.

Managers need to understand both the internal allocation approaches and the way in which emerging technologies are initially financed outside large firms. Good investment decisions result from considering both approaches. By changing their method of internal resource allocation and understanding how to capitalize on the existence of external venture capital markets, managers can create approaches to financing emerging technologies that balance their need to provide sufficient returns to shareholders while investing in the future.

CHAPTER 14

INNOVATIVE FINANCIAL
STRATEGIES FOR
BIOTECHNOLOGY VENTURES

PAUL J.H. SCHOEMAKER
The Wharton School

ALAN C. SHAPIRO
University of Southern California

The costs and benefits of using stocks, bonds, and loans as primary sources of financing are well understood and widely reported on in the business press and financial literature. More innovative forms of financing, such as off-balance-sheet funding, have received far less attention because they are fewer in number, often more complex, and may only succeed in special circumstances. This chapter explores the challenges of information asymmetry and principal-agent conflicts that plague all forms of investment relationships. The authors discuss debt and equity financing, as well as lessons from innovative off-balance-sheet approaches used by biotechnology firms to fund specific research initiatives.

In January 1992, Centocor—a biotech firm specializing in monoclonal antibody technology—offered 2.25 million units of a newly formed paper corporation named Tocor II at $40 per unit. The net proceeds of about $84 million were to be spent on small peptide molecular research aimed at developing therapeutic compounds for rheumatoid arthritis and inflammation, as well as other autoimmune and infectious diseases. Centocor's senior officers managed Tocor II, which had no employees or offices. Tocor II subcontracted all research within its legally

We thank George Day and William Hamilton from Wharton, Stelios Papadopoulos from Paine Webber, and Hubert Schoemaker from Centocor for their helpful information and comments.

307

permitted scope back to Centocor, which in turn had an option to buy all patent and product rights of Tocor II. Centocor's purchase option schedule for Tocor II was as follows: $58 per unit, exercisable any time during 1993; $76 any time during 1994; and $107 during 1995.

Each unit of Tocor II also had a warrant associated with it, which gave the investor a three-year option to buy one share of Centocor common stock at $64.50 starting January 1, 1994 (through December 31, 1996). At the time of the offering, Centocor's common was trading around $52/share. As a protection against Centocor not exercising its purchase option with regard to Tocor II, a callable warrant was issued as well, giving each Tocor II unit holder a two-year option to buy an additional share of Centocor common starting January 1, 1996 at a strike price equal to 124 percent of Centocor's common stock price at the end of 1995. This second, callable warrant would expire automatically if Centocor exercised its purchase option regarding Tocor II. Special provisions covered cases of takeover or sale of substantially all assets.

Tocor II was eventually taken over by Centocor, which fell upon hard times, through a friendly acquisition at a price of $38 per unit. Some investors sued the companies' respective boards for having failed to maximize value. The takeover was prompted by the unusual circumstances surrounding Centocor. During 1992, its stock dropped nearly 80 percent because its major drug against septic shock (called Centoxin) was not approved by the FDA. In 1999, Centocor itself was taken over by Johnson & Johnson for about $5 billion. This valuation reflects the remarkable success of Centocor's second major product (Reopro), an anti-blood clotting drug against heart disease.

FINANCING BIOTECHNOLOGY

Biotechnology—with its voracious appetite for cash over a long and uncertain period of development—requires distinctive financing strategies such as the "stock warrant off-balance sheet research and development securities" (SWORDS) used by Centocor to support its research.[1] SWORDS offer investors an option to bet on the commercial value of a single R&D venture within the larger firm, such as Tocor II. This chapter explores the challenges of financing biotechnology and examines traditional as well as more innovative financing mechanisms. Biotechnology—with its high capital costs as well as long development and approval periods—may represent an extreme case of emerging technology financing. But new insights are sometimes found by studying the extreme outliers, and innovations in biotech may suggest creative strategies for financing other emerging tech nologies.

The distinctive characteristics of biotech have created a demand for new approaches to financing. High-risk, development-stage companies must spend enormous amounts of money on research and development to develop novel technologies and then embody these technologies in salable products. To develop a major new pharmaceutical drug cost traditionally between $100 and $400 million, and the payoffs from the few successful drugs are spread over perhaps 25 years.[2] Figure 14.1 summarizes the various stages and hurdles that must be scaled to develop a new patented drug in the traditional world of pharmaceuticals.

On average, a new drug initiative would require a $129 million investment for discovery and development and another $97 million for clinical trials, followed by $9 million for regulatory review, just to get one compound approved. This figure doesn't even include the early discovery work. Some of the leading independent biotech firms have had to raise and spend over a billion dollars in equity capital before turning a profit. These huge funding needs mean that in addition to the obvious requirements—such as developing a strong organization, targeting appropriate disease groups, and pursuing suitable research strategies— a major challenge for independent biotechnology firms is to raise enough capital to stay the course.

Figure 14.1
Overview of the Drug Discovery and Development Process

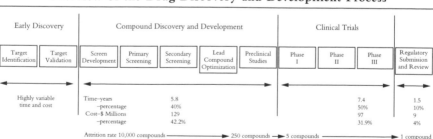

Note: Manufacturing Phase IV, etc., consumer 21.6% or $66 millions, of development expenditures. Total development costs are $305 million.

Source: PARAXEL's Pharmaceutical R&D Statistical Sourcebook, 1997, Malhieu, MO, ed., PAREXEL, Walham, MA, 1997: PYRMA Industry Profile 1995.

The second characteristic of a biotechnology company is that much of its initial value resides in its *growth options*. As discussed by William Hamilton in Chapter 12, these investments are characterized by multistage decisions with downstream embedded options. For example, if research on a new arthritis drug is unsuccessful or strong competitors emerge, managers may decide not to invest in a manufacturing plant or a marketing organization. This ability to alter decisions in response to new information reduces the downside risk while maintaining options on the future upside potential.

Firms with growth options often have few tangible assets in place; the assets consist primarily of specialized knowledge and management skill. Their value lies mostly in the promise of future riches. For example, Amgen, a gene-splicing company, had a stock market value of over $10 billion at the end of 1991 even though earnings for that year were only $98 million, giving it a P/E ratio of over 102. Clearly, the market was valuing Amgen's future ability to capitalize on its research into immune system therapies to treat anemia associated with chronic kidney failure, victims of chronic hepatitis C, and infections in cancer patients undergoing chemotherapy. And indeed, Amgen's market value rose to over $44 billion in 1999 as it started to realize much of this potential. Meanwhile, its P/E ratio has steadily dropped to about 40 after the sixfold increase in its earnings.

Although the recognition of growth options benefited Amgen, the market often has a difficult time valuing biotechnology companies because so much of their value depends upon these growth options, for which there are no obvious comparables. Instead, their valuation must be based on expectations about future profits from yet-to-be-developed products or new market applications for already developed products.[3] Key issues that must be addressed in valuing biotechnology growth options include the therapeutic value of new drugs, patent positions, market potential, and the likely timing of FDA approvals.

In addition to the inherent challenges of predicting the future value of these biotechnology companies, there are two other major challenges that make biotechnology valuations (and all investments, to some degree) difficult:

- *Information asymmetry.* Managers typically know much more about the business at hand than investors do. Investors don't have as detailed an understanding of the technology, the markets, or the true abilities and motivations of managers. This information asymmetry—the fact that one party to a transaction often knows something relevant that the other parties do not know—increases the potential for conflicts or exploitation between managers, investors, and other types of stakeholders.

- *Principal-agent problem.* Given this imperfect and asymmetric information, investors are concerned about how well the incentives of the managers are aligned with their own. This concern is known in economics and finance as the *principal-agent problem.* In particular, the principals (the owners) want to ensure that their agents (the managers) will put in the requisite effort and will spend the firm's resources to benefit the owners and not just themselves. Alignment can be ensured through careful oversight by the owners, but this is usually not practical or desirable. Or the owners can pay bonuses for results, but these could reward managers for short-term gains that undermine long-term results. Instead, many owners make the managers into principals through stock options. But this approach may be expensive since significant stock will be needed to make a difference in the agent's efforts. It is difficult to create incentive schemes that align the interests of principals and agents closely, and nearly impossible to align their interests perfectly.

DIFFERENT FORMS OF FINANCING

The financing options for new ventures include funding from a large established firm, publicly-issued equity, debt, project financing, as well as private placements, venture capital (a special form of private equity), and novel off-balance sheet constructions such as the SWORDS. Traditional forms of financing are examined in more detail by Franklin Allen and John Percival in Chapter 13. We briefly examine below the implications of various forms of financing from the perspective of information asymmetry and principal-agent conflicts before looking in more detail at off-balance sheet approaches.

A Sponsoring Firm

Some biotech companies seek support for the high costs of research and development from established pharmaceutical companies. One strategy, pursued by firms such as Genetics Institute and Centocor, has been to forge strategic alliances with established pharmaceutical companies. In return for financing the R&D, these large firms receive certain commercial rights to any products the research might yield. In the extreme, this route amounts to becoming an R&D division of a major pharmaceutical company or a contract research firm supplying the industry with R&D. A related strategy is to become majority owned by a large pharmaceutical company, but retain an independent culture, research program, and market identity. Genentech and Chiron have pursued this strategy. The benefit of an alliance is that it reduces the need to raise large amounts of equity from the capital markets. However, it may stifle the firm's culture of innovation and thereby restrict its future growth opportunities.

Other biotechnology companies such as Amgen and Biogen have taken the opposite tack and sought to become vertically integrated, from performing R&D, to manufacturing their drugs, and having their own marketing and sales force. This creates tremendous equity needs and the demand for vehicles like SWORDS to help meet them. The desire for vertical integration may be related to reducing operating or transaction costs.[4] It also could be that vertical integration—by enabling the company to retain core capabilities[5] or gain access to markets, as well as direct manufacturing experience—may yield valuable feedback for R&D that is not attainable via joint ventures or arm's length relationships. Also, the ability to attract or retain scientific talent via stock options may be enhanced under vertical integration (as it offers more scope and growth opportunities). Moreover, vertical integration may enable the company to protect unpatentable processes and other trade secrets (see Chapter 11 by Sid Winter).

Equity

Suppose a biotechnology firm needs financing to develop a new treatment for arthritis. If the company goes to the equity market to finance development of the new treatment, information asymmetry may pose a serious problem. How are investors to know exactly how effective the drug will be, what patents it might be infringing on, whether management will be capable of getting it through the FDA, producing it on schedule and at an economic cost, effectively marketing it, and enhancing and supporting it? Given all these uncertainties, investors will expect a sizable risk premium in the pricing of the new equity. The biotech firm would like to delay funding until its situation is more certain, but it knows aggressive competitors may usurp the opportunity. Because they must proceed amid great uncertainty, biotechnology firms generally have to issue equity at a large discount.

Venture capital, which is the most prominent form of financing for technology start-ups, usually commands one-third or more of the young firm's stock. This large ownership stake translates into a high cost of capital for the entrepreneurs, ranging from 30 percent to 70 percent annually, depending on the company's stage of development and its track record. Such high discount rates may have the unintended effect of leading the most competent entrepreneurs to seek alternative sources of risk capital—financial angels who were themselves successful entrepreneurs, corporate partners, or institutional investors such as pension funds—leaving the venture capitalists to finance mostly the second-rate.

This problem, caused by information asymmetry, is the same *adverse selection* or "lemons" problem that leads the buyer of a used car to wonder: If this car is such a good vehicle, why is the owner selling it? The natural fear is that the current owner wants to get rid of the car because he knows that it's a lemon. Fearing that she will get stuck with a lemon, the potential buyer will offer less than she would otherwise for the used car. This price discount in turn may cause owners of good cars not even to consider offering them for sale, thus exacerbating the problem and in effect creating an adverse selection cycle by which increasingly bad vehicles are presented for sale, leading in the extreme to none being sold at all.[6]

One way that venture capitalists deal with the adverse selection problem is to add sufficient value to their deals that the entrepreneurs expect to come out ahead, even with the high cost of capital. In this way, venture capitalists can attract able entrepreneurs. They also carefully check out the entrepreneurs' qualifications and their ideas before investing any money, and then influence decisions as the venture proceeds.

Allen and Percival discuss some of the other key characteristics of venture capital financing in Chapter 13. For our purposes, the key point from venture capital is that the value created by a new business often depends critically on how that venture's financing is arranged. For example, in order to ensure that the founders remain committed to their business, venture capital firms try to structure the deal so that management benefits only if the new company succeeds. This usually means minimal salaries for managers, with most compensation tied to profits and the appreciation in the value of the stock they own. Moreover, venture capitalists typically invest in the form of preferred stock convertible into common shares when and if the company goes public. The venture capitalist, therefore, has a prior claim on the assets of the new business.

One obvious reason for using preferred stock instead of straight equity is to improve the venture capitalist's reward-to-risk ratio. However, this is probably not the primary reason because the founders are unlikely to give something away for nothing; if the founders have to bear more risk, they will raise the price to the venture capitalist of acquiring a given stake in the firm. The more likely reason for using a financial structure that shifts a major share of the risk to the founders is to accomplish two objectives:[7]

1. The venture capitalist is trying to force the founders to signal whether they really believe the promises made in their business plan. If they believe the business will be only marginally profitable, the founders will have little incentive to go ahead with the deal proposed by the venture capitalist.
2. The venture capitalist wants to increase the founders' incentive to make the company succeed. By structuring the financing in this way, the founders will benefit greatly only if they meet their projections.

By their willingness to accept this deal, the founders increase the venture capitalist's confidence in the projections contained in the business plan. The venture capitalist, therefore, is willing to pay a higher price for its equity stake. This particular financial structure also motivates management to work harder and thereby increases the probability that the outcome will be favorable. This example illustrates the twin problems of information asymmetry and agency mentioned earlier and how financing can reduce those problems by forcing the founders to reveal their inside information and giving them incentives to make extraordinary efforts to get their new venture to succeed. Nevertheless, those problems—although somewhat mitigated—remain. If and when the business does go public, it must confront those same problems again, particularly when raising equity capital.

One of the key costs associated with issuing new securities stems from management having inside information about the company's prospects.[8] Recognizing management's ability and incentives to issue overvalued securities, rational investors will revise downward their estimates of a company's value as soon as management announces its intent to issue new securities. The riskier the security being issued, the more important this credibility gap becomes, and the larger the discount applied by investors fearful of buying lemons. Information asymmetry is likely to be pronounced in the case of a high-technology company because management will often have unique information about the technology and the future profitability of undeveloped products and untapped market niches. This increases the opportunity to profit from selling stock when management believes that the company's growth options are overvalued by the market and, thus, increases the amount by which investors will discount the price of new corporate securities to compensate for their informational disadvantage. The natural response—to provide investors with additional information—is often not credible because it is seen as self-serving, or may not be practical because providing outsiders with the necessary information to properly analyze its investments would jeopardize the company's competitive position.

Debt

The discount on debt will generally be much smaller because the cash flows received by creditors are less sensitive to the performance of the firm, excepting the special case of bankruptcy. However, there is a strategic cost to issuing debt that is likely to be particularly great for biotechnology firms. Debt holders generally try to address principal-agent conflicts by requiring detailed covenants to protect themselves against opportunism and potential management incompetence. But these provisions constrain management's choice of operating, financial, and investment policies and reduce its capacity to respond to changes in the business environment. For example, lenders may veto certain high-risk projects with positive net present values because of the added risk they would have to bear without a commensurate increase in their expected returns. The opportunity cost associated with the loss of operating and investment flexibility will be especially high for firms that must respond quickly to continually changing product and factor markets. Even if investors are willing to lend under these circumstances, shareholders will bear the expected agency costs, which are likely to be high. The high costs associated with resolving the conflicts of interest between shareholders and bondholders reduce the desirable amount of debt in a biotechnology firm's capital structure.

In addition, the issuance of debt in amounts large enough to fund a technology company's large appetite for cash will likely exacerbate the conflicts that already exist between a firm's bondholders and stockholders. Because the value of common stock equals the market value of the firm minus the value of its liabilities, managers can increase shareholder wealth by reducing the value of the bonds. This possibility lies at the root of stockholder-bondholder conflicts and limits the use of managerial equity to solve the problem of information asymmetry.

The ability to behave opportunistically stems from bondholders having prior but fixed claims on a firm's assets, while stockholders have limited liability for the firm's debt and unlimited claims on its remaining assets. In effect, stockholders have an option to turn over the firm to the bondholders if things go bad, but to keep the profits if the firm is successful. Consequently, the value of equity rises, and the value of debt declines, as the volatility of corporate cash flows increase. Shareholders (who control the firm) have an incentive to engage in risk-increasing activities—such as pursuing highly uncertain ventures—that have the potential for big returns. Similarly, management can also reduce the value of preexisting bonds and transfer wealth from current bondholders to stockholders by issuing a substantial amount of new debt, thereby raising the firm's financial risk.

Bondholders' fears of being exploited are magnified in the case of high-technology companies. Because growth options involve contingent projects, a large fraction of the investment choices made by high-technology companies cannot be specified in advance. Moreover, other things being equal, the riskier an investment the more valuable is an option on it. Taken together, these factors increase the risk to bondholders of opportunistic behavior on the part of shareholders of high-technology companies with substantial growth options. Another problem for bondholders is that growth options make poor collateral; their value in liquidation is usually nil.[9] In the case of start-ups, whose assets consist primarily, if not exclusively, of growth options, loans are virtually unobtainable (unless personally guaranteed). Large amounts of debt will also increase the probability of financial distress and thereby affect the willingness of non-investor stakeholders, such as customers, employees, and suppliers, to make firm-specific investments. One way to reassure these stakeholders is to maintain a substantial amount of equity on the balance sheet.

Despite the problems associated with debt, many firms consider debt financing to be less expensive than equity financing because interest payments are tax deductible whereas dividends are paid out of after-tax income. However, this comparison is misleading for two reasons.[10] First, it ignores personal taxes. Second, it ignores the supply response of corporations to potential tax arbitrage. Normally, the supply of corporate debt will rise as long as corporate debt is less expensive than equity. As the supply of debt rises, the yield on this debt must increase in order to attract investors in progressively higher tax brackets. In equilibrium, therefore, the tax rate for the marginal debt holder should equal the marginal corporate tax rate, eliminating the tax incentive for issuing more debt. Even if some tax advantage to debt remains, the arbitrage argument implies that only those firms that face the highest effective tax rates are likely to benefit from issuing more debt. Biotech and other emerging technology companies are unlikely to fall into this category.

Most emerging technology companies are unsure of their tax bracket since it is unclear whether they will have net taxable income in any given year. For example, very few biotechnology companies reported any profits at all within the first ten years of their existence. For 1997, publicly traded biotechnology companies as a whole are estimated to have lost $2.7 billion.[11] On average, therefore, the effective tax rate for these companies is significantly below the maximum corporate rate. Moreover, since the variability of profit is likely to be very high for an emerging technology company, there is a lower probability that they will be able to make full use of the interest tax shield, particularly at high levels of debt. Hence, the tax advantages of debt seem less valuable for high-technology companies than for mature companies with fairly stable incomes and few other tax shields.

Although growth companies are unlikely to be able to benefit from the tax advantages of debt, taxes may still play a role in their financing strategy. Specifically, low-tax-bracket growth companies may be able to use financing mechanisms such as leasing in order to transfer certain tax benefits to other companies that can more fully utilize them in return for a lower effective cost of funds. An alternative possibility is to sell R&D partnerships that pass tax benefits along to individuals in high tax brackets, as was often done in the early days of biotechnology.

Convertible Securities

Faced with this unsatisfactory trade off between the steep discount on new equity and the restrictive covenants associated with issuing straight public debt, biotechnology companies are apt to look elsewhere for funds. A more flexible alternative is to issue bonds or preferred stock that are convertible into common stock at the investor's option. These securities offer investors participation in the high payoffs to equity when the firm does better than expected, while simultaneously offering them the downside protection of a fixed-income security when the firm's value falls. If the firm does undertake riskier projects, holders of convertibles or warrants will see the value of their equity claim rise (because stock price volatility increases an option's value), offsetting to some extent a decline in the value of the fixed-income portion. This offset means that the value of a convertible security should be relatively insensitive to the risk of the issuing company.[12] This feature of convertibles also means that the effect of any divergence in risk assessment is much less for convertible than for straight debt: If the market overestimates the risk of a biotechnology company (and thereby undervalues its straight debt), it will overvalue the convertible's call option feature. In this sense, convertible securities are well suited for coping with differing assessments of a company's risk.

A convertible issue may also provide more advantageous financing terms if management believes the market is undervaluing the company's stock. Convertibles provide an indirect means for the firm to sell common stock at a higher price, albeit on a deferred basis, thereby avoiding current market fears that management has chosen to sell equity because it is overpriced. Even if the call option embedded in the convertible security is underpriced due to the credibility gap, once the true information is revealed, and the issue is converted into common stock, the firm will receive a higher price than it would have received had it sold equity directly.

If management truly believes that the stock market is undervaluing the company's shares, the least expensive financing option would be to issue straight debt. Thus, convertibles are the best choice only if straight debt is inappropriate under these circumstances, for instance, because of the added restrictive covenants or the high risk of default.

The problem with convertibles for biotechnology firms, however, is that the very flexibility they afford investors may actually reduce management's financing flexibility. Once issued, a convertible bond is a hybrid security which effectively becomes an equity claim when times are good (and the value of the firm rises) but a straight debt claim when the value of the firm falls. It is, of course, precisely when times are bad that debt can cause problems for biotechnology companies short of tangible assets. On the other hand, the lower coupon payment or preferred dividend when an equity "sweetener" is attached will reduce the debt service burden and, therefore, the likelihood of financial distress. Also, the less restrictive covenants associated with convertibles provide management with more flexibility in responding to unforeseen events than does straight debt.

The high uncertainty of biotechnology ventures increases the challenge of information asymmetry. Measures such as SWORDS that the firm can use to reduce this uncertainty can simultaneously raise the price that investors are willing to pay for its securities and reduce potential conflicts among the various corporate stakeholders.

ALTERNATIVE FINANCING
FOR BIOTECHNOLOGY

An alternative route for those seeking to become major, independent bio-pharmaceutical firms is to solicit equity investments for targeted research projects. Rather than putting the whole company out to investors, these "off-balance sheet financing" strategies, such as Centocor's use of SWORDS, raise funds to finance a particular research initiative.

In the early days of biotech (during the 1980s), two principal forms of off-balance sheet financing were utilized: (1) research and development limited partnerships and (2) stock warrant off-balance sheet research and development securities (SWORDS). These two investment vehicles are largely the same, except that partnerships are sold in a private offering whereas SWORDS are sold in a public offering and trade actively on a national exchange.

Early in the development of biotechnology, numerous private R&D partnerships were formed aimed at well-to-do private investors (e.g., physicians, lawyers). Since 1988, numerous SWORDS have been introduced. The amounts of capital raised per SWORDS varied from $27 million to $84 million. Table 14.1 provides an example of some of the earliest SWORDS used in biotechnology. Figure 14.2 offers a breakdown of the major sources of capital for biotech companies in general over time, whereas Tables 14.2 and 14.3 provides a summary of Centocor's specific financing profile. Although SWORDS (which fall under the category of follow-ons) are not the major source of funding, they are a novel form. We discuss them in some detail next because they illustrate well the various tradeoffs involved in different financing vehicles.

SWORDS, as one form of off-balance sheet financing, target a promising and reasonably well-defined area of research. Typically, the issuing firm manages the investment, with an option to buy out the public investors according to a timetable of preset prices. In return for this limited upside, the investors usually receive warrants or options to buy a share of the issuing firm's common stock at a preset premium strike price and time period. These warrants may trade separately from the original investment unit after some specified period of time. Also, some SWORDS were set up as off-shore companies, so that in case of buy-back a full tax credit would accrue for the foreign asset purchase. In addition to biotech, off-balance sheet financing has been widely used in the movie industry and in oil and gas

Table 14.1
Summary Profile of Nine Biotech SWORDS

Issuing Company	Name of SWORD	Date of Offering	Amount Raised (millions)	Warrants		Buyback Premium		
				Period	Premium	Year 2	Year 3	Year 4
Alza	Bio-Electro Systems	12/12/88	$45.417	12/13/90–12/12/95	34.10%	109%	109%	182%
Centocor	Tocor	07/11/89	24.500	01/01/92–12/31/94	21.60	N/A	100	167
Immunex	Receptech	11/17/89	27.000	02/16/90–01/31/95	10.00	N/A	117	208
Elan	Drug Research Corporation	11/14/90	43.189	11/14/92–11/14/95	25.30	N/A	91	155
Genzyme	Neozyme	10/26/90	47.352	01/30/91–12/31/94	0.00	65	117	196
Genetics Institute	Scigenics	05/23/91	40.000	12/01/92–05/31/96	11.40	52	89	146
Gensia	Aramed	11/27/91	57.500	10/01/93–12/31/96	21.60	61	105	177
Centocor	Tocor II	01/28/92	80.000	01/01/94–12/31/96	24.00	45	90	168
Cytogen	Cytorad	01/31/92	38.500	02/01/94–01/31/97	N/A	N/A	95	164

Figure 14.2
Sources of Capital in Biotechnology Industry

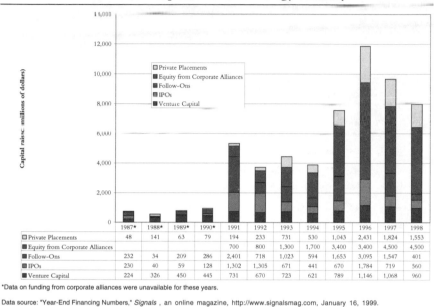

	1987*	1988*	1989*	1990*	1991	1992	1993	1994	1995	1996	1997	1998
☐ Private Placements	48	141	63	79	194	233	731	530	1,043	2,431	1,824	1,553
◼ Equity from Corporate Alliances					700	800	1,300	1,700	3,400	3,400	4,500	4,500
◼ Follow–Ons	232	34	209	286	2,401	718	1,023	594	1,653	3,095	1,547	401
◼ IPOs	230	40	59	128	1,302	1,305	671	441	670	1,784	719	560
◼ Venture Capital	224	326	450	445	731	670	723	621	789	1,146	1,068	960

*Data on funding from corporate alliances were unavailable for these years.

Data source: "Year-End Financing Numbers," *Signals* , an online magazine, http://www.signalsmag.com, January 16, 1999.

exploration. Each of these industries shares with biotech the focused pursuit of low probability—high payoff projects in clearly defined domains (e.g., a disease class, a movie, or geologic field).

Why SWORDS? Within the industry, various arguments have been advanced for using SWORDS to finance R&D projects. The economic logic underlying some of these arguments, however, is suspect if one believes in efficient and rational capital markets. Among the reasons advanced for using SWORDS are:

- *Increasing earnings per share.* One of the most prevalent explanations for the initial popularity of SWORDS is that by shifting research and development expenses to the financing vehicle, this off-balance sheet financing enables parent firms to show higher earnings per share (at least through the R&D stage) and thereby boost their stock prices. There is reason to question this view, however, since the companies issuing SWORDS are also selling off the rights to some of their high-valued growth options. Such a trade-off makes sense only if financial

Table 14.2
Centocor's Financing History

Type	Date	Vehicle	Shares or Units	Ave Price per Share or Unit	Receipts (millions)
a	1979–1981	Venture Capital			$7.415
f	1981–1986	Immunorex Associates			16.444
b	14-Dec-82	Initial Public Offering	1,650,000	14	23.100
c	10-Aug-84	CORP, L.P.	125	40,000	5.000
c	12-Sep-85	CCIP, L.P.	231	100,000	23.100
b	13-Dec-85	Second Public Offering	1,100,000	22.50	24.750
b	24-Apr-86	Third Public Offering	1,050,000	36	37.800
c	13-Dec-86	CPII, L.P.	551	100,000	55.100
c	24-Mar-88	CPIII, L.P.	542	100,000	54.225
d	12-Jul-89	Tocor	2,875,000	12	34.500
e	May-89	Series A Convertible Debentures	10,000	1,000	10.000
e	Dec-89	Series B Convertible Debentures	9,250	10,000	9.250
b	1-Jun-90	Fourth Public Offering	1,495,000	13.25	64.659
e	18-Jan-91	7 1/4 Subordinate Notes	106,650	1,000	106.650
e	16-Oct-91	6 3/4 Subordinate Debentures	125,000	1,000	125.000
d	21-Jan-92	Tocor II	2,250,000	40	90.000
f	Aug-92	Eli Lilly and Company	2,000,000	25	50.000
f	Dec-93	Wellcome plc	2,000,000	10	20.000
g	1987–1993	Warrant exercises	3,873,578	21.49	83.236
g	1994	Warrant exercises	6,151,000	11.65	71.633
f	Oct-94	Wellcome plc	140,000	25	3.500
g	1995	Warrant exercises	728,076	16.59	12.079
g	1996	Warrant exercises	425,731	13.33	5.675
b	Mar-96	Fifth Public Offering	4,025,000	33	132.825
e	Mar-98	4 3/4 Convertible Subordinated Debentures	460,000	100	460.000
		TOTAL FUNDING			$1,525.941

Table 14.3
Key Types Listed in Table 14.2

Type	Type of Financing	Percentage	Amount
a	Venture Equity	.49	$7,415
b	Public Equity Offering	18.55	28,134
c	R&D Partnerships	9.01	137,425
d	R&D SWORDS	8.16	124,500
e	Convertible Debt	46.59	710,900
f	Private Placement Equity	5.89	89,944
g	Exercise of Warrants issued in R&D Partnerships and SWORDS	11.31	172,623
		100.00	$1,525,941

markets systematically undervalue future earnings. The available empirical evidence, however, supports the notion that financial markets take a long-term view of capital expenditures and investments in R&D.[13] Indeed, the valuation of biotechnology firms themselves is perhaps the best demonstration of the market's long-term perspective and willingness to look beyond short-term earnings. At the end of 1991, about ten years or so into its existence, the biotechnology industry had an aggregate market value of over $42.5 billion, even though it produced a combined net loss of over $575 million for the year and discovered very few major drugs in return.[14]

• *Increasing the valuation of future cash flows.* Another common rationale for the use of SWORDS is that by allowing investors to participate in targeted areas of research and development that promise unique risk–return combinations, biotechnology firms may be able to command a higher price for the rights to their future cash flows than if they sold these rights directly in the form of straight equity claims. In general, however, the high elasticity of supply in financial markets means that repackaging a firm's cash flows so that it reallocates risk from one class of investors to another is unlikely to be a sustainable way of creating value. Moreover, as discussed later, biotechnology firms are probably at a comparative *disadvantage* in offering innovative securities like SWORDS. Investors, who are not in a position to properly evaluate the technical merits of the firm's ventures, may fear that the R&D projects bundled in the SWORDS will be lemons. In response, investors are likely to heavily discount these securities, thereby negating the benefits of any innovation.

- *Reducing the cost of capital.* On the other hand, SWORDS make a commitment that can reduce uncertainty and thereby lead to a lower cost of capital. They convey information about the intentions of management (reducing information asymmetry) and limit their actions (reducing the potential for principal–agent conflicts). By reducing uncertainty about the investment, they help lead to a lower cost of equity than less specific investments. Investors are less likely to discount share price based on information asymmetry or principal–agent conflicts. One virtue of SWORDS is that they are irrevocable in their commitment to conduct research; even a board decision could not reverse this unless there were a takeover. SWORDS also offer options for future investments based on the outcome of research, so investors may be willing to accept a smaller ownership share for a given investment. Management benefits from this financing structure because it means giving up a smaller share of ownership for the needed funding.[15]

- *Reducing the problem of information assymetry.* If investors overestimate the riskiness of a specific R&D project because of concerns about information asymmetry, they will overestimate the value of the option on that project. These offsetting effects will make the value of SWORDS relatively less sensitive to the disagreements that are inevitable when trying to create and value novel products and projects. Although the upside value of the SWORDS is capped by the buyback agreements, buyback prices tend to be sufficiently high as to give investors extraordinarily high returns should the projects succeed. The box discusses the analytic aspects of SWORDS in more detail.

- *Reducing the downside risk to the parent.* In contrast to convertibles, if the R&D project turns out to be uneconomic, the parent and its shareholders have no further liability to the SWORDS. Conversely, the parent will exercise its call option and buy back the SWORDS precisely when it is easiest to raise additional equity—when the SWORDS has developed a project worth more than the cost of exercising the buyback option.

- *Using strategic signaling to preempt competitors and attract support.* SWORDS also help to signal the company's commitment to research in a given area, which could serve important strategic purposes. Given limited resources and market sizes for biotechnology, the ability to

Valuing SWORDS

How should Centocor's Tocor II, discussed in the opening of the chapter, (and similar SWORDS) be valued in an efficient market? We define the efficient market view as entailing (1) rational self-interested actors, (2) efficient solutions of agency problems, and (3) valuations based on properly discounted incremental cash flows. Below, we examine the value structure of SWORDS in order to better appreciate their complexities, decision points, information asymmetries, and potential conflicts of interest.

Valuation

A Tocor II unit (which is representative of SWORDS in general) essentially consists of a standard and callable warrant to buy one share of Centocor common plus a repurchase option by Centocor to obtain all of Tocor II's commercial rights. In essence:

$$V(\text{Tocor II}) = \text{Warrant}_1 + p[\text{Buyback}] + (1 - p)\text{Warrant}_2$$

where $V(\text{Tocor II}) = $ value of Tocor II

$p = $ probability of a buyback

$\text{Buyback} = q_1 PV_1(\$58) + q_2 PV_2(\$76) + (1 - q_1 - q_2)PV_3(\$107)$

$q_1 = $ probability of a buyback in year 1, given that a buyback occurs

$q_2 = $ probability of a buyback in year 2, given that a buyback occurs

$PV_i = $ the present value of the buyback price in year i

$\text{Warrant}_1 = $ the present value of the detachable warrant

$\text{Warrant}_2 = $ the present value of the callable warrant

Figure 14.3 diagrams the key decision points for Centocor and the investor. Rationally, since Tocor II is a nondividend paying asset, Centocor should wait until the last moment of 1993 to decide whether to buy back all Tocor II shares at $58. Note that this decision (denoted I in the diagram) should in theory be independent of Centocor's common stock price. It only hinges on the value of obtaining more information about Tocor II's research versus the price of waiting. Since the probabilities of buyback (q_i) should be independent of Centocor's stock price, $V(\text{Tocor II})$ consists of two independent components: two call options (one being conditional) and a probabilistic buyback. Assuming that Tocor's R&D value is a diversifiable risk, investors should only care about its expected value relative to Centocor's

(continued)

325

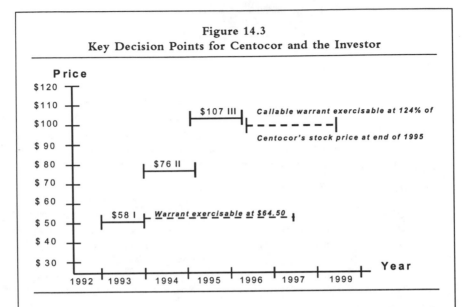

Figure 14.3
Key Decision Points for Centocor and the Investor

Price

$120

$110

$107 III — Callable warrant exercisable at 124% of

$100

Centocor's stock price at end of 1995

$ 90

$ 80 — $76 II

$ 70

$ 60

$58 I — Warrant exercisable at $64.50

$ 50

$ 40

$ 30

Year

1992   1993   1994   1995   1996   1997   1999

buyback prices. The warrants can be priced using the Black–Scholes option pricing model (modified to take into account the effects of dilution). However, considerable uncertainty existed at the time about Centocor's price profile over the ensuing seven years (which in turn would depend partly on the success of Tocor II).*

The more difficult challenge is to assess the probability of buyback. This decomposes into an economic and a reputational issue. Biotech firms wishing renewed financing via SWORDS may exercise their options even if the research results are not quite worth the buyback price. To assess the probability of buyback exclusive of reputational concerns, it may be useful to start with the overall base rate of success in pharmaceutical research and then adjust for (1) the particular class or approach of research and (2) possible adverse selection. The more narrowly defined the class, the better the base rate estimate will be but at the price of a higher standard error (due to the smaller sample). Within the capital asset pricing model (CAPM), only the expected probability matters; however, investors may in fact be ambiguity averse. Experimental evidence suggests that most people prefer crisp over diffuse priors, even if the expected probabilities are the same.†
Regarding adverse selection, a downward adjustment may be needed in the base rate to reflect that biotech firms might offer the less desirable or more

risky R&D projects disproportionately for SWORDS financing. Estimates of this effect may be obtainable from the venture capital literature, as the less promising ventures typically seek outside funding.[‡] This is the classic lemons problem.[§]

Pricing Efficiency

Yet another problem in analyzing SWORDS is assessing whether they are efficiently priced, that is, whether they are undervalued or overvalued. Ordinarily, one would address this question by looking at the pattern of returns to see if they are serially correlated. Although autocorrelated returns typically suggest an inefficiency in pricing, this is not necessarily the case here because of what is known as the "peso problem." The peso problem stems from an external event whose outcome is unknown (such as a government decision to devalue the currency) but where the passage of time itself reveals important information about the likelihood of that event occurring. In such a case, because the probability of devaluation depends on the amount of time that has already elapsed, returns could be serially correlated without there being an arbitrage opportunity ex ante. The peso problem here could stem from something like uncertainty over FDA approval of an important new drug, with every additional day that passes without approval reducing the probability of future approval. Moreover, high (low) returns ex post could just mean that the drug was approved earlier (later) than expected, not that the SWORDS was mispriced initially. For SWORDS in general, high (low) returns ex post could just mean that the FDA changed its approval process in a way that was not anticipated ex ante. Hence, the complexity of SWORDS makes even ex post analysis of their returns an insufficient indicator of pricing efficiency.

---

[*] The original Black-Scholes formula assumes that the number of shares remains constant. However, if a warrant is exercised, the number of shares will increase, and the firm's earnings and assets must be divided among more units of stock.

[†] H. Einhorn and R. Hogarth, "Decision Making Under Ambiguity," *Journal of Business,* *59* (4), Pt. 2 (1986), pp. S225–255.

[‡] See R. Amit, L. Glosten, and E. Muller, "Venture Formation," *Management Science,* October 1990, pp. 1232–1245.

[§] G. Akerlof, "The Market of 'Lemons': Qualitative Uncertainties and the Market Mechanism," *Quarterly Journal of Economics, 84,* 1970, 488–500.

scare off competitors may become important. This will require past success, a strong R&D reputation, persistence, deep pockets, and a willingness to fight vigorously (in court as well as in the market place). The issuance of SWORDS can reinforce these other factors by serving as an irreversible commitment. If a firm can credibly signal that it intends to pursue and dominate a given disease group, it may keep competitors at bay and enjoy near-monopoly rents. The downside of the early signaling strategy, however, is that it eliminates surprise and diminishes potential first-mover advantages. Instead of scaring off competitors, disclosure of intent and approach via a SWORDS may in fact accelerate and intensify competition. Financing via bonds or common stock does not entail these costs and benefits. A game theoretic model would be needed to judge whether, on signaling grounds alone, SWORDS are strategically more desirable than other forms of financing. Biotech firms could always signal their intentions through press releases and other communications, but such pronouncements are less credible because they are a revocable commitment to conduct research.[16] Likewise, SWORDS may help develop crucial external linkages with universities and other research-intensive firms, due to their credible commitment.[17] For example, outsiders may reveal interest or provide access to know-how in ways the firm might not otherwise have discovered.

- *Motivating employees.* The irreversibility associated with SWORDS may also bear on certain intra-organizational issues. A SWORDS project has clear identity and performance measures. Management could offer key scientists bonuses depending on whether the firm exercised its option to buy back the commercial rights. This may incentivize key researchers to try harder because their bonus is less discretionary or arbitrary than with other R&D projects that are judged on patents filed, general impressions, and other soft criteria. Management in turn is unlikely not to exercise its buyback option just to save on the promised bonuses. Since the buyback schedule of SWORDS entails two to five years, they offer alternative binding mechanisms (in addition to stock options) for key scientists.

- *Avoiding public offering expenses.* Companies also point to the fact that the exercise of these warrants generates additional equity capital without having to go through a time consuming and expensive common stock offering (usually entailing 5–7 percent commissions). However, conversion will occur only if prices rise beyond the warrant's exercise price. If the stock price does *not* rise above the exercise price, the company may be forced to raise additional equity anyway, at the worst possible time—namely when the company's prospects, as reflected in its stock price, have already dimmed and caused concerns about additional, but hidden, problems. The resulting discount on the sale of new equity will likely more than wipe out any expected savings from using warrants to issue new equity. In addition, the greater complexity of SWORDS will likely lead to higher marketing expenses and, hence, higher issuing costs.

CUTTING BOTH WAYS

Although SWORDS appear to solve certain agency and strategic problems, they may create other problems. The commitment of these SWORDS can cut both ways because commitment reduces flexibility. If a biotech firm encounters hard times, for instance owing to research setbacks or regulatory delays, it cannot divert SWORDS funds to support its general operations. A dramatic episode in Centocor's evolution illustrates this dilemma. Approval of its flagship product Centoxin was unexpectedly delayed by the FDA on April 15, 1992, resulting in an overnight loss of 44 percent in overall market value. The company had built a 200+ person sales force in anticipation of its first major U.S. drug. With a cash balance of about $200 million and a quarterly net cash outflow ("burn rate") of $40 million, Centocor had less than a year and a half before drastic actions had to be taken. If the $84 million obtained from the Tocor II SWORDS had instead been raised via straight equity, Centocor would have gained an additional half year.

Further delay of FDA approval would present Centocor with three unattractive strategic options: (1) Raise more equity capital under the worst possible circumstances, (2) sell off future product rights, or (3) give up independence. Centocor eventually abandoned its research on Centoxin and pursed option (2) through a $100 million deal with Eli Lilly and finally elected option (3) in its $5 billion merger with Johnson & Johnson. It is hard to judge in hindsight whether the SWORDS financing route was smart. It did provide an extra financing buffer to Centocor, which might have been very costly via straight equity or debt, but sacrificed flexibility in the use of these funds. Industry observers feel that it was Centocor's innovative financing strategy—and the resulting financial cushion it created—that enabled it to survive the Centoxin debacle. The turnaround after the stock had tumbled from $60 per share to below $6 was an arduous route, prompting skeptics to refer mockingly to the company as "Cento-corpse." The cost of SWORDS financing is clearly high in hard times, when it offers the company less operating flexibility than straight equity. The knowledge that the company will run out of cash in the relatively near future also has a powerful motivating effect on management. Its energies are focused on getting the most out of limited resources. By treating cash as a scarce resource, management will behave in a way that benefits itself and the shareholders.

Another problem with SWORDS is potential conflicts of interest. Since the typical R&D investment vehicle underlying SWORDS is a paper company, totally dependent on decisions made by its parent, there are obvious conflicts of interest. For example, the prospectus of Centocor's Tocor II contained over a page of small type describing the various conflicts of interest that can arise because of its arrangement with Centocor—ranging from transfer pricing to allocation of scarce Centocor resources to Centocor's acquisition of valuable expertise without paying for it. The net result is that the project's benefits may unduly flow to Centocor. This possibility is mitigated somewhat by granting warrants on Centocor's stock: Any below-market transfer of value from Tocor II to Centocor will simultaneously depress the value of Tocor II and increase the value of the option to acquire Centocor stock. These offsetting effects will reduce the sensitivity of the Tocor II units to conflicts of interest. In addition, the board of the SWORDS and standard auditing committees can, in theory, allay much of the transfer pricing risk. Any mistreatment of Centocor's current SWORDS investors will also be constrained by its desire to issue additional SWORDS in the future.

Management also could play games regarding its exercise of the buyback option. In a one-period model, the investor runs the risk that management would not exercise the buyback option even though the commercial rights are worth the price. Either the company figures it can obtain the scientific benefits by other means, since it performed all the relevant research, or it assumes that the SWORDS lack the knowledge and financial means (once all funds are spent) to contest improper appropriation of research and commercial rights. Several factors counter such short-term opportunism. First, the company plays a multiperiod game, with other SWORDS to be considered (present and future). Second, legal recourse and remedies exist for wronged investors. Third, the senior managers individually are unlikely to benefit the company at their own ethical, legal, and reputational expense.

CONCLUSIONS

To reassure its stakeholders, reduce the costs of financial distress, and take advantage of growth options at opportune times, biotechnology companies should maintain substantial financial resources in the form of unused debt capacity, large quantities of liquid assets, excess lines of credit, and access to a broad range of fund sources. The desire for financing flexibility requires firms to perform a balancing act, which can be viewed in the context of the financial pecking order: internal funds, issuing debt, and issuing equity.[18] Assuming that information asymmetry and transaction costs make it more expensive to use financing sources at the bottom of the pecking order (equity), a firm that needs to raise funds faces an inter-temporal tradeoff. If sources high on the pecking order (internal funds and debt) are used this period, current financing costs will appear to be low, but the firm faces the hidden opportunity cost of being pushed down the pecking order (by being forced to issue more-costly equity in the future). On the other hand, if new equity is issued this period and the funds are held as cash, current costs are high, but the option to move up the pecking order in the future (to internal funds and debt) may actually provide the firm with a cheaper source of funds overall.

Investments in emerging technologies often entail growth options that may require different financing and evaluation approaches than more traditional investments in the mainstream of the business. Significant information asymmetry and agency concerns typically surround emerging technology initiatives, which in turn may complicate investments in growth options beyond the ordinary. The use of debt as opposed to equity is often less attractive in such investments. SWORDS constitute one (partly successful) mechanism to overcome the challenges of financing biotech; we examined it to elucidate the issues posed by information asymmetry and principal-agent issues.

New forms of off-balance sheet R&D financing in biotechnology are an innovative means to raise large sums of equity funding without losing independence or too much operating flexibility. The use of SWORDS seems especially appropriate for firms striving to become independent, vertically integrated biopharmaceutical companies. The downsides of

SWORDS, however, are several. First, they substantially increase the risk of the parent firm in that additional finances might be needed when the firm is least able to raise it. Second, SWORDS entail serious agency conflicts (due to information asymmetry) that may unduly lower their value to investors. Third, SWORDS are complex securities that may involve higher underwriting expenses. This may explain why their popularity waned.

Although our focus has been on the external financing options available to independent emerging technology firms, established companies face similar tradeoffs regarding internal new ventures. The problems of information asymmetry and principal-agent conflict may be somewhat less acute, however, in established firms. First, information flows more freely within the boundaries of a large firm, so that in theory there should be less concern about violations of confidentiality or the strategic withholding of information. Second, a large firm may already have resolved the main incentive problems among key internal stakeholders, through compensation plans and perhaps by developing a culture of collaboration and mutual support. Nonetheless, with respect to external stakeholders, established firms may encounter the same problems that upstarts do. As such, the observations in this chapter should be of value to them as well.

PART V

RETHINKING THE ORGANIZATION

Is Jack Welch concerned that the disruptive technology of the Internet may play havoc with his finely tuned organization? "This is about as good as it gets," the CEO of General Electric Corp. said in an interview with *Business Week.* "This takes everything that's mundane, everything that's slow, and it takes them out of the game. You're not pulling an organization in this one. You're letting it grow like flowers. You have a group of people dying to get into the Internet. That's so much easier to manage. You're asking them to break the mold. Everybody is excited. They want a piece of this too. And we can give it to them. It just doesn't get much better than this."[1]

Whether or not executives embrace these changes as enthusiastically as Welch, emerging technologies have a tremendous impact on organizations and the people within them. Managers are challenged to rethink the organization. This section explores some of these challenges and new approaches to managing interorganizational knowledge networks, developing alliances and other partnerships, creating new organizational designs, and redesigning relationships with employees.

A BROADER VIEW

To deal with the uncertainty and complexity of emerging technologies, managers need to take a more expansive view of the firm. While the unit of analysis in the past has been a single corporation, the competitive reality in emerging technologies is that success depends upon whole constellations of firms. The first two chapters in this section explore the management of networks and alliances.

In Chapter 15, Lori Rosenkopf examines the impact of informal and formal "knowledge networks" that are crucial to the development of new knowledge and the spread of technological standards. Although these networks play a critical role in the creation and diffusion of new technologies, many companies are largely unaware that they exist. Rosenkopf offers strategies for evaluating network advantages and strengthening these networks.

Because of the high cost and uncertainty of emerging technologies, companies often choose to participate through alliances and other partnerships. But managing these alliances presents its own challenges, as discussed by Jeff Dyer and Harbir Singh in Chapter 16. They point out that the role of these alliances depends upon when they are undertaken. In early-stage technologies, alliances are often used as a "window" to allow firms to explore a broad spectrum of technologies. Later in the process, alliances are typically used to place strategic bets, creating options for the future. The authors discuss the strategies needed to build successful alliances and the underlying core capabilities required for successfully managing emerging technologies.

In Chapter 17, Jennifer Herber, Jitendra Singh, and Michael Useem examine emerging organizational models that are replacing the hierarchical designs of the past. They discuss how organizational innovations—such as virtual organizations, network firms, spinouts, ambidextrous firms, front/back designs, and sense-and-respond models—can offer companies the adaptability and innovation needed to successfully manage emerging technologies.

Knowledge is central to emerging technology firms, and this knowledge often is held by employees. The rapidly changing demands of emerging technologies mean that companies need greater flexibility from employees. At the same time, employees are increasingly seeking more independence and control over their work and personal lives than can be found in traditional, one-size-fits-all job designs. In Chapter 18, John Kimberly and Hamid Bouchikhi examine the human resource challenges of emerging technologies. They explore how the concerns of companies and employees can be brought together into a customized workplace where workers and employers jointly determine the conditions of employment on a case-by-case basis.

Even outside of emerging technology firms, alliances, organizational forms and workplaces are changing. But in the demanding environment of emerging technologies, companies have a greater appetite and need for these innovations to help solve their sticky human resource challenges. Radical technologies are especially disruptive to organizations, but whether this disruption leads to dysfunction or to greater flexibility and creativity depends much on how managers respond. With the right approaches, these changes can drive the organization to higher levels of commitment and performance. It can be, in the words of Jack Welch, "as good as it gets."

CHAPTER 15

MANAGING DYNAMIC
KNOWLEDGE NETWORKS

LORI ROSENKOPF
The Wharton School

Emerging technologies are not developed and commercialized by individuals or single firms. They are developed in networks. As the complexity and required resources for developing emerging technologies continue to increase, the development and management of these knowledge networks become a central strategic issue. How are these knowledge networks shaped? How do the influence the progress of emerging technologies? Many of these networks are very informal, so how do companies begin to map and understand their existing networks? What can managers do to build and strengthen them? This chapter explores the role of knowledge networks and strategies for managing them.

By the mid-1970s, two alternative standards for flight simulators used to train military and commercial pilots were engaged in a dogfight. The first model was the Full Flight Simulator (FFS), with all the bells and whistles, as well as a $15 to $20 million price tag. These systems offered integrated cockpit instrumentation, full motion and visual capability. The second primary option was a more bare-bones Flight Training Device (FTD) for just $1 to $3 million, based on the principle that faithful replication of the flight experience is not necessary for effective pilot training. Despite the difference in pricing and features, it would not be cost, technology, or effectiveness alone that would decide who would triumph in these virtual skies. It would be networks.

Thanks to George Day and Paul Schoemaker for many helpful comments, as well as to Anca Turcanu for her careful analyses of cellular industry data. I also appreciate the assistance of Sonya Ahluwalia, Joan Allatta, and Anjana Pandey. I am grateful to Elaine Baskin, editor of *Communications Standards Review*, for providing technical committee participation data.

The Full Flight Simulator was supported by commercial airlines, regulatory bodies, and aircraft manufacturers. Ex-pilots who ran the training programs at commercial airlines liked the realism of the simulators. Manufacturers liked the integration of instruments, which generated a larger market for their instruments, aeronautical models, and flight test data. This standard was supported by several strong professional organizations, including the Air Transport Association's Training Committee and Federal Aviation Administration's Advanced Simulation Plan Working Group. Lobbying by these organizations helped to make FFS the regulated norm for commercial aircraft training and certification in 1980.

On the other side, the Flight Training Device standard was supported by academic and military researchers who believed that faithful replication of the flight experience was not necessary for effective training. The FTD systems were also used by regional and general airlines and flight schools that could not afford the more expensive systems. The FTD standard was not supported by strong organizations until the 1980s. It was only after this network of organizations was developed that the flight simulation community began to actively consider the increased use of FTD-based training.

The structure of the networks that were formed before 1980 made the success of the FFS standard almost inevitable. As shown in Figure 15.1, the

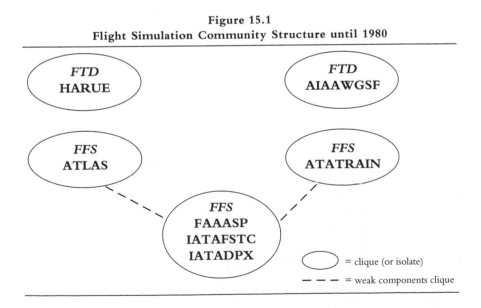

Figure 15.1
Flight Simulation Community Structure until 1980

cooperative technical organizations that supported the FFS standard were in the interlinked organizations shown at the bottom of the figure. The three organizations in the bottom oval share the same membership to form a powerful clique supporting the FFS standard. The supporters of the FTD standard, as shown by the two detached ovals at the top of the figure, did not have these connections. With the establishment of the FFS standard in 1980, this supporting network became even stronger and more locked in, as shown in Figure 15.2.

It was only in the late 1980s that the FTD supporters woke up to the power of networks. A period of technological ferment that resulted from the rise of modular simulation approaches led to a rapid growth of cooperative technical organizations and an increase in membership. This change reshaped the networks and allowed FTD supporters to begin to link with one another to develop cliques that could garner enough support to make the technology a viable alternative, as illustrated in Figure 15.3.[1] The FTD approach (new guard) could now compete effectively with FFS products (old guard). One sign of the acceptance of the FTD was that most of the traditional FFS manufacturers added high-end FTD lines. Some organizations have aligned themselves with one or the other standard, but some of

Figure 15.2
Flight Simulation Community Structure 1981–1986

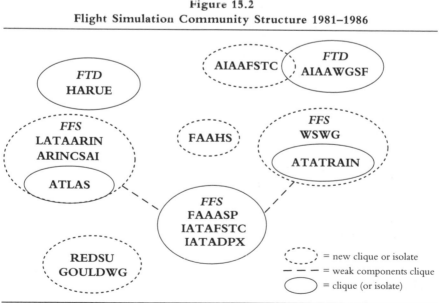

Figure 15.3
Flight Simulation Community Structure 1987–1992

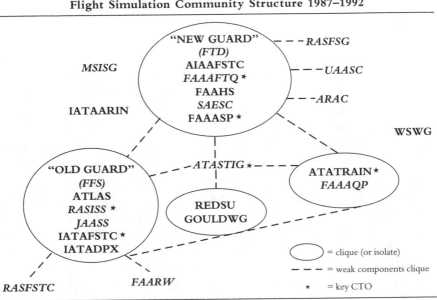

the key organizations in the industry are connected to both groups (as shown by the circles outside the old and new guards). The reshaping of these networks has helped make FTD more successful as an alternative to the high-cost full flight simulator.

As illustrated in the flight simulator example, networks can make or break a new technology. These networks play a role in research, product development, and commercialization. Yet, because they are often complex, undefined, and run across organizational borders, managing them is a daunting challenge. Even recognizing these networks can be a test of corporate perception.

A multitude of ties facilitate the flow of knowledge between organizations; these ties and organizations constitute *interorganizational knowledge networks*. Such networks influence both the fates of competing technologies and the fates of competing firms. In this chapter, the managerial implications of these networks and strategies for network development are explored.

THE POWER OF NETWORKS

Networks play a central role in the management of emerging technologies. A multitude of players generate and circulate relevant knowledge, yet it is extremely difficult to ascertain which knowledge is most critical for success. Research suggests that the choice of new technologies is intimately tied to sociopolitical considerations, and so the structure of these knowledge networks can strongly influence the development of technologies and the success of firms.[2] Exploring knowledge flows in interorganizational networks, therefore, can help managers think systematically about targeting useful external sources of knowledge as well as mechanisms for obtaining this knowledge.

In particular, because emerging technologies are built upon emerging knowledge, *knowledge networks* are central. The webs of relationships that companies form determine their success. Successful firms must capture the knowledge circulating through these webs and use it to their advantage. By placing themselves at the center of these connections, companies can gain access to new knowledge percolating through the system and ensure that the company's knowledge and standards gain widespread support.

Consider these examples of the intricate webs of relationships that connect competitors and related firms:

- Telecommunications firms cooperate to develop standards in technical committees sponsored by organizations like the Telecommunications Industry Association (TIA). Traditionally, employees of AT&T (and its related entities) have maintained a strong presence in these types of committees. This presence reinforced their reputation for expertise throughout the industry and allowed them to control many aspects of standards-building for decades. In the wireless communications arena, as Qualcomm developed the digital cellular technology called CDMA, it adopted AT&T's approach and devoted significant resources to standards body activity. This effort created many informal links with engineers throughout the industry, and Qualcomm also developed alliances that created additional links with the firms these engineers represented.
- Microsoft's continued success is frequently attributed to its ability to integrate other firms' technological innovations into its own strategy. The company has learned about and marketed new technologies through the use of strategic alliances and acquisitions. Its numerous

relationships have enabled flexibility and adaptability in the face of rapid technological change even as the firm has become larger and more bureaucratic.[3]

The prestige and power of organizations like AT&T, Qualcomm, and Microsoft may well be attributed to their ability to create and navigate these webs of relationships. In emerging technologies, knowledge about technological possibilities and strategic intentions circulate through these webs. Since one of the requirements of managing emerging technologies is the ability to synthesize new knowledge with the existing knowledge base of the firm, successful firms must capture the knowledge circulating through these webs and use it to their advantage.

ANALYZING THE NETWORK

To chart the shape and flows of networks, network theory draws upon a branch of mathematics called graph theory. Graph theory has been used to find the best way to move petroleum between refineries or telephone calls between switches ("the maximal flow problem") and find optimal routes for traveling salesmen ("shortest-path problem"). Graphs are composed of nodes and arcs, as shown in Figure 15.4, where arcs (arrows) carry quantities of items between nodes (circles) that serve as repositories for these items.

The most relevant application of graph theory for our purposes has been in the sociological study of communication networks. Figure 15.4 can be thought of as a communication network by assuming that each node is a person or firm and understanding that an arc between two nodes means that those two nodes communicate with each other. The arcs have arrowheads because flows of information between any two people or firms may be reciprocal or unidirectional.

In this communication network, then, we might imagine that Firm A, at the center of many arrows, has advantages because it has access to more information than anyone else in the network. Or we might imagine that if Firm C were to try a powerful new practice, we might see a chain of adoptions due to this communications network. Firm C communicates to Firm F, which passes the idea to Firms A and G, who relay it to B and E. Since H, J, and K do not communicate with the larger group, they will not hear about the new approach. Neither will Firm D; although it is linked to the larger group through A, information flows only one way. Firm D shares information with A, but A does not return the favor.

Figure 15.4
Hypothetical Network

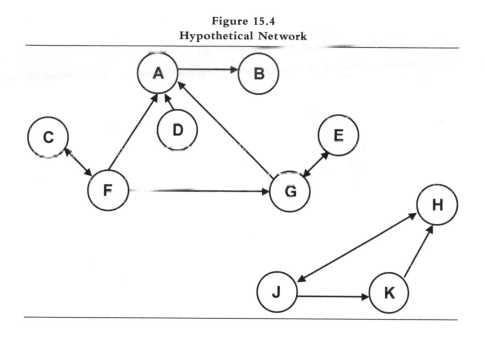

Identifying the Network

The first challenge in managing the knowledge network is to identify it. Formal alliances make the knowledge transfers fairly easy to see (although even here there is much more to an alliance that what is on paper). But knowledge flows in and out of firms through many other channels. Identifying all the many actors in the network and the links among them is a considerable challenge.

The design of the network is shaped by the types of nodes, or network actors, and arcs, or network ties. To understand knowledge networks, managers need to identify the actors involved in the network and the mechanisms by which they are linked.

Identify the Network Actors.   The set of actors creating and disseminating knowledge about emerging technologies may be quite broad.[4] The biotechnology community, for example, is far more diffuse than biotechnology firms alone. Other actors contributing to the development of the technology include universities, research labs, and pharmaceutical firms linked together in "networks of learning."[5]

Likewise, the community of actors involved in the development of laser technology is not limited to manufacturers, universities, and research labs, but also includes government and military bodies. Specifically, patents in the optical disc arena are held by a diverse set of actors—while companies such as Sony, Philips, and Matsushita are key players, research laboratories such as Battelle and government military organizations also hold patents in the area.[6]

Since the underlying technology for optical disc components is useful in many different types of products, the manufacturers are not merely optical disc firms, but also come from the health care and semiconductor industries.

Determine the Network Ties. Although the network represented in Figure 15.4 designates connections with simple arrows, there are actually a wide variety of formal and informal links between actors in the network. The arrows convey one important characteristic of the links between organizations—the direction of knowledge flows. But there are many other characteristics of these links that affect the functioning of the network.

Interfirm linkages may be through formal or *intended* networks, such as written contracts, or through informal or *emergent* networks, such as exchanges of knowledge in chat rooms between engineers from competing firms. These latter networks are much less visible, and so are quite difficult to manage. Among the places to look for these linkages are:

- *Alliances.* The role of alliances in knowledge-sharing has been well documented.[7] Alliance relationships range from licensing agreements to equity joint ventures, but all are formalized through contracts. The intensity of knowledge-sharing through alliance relationships tends to deepen over time as players in a satisfying relationship formalize additional arrangements and increase their capacities to learn from one another.[8]
- *Participation in cooperative technical organizations.* Another link between firms is common participation in professional societies, trade associations, standards bodies, technical committees, and other "cooperative technical organizations."[9] In most industries, there are numerous "supra-organizations" that bring together representatives from many firms. These organizations provide many venues for informal know-how trading among the firms that send their engineers and managers to participate in the activity.[10] In the wireless communications

arena, for example, firms such as Lucent, Motorola, and Qualcomm all send representatives to attend technical committee meetings sponsored by the Telecommunications Industry Association. Whether or not these firms have formal contractual agreements, common participation in this association facilitates knowledge transfer by informal means. Research also suggests that the development of informal relationships through CTO activity may be associated with the formation of alliances. So the informal knowledge exchange that occurs in the technical committee venues may enable identification of joint product development opportunities.[11]

- *Joint authorship.* Authorship of technical papers with partners from other organizations or firms is another way researchers create links across different organizations. Technical papers demonstrate tangible collaborative R&D activity. Research focused on the biotechnology industry suggests that firms that collaborate with universities on R&D efforts as demonstrated by co-authorship find that these boundary-spanning relationships enhance learning and flexibility.[12]

- *Board interlocks.* By "board interlocks," we mean that two firms have the same person serving on each of their boards. This common board member's accumulated expertise and contacts will be available to each of the two firms, and information about one firm's activities may be more transparent to the other. Studies have shown poison pill strategies[13] and M&A activity[14] have been diffused through these board interlocks and that firms linked through board interlocks are more likely to form alliances.[15]

- *Job changers.* Knowledge also moves between firms due to the career mobility of engineers and managers. Firms are more likely to build on the knowledge of other firms when an engineer has moved from the built-on firm to the builder.[16] Other research has found that the more previous jobs managers have held, the bigger their set of opportunities for alliance formation, because they know more about potential partners.[17] Managers pursuing particular product strategies tend to imprint these same strategies onto new firms when they move.[18] A study of semiconductor start-ups suggests that the transfer of knowledge through staff mobility may reduce the likelihood of formal alliances with the same firm. Presumably the new staff brings the knowledge to the firm, so the formal alliance is no longer necessary.[19]

- *Electronic communication.* With the rapid expansion of electronic communication, employees do not have to change jobs or attend professional meetings to exchange information. There is an emerging set of computer-assisted interactions between engineers of different firms including conversations in chat rooms, visits to web sites, or direct e-mail communication. While these links are still largely invisible, technology to enable the systematic collection and monitoring of these activities is in various stages of development.

Proxy Measures for Knowledge Flows.   These ties can be directly mapped or systematically studied but it is sometimes unwieldy to accumulate systematic data on network ties. From a practical standpoint, it is difficult for a firm to track the knowledge that is transferred in a chatroom or at a cocktail party at an association meeting. Another way to identify possible knowledge flows among firms is to consider proximity measures, such as how geographically or technologically close the firms are. While these measures do not explicitly identify the mechanisms of knowledge transfer, they help to predict the likelihood of knowledge transfers:

- *Technological similarity.* Two firms that are similar technologically are more likely to pursue intelligence about each other, to participate in the same associations, and to have engineers communicating electronically. What constitutes a technologically similar firm? Researchers have used patent data to determine whether firms patent in similar areas[20] or whether the firms' patents cite similar work.[21] Similarly, citations in the same technical publications[22] or actual product data[23] may be compared.
- *Geographic similarity.* Geographic proximity also can serve as a proxy for various informal mechanisms that transfer knowledge. The infamous stories of engineers trading know-how over beers at the local watering holes, as well as the movement of engineers between firms in a region, both support this notion.[24] Studies also have shown that the tendency of knowledge to cluster locally can be attributed to the mobility of inventors within regions.[25]
- *Patent citations.* In addition to looking for similarities in patent citations, examining networks of related patent citations offers fascinating insights into the paths of search and exploration firms follow.[26] Ties among firms can also be observed in overlapping patent citations to the scientific literature.[27]

There are many links that facilitate the flow of knowledge among firms. In understanding their own knowledge networks, managers can explore each of these ties and also proxy measures that might indicate the potential for linkages, even if they are not apparent. As can be seen, because these links are not limited to formal ties that are intentionally created, identifying all the links can be challenging.

Assessing Network Advantage

After identifying the relevant actors and linkages in the knowledge network, the next challenge for managers is to assess what this means for the firm. Once they have drawn a map of their networks similar to Figure 15.4, how can they assess their relative network advantage? What does this network imply for the fate of each actor? And how can they reshape the network for greater advantage? For example, if the FTD flight simulator firms had done this analysis, they might have recognized their disadvantage as a result of a far weaker network and could have begun more quickly to put together the network needed for success.

Several types of measures have been used to quantify an actor's position in a network, including measuring the ties of each actor (degree centrality), the centrality of the actor in conveying knowledge throughout the network (betweeness centrality), and identifying cliques.

Degree Centrality. Most simply, we can count the number of ties each actor maintains. In network parlance, this measure is known as *degree* or sometimes *degree centrality*. Considering the network depicted in Figure 15.4, calculating degree requires consideration of the directionality of the network links. While one could calculate degree simply by counting the number of arcs attached to each node without regard to directionality, many researchers would distinguish between "in-degree" (the number of arcs coming in to the node) and "out-degree" (the number of arcs leaving the node). For our purposes, in-degree measures the amount of information available to the node and is more appropriate. Figure 15.5 displays the in-degree centrality measures for each firm of the network. Firm A has the highest in-degree (3) and would be expected to reap the greatest information benefits. In contrast, Firm D has the lowest in-degree (0) and would be expected to suffer from the lack of information.

Weighted Degree. While degree centrality is an easily measured quantity that gives a intuitive measure of how much knowledge an actor might access, we must recognize that knowledge from certain prestigious firms may be more valuable than knowledge from run-of-the-mill firms. So one

Figure 15.5
Using In-Degree Centrality to Calculate Network Advantage

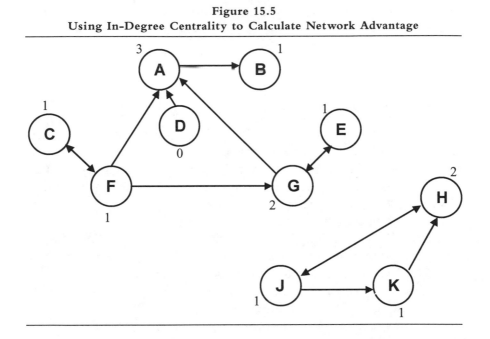

might use a weighted degree score that incorporates the prestige of each actor to whom the focal actor is connected. For example, assume that Firm A is high-prestige and that Firm C is low-prestige. We might weight links to Firm A more heavily and discount links to Firm C. More specifically, assume we double the impact of links to Firm A and halve the impact of links to Firm C. Then our predictions about knowledge benefits for firms will differ from the straight degree calculations: While Firms B and F each had the same in-degree (1), once we weight their links, B's weighted in-degree is 2 and F's weighted in-degree is ½, reflecting our belief that B's link generates more useful knowledge than F's.

Betweenness Centrality. But knowledge flows not just between two points but through the entire network. As we discussed earlier, the knowledge held by C, is ultimately transferred to A, B, E, F, and G. In contrast, knowledge from D is only passed along to A and B. There are other more complex measures such as "constraint" and "structural autonomy" that refine the associations to knowledge collection and potential, but these can be fairly complicated to calculate.[28] These measures consider whether

multiple linkages really provide access to additional, non–redundant knowl–
edge, or whether they duplicate the same knowledge. Thus, one measure
that focuses on the actor's value in the overall network is "betweenness
centrality," which loosely measures how many pairs of actors rely on a par–
ticular actor to transmit information between them.[29]

Specifically, to determine the betweenness centrality of actor k, we
calculate how many paths link actors i and j, and then calculate the pro–
portion of these paths that k is on. This fraction suggests how "between"
i and j actor k is, and then we sum this figure for all pairs (i,j) to get k's
overall betweenness centrality. Figure 15.6 displays the betweenness cen–
trality values for our hypothetical network. Once again, A is the most
central actor. Note that F and G obtain the same values of betweenness
centrality, even though their in–degree values were different. This suggests
that while G may receive more information than F, G has no more control
over the flow of information between other players than F does.

Cliques.   Another type of network concept that yields insight is the
clique. A clique is a set of actors who are all connected to one another.

Figure 15.6
Using Betweenness Centrality to Assess Network Advantage

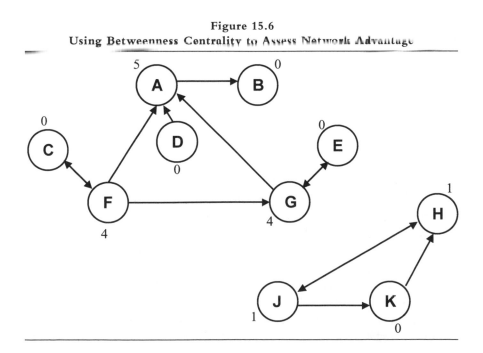

Cliques represent coalitions that typically work toward common goals. Actors who are not linked into cliques may be too peripheral to influence or profit from technological development. In Figure 15.3, we can observe two distinct cliques: Firms A–G are in one clique, while firms H–K are in the other clique.[30] Very different knowledge may circulate in each clique since it does not pass from one clique to the other.

Analyzing Cellular Networks

How can these concepts of degree centrality and betweeness centrality be used to analyze actual networks? Let's consider an example from the cellular industry. In 1995, 59 firms participated in the technical committees sponsored by the Telecommunications Industry Association. We can construct a network of informal knowledge flows by making our network tie value equal, for any pair of firms, the number of meetings to which they both sent representatives. Such a network is very dense, in contrast with the hypothetical network we examined earlier. This is because any firm that sends representatives to at least a few meetings is likely to come into contact with many of the other firms. So few of the tie values are zero-valued, and many of the tie values are much greater than 1. We list the top firms by degree centrality and betweenness centrality in Table 15.1. Note that the same firms dominate both lists, although there are subtle differences in the rankings.[31]

For example, NEC is tied for first on betweenness centrality although its degree centrality score is substantially lower than AT&T or Motorola. This means that even though NEC has fewer interactions with other firms, it still reaps the same benefits from controlling information—an efficient strategy.

As shown in Table 15.2, the top firms tend to congregate into one of two strong cliques. AT&T, BellSouth, Nokia, Bell Atlantic, and Mitsubishi Electric form one clique, where all firms have participated in many of the same CTOs, thus circulating common knowledge. Likewise, Motorola, Qualcomm, NEC, and SBC Communications also interact strongly. Of our top firms, only Ericsson, GTE, and Northern Telecom maintain more independent, unique participation patterns.

This analysis has important implications for the firms involved. A firm with high centrality might question whether too much knowledge may be floating out of the firm (particularly if it looks at out-degree centrality). A firm with low centrality might ask whether it has sufficient connections to

Table 15.1
Top Firms by Degree and
Betweenness Centrality

Firm	Degree Centrality
1) AT&T	677
2) Motorola	663
3) Ericsson	641
4) Northern Telecom	592
5) Qualcomm	499
6) NEC	497
7) (tie) BellSouth	443
7) (tie) Nokia	443
9) SBC Communications	430
10) GTE	425
11) Bell Atlantic	390
14) Mitsubishi Electric	383
Minimum value in 59-firm network	11

Firm	Betweenness Centrality
1) (tie) AT&T	58.7
1) (tie) Motorola	58.7
1) (tie) NEC	58.7
4) Ericsson	54.3
5) Northern Telecom	49.8
6) Nokia	47.1
7) Bell Atlantic	39.8
7) Qualcomm	39.0
9) SBC Communications	36.8
10) Mitsubishi Electric	30.2
11 (tie) BellSouth	27.8
11(tie) GTE	27.8
Minimum value in 59-firm network	0

Table 15.2
Top Firms Tend to Congregate

Clique 1	Clique 2	Unaffiliated with Other Top Firms
AT&T	Motorola	Ericsson, GTE, Northern Telecom
Bell Atlantic	NEC	
BellSouth	Qualcomm	
Mitsubishi Electric	SBC Communications	
Nokia		

obtain the knowledge it needs to succeed. Managers might also use the framework to assess whether the company is in the right cliques to carry the business forward. Based on this analysis, managers might choose to reshape the ties that define the network.

THE IMPACT OF KNOWLEDGE NETWORKS

Why is this analysis of networks important? As indicated in the flight simulator example, these knowledge networks have an impact on the evolution of technology and the fate of individual firms. Understanding this impact and actively managing the knowledge networks can provide an advantage to firms.

Networks Influence Technological Outcomes

While traditional thinking on technological evolution recognizes the firm as the locus of innovation, more recent work acknowledges the many situations in which technological selection is the result of many, varied, interdependent organizational actions. Dominant designs are the result of community-level processes.[32] In other words, as shown in the flight simulator example, technologies that capture the majority of the market do not emerge victorious due to simple technical and economic considerations. Rather, social and political activities such as coalition-building take place in community networks and determine technological outcomes. The influence of the community is all the more pronounced when uncertainty is high, investment is large, products are systemic, and knowledge is paramount—all conditions that often characterize emerging technologies.

Many historical case studies suggest the power of coalitions and networks in shaping technological outcomes. The cancellation of the TSR 2 British military aircraft project has been attributed to incomplete network linkages among critical actors,[33] and the differences in electrical supply distribution systems in Germany, the United Kingdom, and the United States can be traced to different political systems that shaped the actions of firms.[34] Similary, numerically controlled machine tooling systems may have dominated the much more cost-effective record-playback tooling systems because of the coordinated actions of GE, MIT, and the U.S. Air Force.[35]

Networks Affect the Fates of Firms

Networks affect not only technology but the fate of individual firms. Studies have indicated that the firm's position in the network is associated with success (or the lack of it). A study of network of alliances among dedicated biotechnology firms showed the importance of a "network of learning."[36] The price of admission to this network is acceptance (for example, a contractual alliance) by an incumbent member of the network. Researchers found that firms in the main interconnected component of the network achieve higher growth rates, and that firms more centrally located in the overall network do the same. Thus, learning through network ties is associated with a performance outcome (growth).

Several studies in technology-based industries replicate and extend these findings. One found that venture-backed biotechnology firms go to IPO faster, and at higher valuation, when their alliance partners and equity investors are more prominent (that is, they focus on weighted degree centrality to predict these IPO outcomes).[37]

As noted, knowledge flows seem to support alliance formation and alliance formation supports knowledge flows. In the semiconductor industry, research found that firms with highly-cited patent portfolios form more alliances.[38] The study also found that firms in crowded regions of technology space (where the network is defined by common patent citations by pairs of firms) also form more alliances. A study of the chemical industry, demonstrated that innovativeness (as defined by patent output) is associated with the number of alliance partners, the number of the partners' alliance partners, and structural autonomy in the alliance network.[39]

MANAGING NETWORKS FOR ADVANTAGE

Given the importance of these knowledge networks for technological and performance outcomes, it is surprising how little attention they are given by managers. In conversations with managers in the telecommunications area, where participation in the voluntary standards committees is so critical, I've been told two surprising things. One, that there is little centralized oversight of the patterns of participation in technical activities—decentralized work groups tend to send representatives at their own discretion, and the knowledge they collect is not shared with many others from the sponsoring firm. How can managers determine whether they are allocating these resources wisely?

Second, managers said they would find information about the participation patterns of other organizations valuable, but that they have not attempted to collect this information systematically. Understanding the overall structure of knowledge flows across the community not only helps firms to assess useful targets for creating new network links, but also can help firms assess their future competitors. That is to say, information about knowledge overlaps in patenting behavior or standards body participation suggests the firms you will compete with in the future (in contrast to product information which shows the firms you compete with today).

When they understand these networks, companies can reshape them to their advantage. For example, Mitsubishi reshaped its knowledge network to its advantage. Patent citation data shows that Mitsubishi had a peripheral network position among top Japanese semiconductor manufacturers in 1982.[40] Through the strategic use of alliances, Mitsubishi developed its technological capabilities and occupied a much more central network position by 1992. This change in network position was accompanied by growth in not only patents and patent citations, but also market share.

How can managers begin to more actively manage these knowledge networks? There are several steps:

Envision Your Network.   Who are the relevant actors? An interesting question for managers is how to draw the boundaries of a network. Most academic studies constitute networks by including all the firms in an industry. Yet there are two practical concerns here. First, for emerging technologies, we frequently find that industry boundaries are blurred, since emerging technologies often come from the convergence of industries. In this case, it is useful to look at interfirm relationships, such as alliances, to cast your net beyond the standard industry boundaries and to think about the types of partners other players are engaging. Second, while we should include all firms in the industry to constitute our networks, we need to realize that our *comparison set* may be smaller than the entire industry. For example, in cellular networks, Omnipoint is a successful firm that focuses on PCS. If we compared their network measures to those of generalists such as AT&T, they would look inferior. Specifically, Omnipoint's 1995 degree centrality score was only 136 (about a fifth that of AT&Ts), and its betweenness centrality score was just 2.4 (just 4 percent of AT&Ts). Its position, however, is better than these assessments might indicate. Omnipoint competes with a different set of players and should compare network measures with them, while striving for long-term positions like the giants.

What are the most critical types of ties to examine? As discussed earlier, a multitude of ties may be used to generate the networks. The question of which ties to examine depends upon the specifics of the industry and the availability of data. In standards-oriented industries such as telecommunications, the technical groups are both critical and well-documented. Alliancing data are easy to obtain, particularly for public firms. Patent and publication data are also useful for knowledge-based industries.

Calculate Your Network Position. With the image of the network in mind and specific data on hand, you can calculate network measures associated with your position as well as those of your competitors. It is relatively easy to assess measures like degree centrality, and useful to weight this measure in accordance with the status of the firms to which you are connected. Software is available to help calculate the more complex measures as well.[41] How does your firm stack up?

Look for Unique Combinations of Knowledge. Does the set of ties that your firm maintains provide access to many different sources of knowledge, or does it duplicate the same sources? The tendencies for knowledge to circulate among firms in the same geographic region, and also among firms that are technologically similar, is becoming well-documented. A study of knowledge flows in the semiconductor industry, found that startups tend to devote resources to creating alliances with other firms that are in the same region, and are technologically similar.[42] These startups also tend to hire engineers from similar firms. Are these startups devoting their resources in arenas where they can already access information? Might they do better to devote these limited resources to learning something more unique from a more distant firm? Calculating measures like betweenness centrality can help you assess the uniqueness of your knowledge position.

Look for Ways to Improve Your Network Position and Benefits. Companies have limited resources to devote to accessing knowledge. The question then becomes, how can you deploy these resources in a fashion that maximizes your benefits? With the network images and measures in hand, you can target sources of non-redundant information as well as the types of ties that might be effective. Look to increase betweenness centrality by bridging unconnected groups.

As an example, consider our hypothetical network from Figure 15.4. Firm G can increase its centrality dramatically by developing a tie with Firm J. If we assume that the new G-J link is bidirectional, we find that G and J's in-degree centrality each increase by 1, while all other firms stay the same. The change in betweenness centrality is even more dramatic, as shown in Figure 15.7. The betweenness centrality values increase for several firms (A, F, G, J, and H), but most dramatically for G and J, as they control information flow between the two previously unconnected cliques (see Figure 15.7). Both G and even J ultimately surpass A in their ability to control and access information in the new network.[43]

Figure 15.7
G and J Improve Their Network Position by Adding Link

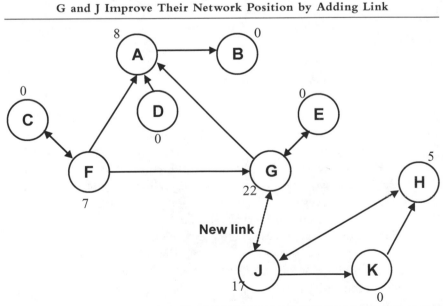

CONCLUSIONS

Emerging technologies are characterized by massive uncertainty. Whatever form these technologies ultimately take emerges from the coordinated action of multiple actors, so the importance of social networks and political coalition-building is paramount.

Managers need to pay more attention to systematically understanding knowledge networks and managing them. While numerous authors stress the importance of keeping knowledge proprietary, and protecting against knowledge spillovers, the many forms of informal linkages discussed in this chapter indicate that this is a difficult, if not impossible, task. Assuming that knowledge flows between firms, the challenge is not to prevent these flows but to manage them.

Flows of knowledge through various types of networks have been demonstrated to influence both technological outcomes and firm-specific outcomes. So it behooves managers to envision potential actors in their technological communities, to evaluate the network connections between these actors, and to assess their firms' network positions in these structures. Only through this activity can managers improve their deployment of resources toward networking activity to influence overall technological development and their firms' roles in this process.

Managing networks is an ongoing process. As the company shifts its network linkages, other players may also be shifting theirs. Many players in a network will seek to improve network position, so strategic links may increase or decrease in value. Managing networks requires ongoing, systematic reevaluation as actors and linkages evolve.

CHAPTER 16

USING ALLIANCES TO BUILD COMPETITIVE ADVANTAGE IN EMERGING TECHNOLOGIES

JEFFREY H. DYER
Marriott School of Management

HARBIR SINGH
The Wharton School

Alliances play a central role in the success of emerging technology businesses. With high uncertainty and high costs to develop technologies, alliances offer a way to share resources and spread risk. Yet they are very difficult to manage and many alliances fail to live up to their promise. In this chapter, the authors examine strategies for creating and managing successful alliances. They explore the different uses of alliances as the technology evolves—from providing early windows on technologies, to creating options, to achieving positioning as the technology matures. They also examine four key factors that determine the advantages created by alliances: knowledge sharing, finding complementary partners, creating and managing specialized assets unique to the partnership, and establishing effective governance systems.

Pharmaceutical company Rhone Poulenc-Rorer (RPR) developed a shifting web of alliances with nineteen partners to learn more about gene therapy research. This network of alliances served as a nexus of multiple research efforts, each providing access to a new technological area for RPR. These alliances were highly flexible. Partners were moved in and out of the network based on their achievement of milestones, or based on RPR's strategic priorities. But is this an effective way to structure alliances? Does it provide advantage? At this time, it is unclear whether RPR's set of relationships has yielded large profits, but it illustrates the attention and creativity companies are focusing on alliances in emerging technology businesses.

Alliances are so important to high-technology companies that firms such as Hewlett Packard, Xerox, and Microsoft have created new positions of "director of strategic alliances" to better identify and evaluate alliance partners. These new offices, which act as a focal point for learning and leveraging lessons from prior and ongoing alliances, signify that alliance creation and management skills are key competencies for emerging technology firms.

During the past two decades, there has been an extraordinary increase in interfirm alliances, particularly in emerging technologies industries. Between 1980–1989, firms in the high-tech industries of biotechnology and new materials technology reported 1,277 alliances compared to only 99 alliances in lower-technology industries, food and beverage and consumer electronics.[1] In recent times, estimates of the numbers of alliances are even higher for technology-intensive industries.

Why are alliances so important to high-tech firms? Alliances offer a variety of opportunities to enhance the competitive position of high-tech firms by providing:

1. Opportunities to learn and acquire new technologies.
2. Access to complementary technological resources and capabilities that reside in other firms.
3. Access to new markets.
4. Access to resources that can enhance the competitive position of the firm (e.g., through minimizing costs).
5. Opportunities to influence or even control technological standards.

These potential advantages have compelled firms to develop a portfolio of alliances as part of a strategy to effectively compete in the marketplace.

As important as they are, however, alliances are very difficult to manage. A large percentage of alliances fail to live up to expectations. In fact, some studies indicate that approximately half of all alliances fail.[2] Reasons for failure include lack of organizational fit (e.g., cultural clashes, poor conflict management, lack of effective coordination mechanisms), lack of strategic fit (partner lacks viable complementary resources, assets, knowledge), or changes in strategic objectives on the part of one or both firms. Thus, while alliances can create economic value, they are also fraught with risk. How can firms manage alliances to maximize the probability of success?

In this chapter, we describe the various key motives for alliances and discuss how the focus of a firm's alliance strategy will vary depending on the evolutionary stage of the emerging technology. We also describe how the key skills required for alliance success change as the technology (industry) evolves. These insights can help managers create more successful alliances.

KEY FEATURES OF
STRATEGIC ALLIANCES

A strategic alliance is "a *cooperative* relationship between two or more *organizations,* designed to achieve a *shared strategic goal.*" Each of the italicized keywords is important: the relationship must be cooperative in nature to constitute an alliance. In addition, we use the term *organizations* broadly, to include firms, universities, or governmental agencies. The idea of the shared strategic goal has a twist to it: The partner organizations may have several goals related to the alliance: some shared and others unshared. This leads to one of the tensions associated with alliances: Do the benefits associated with the shared goals offset the costs associated with conflict between unshared goals? In forming an alliance, this tension should be addressed.

Alliances are unique organizational forms. They tend to have a great variety of structures, entail simultaneous cooperation and competition, and they are often temporary. Alliances have a great variety of structures, starting with equity versus non-equity alliances. Non-equity alliances tend to be governed by a contract, which delineates the roles and responsibilities of each party involved. Equity-based alliances often take the form of joint ventures, which have a separate organizational structure from the parent firms. For the purposes of our discussion, we refer to the entire spectrum of alliances, unless explicitly stated otherwise.

The co-existence of cooperation and competition is a hallmark of strategic alliances, so much so that a new term, *co-opetition,* has been coined to describe it.[3] This challenge is exacerbated when the firms are traditional rivals in their industries, and the alliance is in a segment of the market where the firms have chosen to cooperate. Keeping the boundaries of cooperation and competition well-defined and easily discernible is a significant challenge. Many a well-conceived alliance has fallen prey to the tension between cooperative forces and competitive forces.

One of the features of the relationally skilled organization is the ability of its managers and employees to effectively walk the line between cooperative behavior, where needed, and behavior that protects the firm's proprietary assets. The key desired behavior in a cooperative mode is to maximize joint competitive advantage, while in the competitive mode, it is to maximize the firm's individual competitive advantage. Similarly, in terms of proprietary resources, such as key technologies, cooperative behavior would entail sharing such resources, while competitive behavior would entail protecting them while trying to absorb as much as possible of the partner's resources.

A STRATEGIC PERSPECTIVE ON ALLIANCE FORMATION

Why are firms in emerging technology industries more likely to enter into alliances than firms in more technologically mature product industries? The fundamental reason is *market uncertainty,* a combination of uncertainty about customer demand and technological development.

This combination of demand and technological uncertainties in emerging technology industries naturally increases the risks associated with doing business. In particular, these market uncertainties raise the risks associated with internal development of assets, resources, or capabilities. The risks associated with internal development are higher when there is a great deal of market uncertainty for two primary reasons. First, internal development takes time. Important and valuable resources cannot be developed overnight because they are often subject to time-compression diseconomies.[4] This presents a particular problem in high-tech industries because speed is especially critical when uncertainty is high—in part because there are powerful first-mover advantages associated with being first to market with a new technology (e.g., building an installed base or becoming the industry standard). An inability to quickly respond to market opportunities can result in the firm being outmaneuvered by its competitors in a particular technological arena.

Second, internal investment typically requires a commitment of resources and assets to a particular technological direction or trajectory. Many of these commitments involve investments in specialized assets that increase sunk costs and lead to high exit barriers, thereby reducing the overall flexibility of the firm in responding to market uncertainties.[5] This may explain why studies to date have typically found a negative relationship between technical uncertainty and vertical integration.[6] "A highly volatile industry characterized by frequent technological changes, therefore, will be unattractive for high levels of [vertical] integration."[7] Rather than make these direct investments, companies turn to alliances to decrease their sunk costs and increase their flexibility. Alliances offer greater speed and flexibility in responding to market uncertainties.

USING ALLIANCES TO CREATE ADVANTAGE

Although market uncertainty explains why alliances are important to competitive success in emerging technology industries, it does not explain the specific role of alliances or how they help firms achieve competitive advantage. Alliances serve different roles during the evolution of the technology—from an initial stage characterized by a high degree of market uncertainty to a mature stage characterized by a much lower degree of market uncertainty, as illustrated in Figure 16.1. Alliance strategies related to each stage of this evolution are shown in Table 16.1. During the early stage of technology development, alliances offer windows to learn about

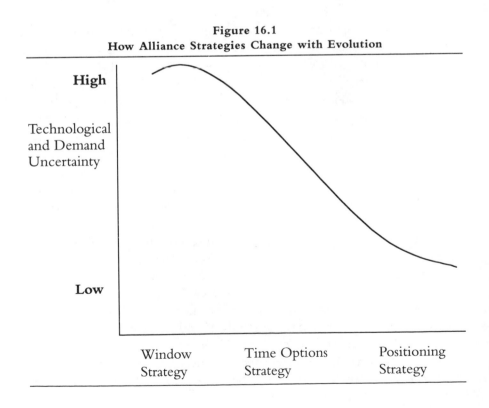

Figure 16.1
How Alliance Strategies Change with Evolution

Table 16.1
Summary of Elements of an Alliance

	Window Strategy	Options Strategy	Positioning Strategy
Strategic objectives	Learning Monitoring	Building platforms	Scale-based advantages
Key success factors	Effective tracking	Scalability	Scale, operational effectiveness
	Knowledge absorption	Ability to evaluate technologies	Ability to identify complementary resources
Key difficulties	Leakage of knowledge	Value of option	Speed and responsiveness (Partner dependence)

diverse technological streams. As promising technologies begin to emerge, alliances are used to create options for future investments in them. Finally, as the promise of these technologies becomes clearer, alliances are used to position the company in the new emerging industries.

At the beginning of research and development into new technologies, there are many unanswered questions, many possible technological directions the firm can take, and great uncertainty as to whether the technology will work as envisioned. Over time, as development of the spectrum of related technologies continues (both internally and perhaps with the assistance of alliance partners), uncertainty is reduced. The firm can, with greater certainty, predict which technologies will work and whether a market will materialize for a particular technology. Finally, as new technologies become embodied in commercialized products or services, uncertainty is further reduced because the firm is able to acquire further knowledge by monitoring market acceptance of the technologies.

At each of these stages in the development of the technology, alliances serve a different purpose:

- *Opening Windows.* During the initial phase, when market and technical uncertainty is the highest, firms may enter alliances using a "window strategy," using alliances as windows to monitor a spectrum of related or complementary technologies. The primary reason for such alliances is to acquire knowledge that will help the firm further develop the technology and/or reduce the uncertainty about the strengths of the technology relative to possible substitute technologies.

- *Exploring Options.* As the technology (and related and complementary technologies developed by other firms) matures, the firm may still have some uncertainty regarding which technologies will be the future winners, but it may have enough information to place some calculated bets. There may be a number of competing fledgling technologies in the marketplace. Thus, alliances may be used as an "options strategy" or as an opportunity to access future winning technologies.
- *Gaining Position.* Finally, as particular technologies become embodied in products and services that compete in the marketplace, alliances are more likely to be used to enhance the competitive position of the firm and its alliance partners. Thus, alliances are more likely to be used as a "positioning strategy" to further lower costs, influence market structure or gain access to markets, or to better differentiate a product offering.

Alliances at each stage have different strategic objectives, key success factors, and potential problems, as summarized in Table 16.1.

Opening Windows

In the early stages of technological evolution of an emerging technology, there are very high levels of both technological and market uncertainty. The formation of alliances in this stage of the technology's evolution will be guided by a window strategy, in which the firm will form alliances to open windows on the technological activities in the field. At this stage, the emphasis is on effective tracking of technologies and learning about them. Firms that have a market or technologies affected by the emerging technology will seek to form alliance relationships with firms that possess technologies that show promise. For example, Rhone Poulenc Rorer's network of alliances with nineteen partners was used to gain windows on multiple streams of technology that might be needed for its gene therapy research. The company moved partners in and out of this web as their technological potential proved itself or failed to materialize. An important success factor is knowledge absorption: the ability to absorb the technological know-how being followed and to develop a presence in it over time. An important difficulty in the windows stage of technological evolution is the potential leakage of the firm's technologies as it seeks to open windows on its partner's technologies. The biotechnology field has a very large number of alliances, and basic research alliances tend to be the most frequent type of partnership.

This stage often involves managing shifting webs of alliances. The technologies that will be successful in producing strong market applications are very uncertain. Companies often use alliances to pursue multiple paths. Studies of innovation in biotechnology have revealed that the locus of innovation is the network, rather than the individual firm. This is in part due to the use of alliances as windows into the research of other firms.

Exploring Options

As emerging technologies evolve, technological and market uncertainty are reduced. At this stage, the role of alliance shifts to the creation of real strategic options for the firm. The fog of uncertainty has lifted a little, but there are still multiple paths to advantage in the future marketplace. In such situations, there are difficult choices: Committing prematurely to a technological path may be unwise if the technology proves unsuccessful. Yet, committing to large numbers of projects would not be feasible in a resource-constrained environment. In such a situation, the use of multiple alliances, each representing a possible path to success in the future, is an appropriate response.

To illustrate, Intel is a company that has used an options approach to many of its alliances, particularly with companies involved in developing Internet technologies. Intel views certain new Internet technologies as having high potential, as well as posing a potential threat because they may allow individual users to tap into applications, data, and processing power through relatively simple and inexpensive desktop devices. These technologies could be devastating to Intel's microprocessor business and, therefore, have been identified by Intel CEO Andrew Grove as having the potential to "represent a '10X' change for Intel's business." To hedge its risk, Intel has invested more than a half billion dollars in venture capital, taking equity positions (buying options) in over 50 companies, many of which are involved in developing Internet technologies.[8]

Gaining Position

Over time, technological and demand uncertainty declines sufficiently to provide the opportunity to pursue *positional* advantages in the marketplace.

In such situations, alliances are used to achieve scale or scope-based advantages and to enhance the participant firms' market positions. Firms find partners with complementary capabilities at this stage, attempting to create a combination of firms with the best capabilities in the industry. One study of alliances in medium- to low-tech industries found that the single most important motive for forming alliances was market access and positioning.[9] This finding contrasted with that of high-tech industries where alliances were used more frequently to access complementary technologies, reduce the innovation period, or for basic research.

An example of the use of alliances for positioning and commercializing more developed technologies is the joint venture between Xerox and Fuji in Japan, created in 1967, to manufacture and market photocopying equipment in Japan. Xerox had the patents on the technology, but faced acute market uncertainty in pursuing the Japanese market alone. It had no experience manufacturing in Japan (and had relatively weak manufacturing skills relative to Fuji) and had little access to, or experience with, complicated Japanese distribution networks. A partnership with Fuji gave Xerox the complementary skills and resources it needed to pursue the Japanese market aggressively and gain positional advantages in the market by having high quality copiers that catered to the medium and high priced end of the market.

The objectives and approaches to forming alliances vary according to the stage of development of the technology, the capabilities of the firms, and the competitive environment. Whatever the purpose of the alliance, however, there are a set of superior "relational skills" that can help companies form, and generate value from, successful alliances. These are the focus of the following section.

STRATEGIES FOR SUCCESSFUL ALLIANCES: BUILDING RELATIONAL ADVANTAGES

As noted at the opening of the chapter, many alliances fail to live up to their potential. What can companies do to manage their alliances more effectively? Four key strategies contribute to the success of a given alliance and build relational advantage:[10]

1. Creating knowledge sharing routines.
2. Choosing complementary partners.

3. Building and managing co-specialized assets.
4. Establishing effective governance processes.

In the context of high-tech industries, these four mechanisms are used by firms to obtain relational advantage, or develop competitive advantage via alliances. The strategic objective of the alliance influences the relative importance of each of these mechanisms as well as the skills that firms need to effectively realize the full potential of their alliances (e.g., knowledge-sharing routines are relatively more important when a firm is using an alliance as a window into new technologies).

Creating Interfirm Knowledge-Sharing Routines

As the alliance evolves, there is a need to develop effective ways of sharing knowledge across organizational boundaries. Research shows that knowledge resides in the management routines used by firms. Building effective knowledge-sharing routines is an important element in obtaining returns from alliances. Knowledge sharing and acquisition are likely to be particularly important when the alliance is used as a window on new technology.

Interorganizational learning is critical to competitive success because firms often learn by collaborating with other firms.[11] For example, in some industries (e.g., scientific instruments) more than two-thirds of the innovations could be traced back to a customer's initial suggestions or ideas. Transferring this knowledge from customers to the firm is therefore crucial to success. A production network with superior knowledge-transfer mechanisms among users, suppliers, and manufacturers should be able to "out-innovate" production networks with less effective knowledge-sharing routines.[12]

Beyond simply arguing that alliance partners can benefit through knowledge-sharing routines, it is important to understand *how* partners create knowledge-sharing routines that result in competitive advantage. Knowledge can be divided into two types: (1) *information,* and (2) *know-how.*[13] Information is defined as easily codifiable knowledge that can be transmitted clearly, including "facts, axiomatic propositions, and symbols."[14] By comparison, know-how involves knowledge that is tacit, "sticky," complex, and difficult to codify.[15]

Know-how is also difficult to imitate and transfer. However, these properties also suggest that compared to information, know-how is more likely to result in advantages that are sustainable. As a result, alliance partners that are particularly effective at transferring know-how are likely to outperform competitors who are not.

The ability to exploit outside sources of knowledge is largely a function of prior related knowledge or the "absorptive capacity" of the recipient of knowledge. Absorptive capacity is "the ability of a firm to recognize the value of new, external information, assimilate it, and apply it to commercial ends."[16] This definition suggests that if a firm has absorptive capacity, it is equally capable of learning from all other organizations. Although Cohen and Levinthal focus on the absolute absorptive capacity of individual firms, the concept is particularly useful in thinking about how alliance partners may systematically engage in interorganizational learning. Thus, *partner-specific absorptive capacity* refers to the idea that a firm has developed the ability to recognize and assimilate valuable knowledge *from a particular alliance partner*. This capacity would entail implementing a set of interorganizational processes that allow collaborating firms to systematically identify valuable know-how and then transfer it across organizational boundaries. Partner-specific absorptive capacity is a function of: (1) the extent to which partners have developed overlapping knowledge bases, and (2) the extent to which partners have developed interaction routines that maximize the frequency and intensity of their interactions. The ability of a receiver of knowledge to assimilate knowledge is largely a function of whether or not the firm has overlapping knowledge bases with the source.[17] Thus, this is a critical component of partner-specific absorptive capacity. In addition, partner-specific absorptive capacity is enhanced as individuals within the alliance partners get to know each other well enough to know *who knows what* and *where critical expertise resides* within each firm. In many cases, this knowledge develops informally over time through interfirm interactions. However, it may be possible to codify at least some of this knowledge.

For example, Fuji and Xerox have attempted to codify this knowledge by creating a "communications matrix" which identifies a set of relevant issues (e.g., products, technologies, markets) and then identifies the individuals (by function) within Fuji-Xerox, Fuji, and Xerox who have relevant expertise on that particular issue. This matrix provides valuable information with regard to where relevant expertise resides within the partnering firms. This example illustrates that alliance partners can increase partner-specific absorptive capacity by designing routines that facilitate information. These types of routines are particularly important since know-how transfers typically involve an iterative process of exchange, and the success of such transfers depends on whether personnel from the two firms have direct, intimate, and extensive face-to-face interactions.[18]

Finally, to generate advantage through knowledge-sharing routines, alliance partners need incentives that encourage them to be transparent, to transfer knowledge, and not to "free ride" on the knowledge acquired from the partner. In particular, the transferring firm must have an incentive to devote the resources required to transfer the know-how since it typically incurs significant costs during the transfer that are comparable to those incurred by the receiving firm. Thus, the mechanisms employed to govern the alliance relationship (discussed below) must create appropriate incentives for knowledge sharing. These may be formal financial incentives (e.g., equity arrangements) or informal norms of reciprocity. Equity arrangements are particularly effective at aligning partner incentives and therefore promote greater interfirm knowledge transfers than contractual arrangements.

Choosing Complementary Partners

There is a priority on seeking partners whose assets complement those of the firm seeking the partnership. The greater the complementarity of the assets involved, the greater are the returns from combining the assets under the rubric of a strategic alliance. This is especially important when a firm is choosing alliance partners for their option value as part of an options strategy. Information about potential partners is a key resource and the ability to assess and predict the potential value, and complementarity, of alliance targets' resources is critical.

We define complementary resource endowments as distinctive resources of alliance partners that collectively generate greater competitive advantage than the sum of those obtained from the individual endowments of each partner. For these resources to generate advantage through an alliance, neither firm in the partnership can purchase the relevant resources in a secondary market. Also, these resources must be indivisible, thereby creating an incentive for each firm to form an alliance to access the complementary resources.

Apple Computer Inc.'s alliance with Sony Corporation to manufacture Apple's Powerbook computers is an example of how competitive advantage can be created by combining complementary resource endowments. The Apple-Sony alliance linked Apple's capability at designing easy-to-use computer products with Sony's miniaturization capability, including the manufacturing know-how necessary to make compact products. Neither firm had the capability to develop the Powerbook individually.

A study of global strategic alliances in biotechnology found that complementarity of both firm- and country-specific resources between domestic and foreign firms was a key factor in the formation of alliances.[19] The complementarity in this case consisted of linkages between the strong basic research capabilities of U.S. firms with the unique local knowledge and distribution capabilities of their partners in overseas markets.

In each of the cases described, the alliance partners brought distinctive resources to the alliance which, when combined with the resources of the partner, resulted in a synergistic effect whereby *the combined resource endowments were more valuable, rare, and difficult-to-imitate than they had been before they were combined.* Consequently, these alliances produced stronger competitive positions than those achievable by the firms operating individually.

Not all the resources of a potential alliance partner will be complementary. In assessing the relational advantages provided by a given alliance, potential partners should examine the resources that are synergy-sensitive with their own resources. As the proportion of synergy-sensitive resources of the potential partners increases, so does the potential for earning relational advantage by combining the complementary resources.

There are several challenges faced by firms attempting to generate relational advantage through complementary resources. First, they must find each other and recognize the potential value of combining resources. If potential alliance partners possessed perfect information, they could easily calculate the value of different partner combinations and then rationally form an alliance with the partner(s) that generated the greatest combined value. It is often very costly and difficult (if not impossible) to place a value on the complementary resources of potential partners. In fact, firms vary in their ability to identify potential partners and value their complementary resources for three primary reasons:

1. *Differences in prior experience.* Firms with higher levels of experience in alliance management may have a more precise view on the kinds of partner/resource combinations that allow them to generate supernormal returns. Prior alliance experience results in more opportunities to enter into future alliances, presumably due to the development of alliance capabilities and reputation.

2. *Differences in internal search and evaluation capability.* Many organizations are developing ways to accumulate knowledge on screening potential partners by creating a "strategic alliance" function. For example, firms such as Hewlett Packard, Xerox, and Microsoft have appointed a Director of Strategic Alliances with his or her own staff and

resources. The role of these individuals is to identify and evaluate potential alliance partners as well as monitor and coordinate their firm's current alliances. The creation of the role of alliance manager ensures some accountability for the selection and ongoing management of alliance partners. It also ensures that knowledge on successful partner combinations and on effective alliance management practices will be accumulated. An opportunity exists to codify some of this knowledge as illustrated by the fact that some firms, such as Hewlett Packard, have created manuals which attempt to codify alliance-specific knowledge. (Hewlett Packard has 60 different tools and templates for managing alliances included in a 300-page manual.) Lotus Corporation has created "35 rules of thumb" to manage each phase of an alliance, from formation through termination. Other firms such as Xerox, SmithKline Beecham, and Oracle have followed a similar approach.

3. *Differences in the firm's ability to acquire information about potential partners due to different positions in their social/economic network(s).* Finally, the ability of a firm to identify and evaluate partners with complementary resources depends on the extent to which the firm has access to accurate and timely information on potential partners. An investment in an internal alliance function will likely facilitate the acquisition of this information, but it also depends on the extent to which the firm occupies an information-rich position within social/economic networks. Firms that occupy central network positions with greater network ties have superior access to information and are thus more likely to increase their number of alliances in the next time period.

Thus far, our discussion has focused on the benefits from combining resources with *strategic complementarity*. However, once a potential partner with the requisite complementary strategic resources has been identified, another challenge is developing *organizational complementarity,* or organizational mechanisms to access the benefits from complementary strategic resources. The ability of alliance partners to realize the benefits from complementary strategic resources is conditioned on compatibility in decision processes, information and control systems, and culture.[20] Although complementarity of strategic resources creates the *potential* for competitive advantage, this advantage can only be realized if the firms have systems and cultures that are compatible enough to facilitate coordinated action. Research suggests that a primary reason for failure of both acquisitions and alliances is *not* that the two firms did not possess strategic complementarity of resources, but rather they did not have compatible operating systems, decision-making processes, and cultures.[21] Managers need to distinguish between initial complementarity (strategic complementarity) based on potential combinations of resources and revealed complementarities (organizational complementarity) based on the realized results of cooperation between the firms involved in the partnership.[22] Both strategic and organizational complementarity are critical for realizing the potential benefits of combining complementary strategic resources.

Building and Managing Co-Specialized Assets

Although the complementary assets that partners bring to the table are important at the start of the alliance, over time the new assets that are created as a result of the partnership become increasingly important. Successful alliances need a process for managing the assets and capabilities each partner develops that are distinctly linked to the partnership. Noted economist Oliver Williamson (1985) identified three types of assets that firms can make that may be co-specialized with the assets of an exchange partner: site, physical, and human asset specificity.[23] *Site specificity* or physical proximity refers to the situation whereby successive production stages that are immobile in nature are located in close proximity to one another. Site-specific investments can substantially reduce inventory and transportation costs and can lower the costs of coordinating activities.[24] *Physical asset specificity* refers to transaction-specific capital investments (e.g., in customized machinery, tools, dies) that tailor processes to particular exchange partners. Physical asset specialization allows for product differentiation and may improve quality by increasing product integrity or fit.[25] *Human asset specificity* refers to transaction-specific know-how accumulated by transactors through long standing relationships (e.g., dedicated supplier engineers who learn the systems, procedures, and individuals that are idiosyncratic to the buyer). Human co-specialization increases as alliance partners develop experience working together and accumulate specialized information, language, and know-how that allow them to communicate efficiently and effectively. Human asset specificity reduces communication errors, thereby resulting in higher quality and faster product development cycle times.[26]

These co-specialized assets are likely to be particularly important for positioning strategy or achieving operational efficiencies after some of the market uncertainty has been resolved. These investments typically take place over time but can improve alliance performance as illustrated by alliances formed by many Silicon Valley firms in the late 1980s. Hewlett Packard and other Silicon Valley firms greatly improved performance by developing long-term partnerships with physically proximate firms.[27] This proximity in high-technology industries "greatly facilitates the collaboration required for fast-changing and complex technologies which involve ongoing interaction, mutual adjustment, and learning."[28] As Sun Microsystems Materials Director Scott Metcalf observed, "In the ideal world, we'd draw a 100-mile radius and have all our suppliers locate plants into that area."[29] Physical proximity created through firm-specific investments facilitates interfirm cooperation and coordination, thereby enhancing performance. An advantage stemming from a unique customization of assets is that it is often difficult for rivals to imitate because they take time to put into place, are often costly, and may involve irreversible asset investments.

This means that firms involved in alliances tend to customize their assets to each other, such that the combination of assets is uniquely valuable. An advantage stemming from a unique customization of assets, called *asset specificity,* is difficult for rivals to imitate and compete against.

In addition, the ability to substitute special-purpose assets for general-purpose assets is influenced by the total volume of transactions between the alliance partners. Just as firms that achieve production economies of scale are able to increase productivity by substituting special-purpose assets for general-purpose assets, alliance partners are also able to increase the efficiency associated with interfirm exchanges as they increase the volume of transactions between the alliance partners.

Establishing Effective Governance Processes

As alliance partners create more co-specialized assets, they also increase their risk. The more specialized these resources become, the lower their value in alternative uses. This increases the risks of opportunism by alliance partners.[30] For partners to make this kind of commitment to the alliance, an effective system of governance is critical. This mechanism for governing the relationship helps minimize transaction costs, thereby enhancing efficiency.[31]

Firms use a combination of both formal and informal mechanisms to govern alliance relationships. Formal mechanisms include the use of legal contracts (with tight contracting language) the composition of boards that would oversee the operation of the alliance as it evolves, and/or financial investments in either equity or other collateral bonds. Informal mechanisms include the building of interpersonal trust between the partners, and the reputation of the firms in the marketplace. (A firm is less likely to behave opportunistically if such behavior will result in its reputation being damaged by an alliance partner.)

In general, in high-tech industries, informal mechanisms and financial bonds (which align the incentives of the parties) are superior to contractual arrangements (though contracts are likely to be used in addition to informal mechanisms). There are two main reasons that informal mechanisms are superior as alliance governance arrangements. First, informal mechanisms are "self-enforcing agreements" that are more flexible and allow alliance partners to adjust the agreement to respond to unforeseen market changes.[32] If alliance partners must rewrite a contract to respond to market changes, it will slow their ability to respond. Alliances governed by informal mechanisms respond to market uncertainty with more flexibility. Second, a governance arrangement that involves equity participation aligns the incentives of the alliance partners. Contractual agreements typically only identify the minimum requirements of a partner, but do not provide incentives for maximum effort and resources.

The appropriate governance arrangement will vary somewhat with the alliance strategy. For example, in both windows alliances and options alliances, there may be tensions between sharing knowledge, and protecting one's own proprietary knowledge. Alliance partners may be unwilling to share valuable, proprietary knowledge if they are not assured that this knowledge will not be readily shared with competitors or otherwise used inappropriately. The willingness of firms to share knowledge and combine complementary strategic resources may also hinge on credible assurances that the trading partner will not attempt to duplicate those same resources, thereby becoming a future competitor.

Contracts typically cannot anticipate or specify all of the ways that knowledge may be misused to the detriment of the alliance partner. Alignment of the incentives of the partners is often a more effective mechanism for ensuring that knowledge transfers are not abused. A financial bond, such as equity sharing, is a superior governance mechanism for facilitating knowledge transfers.[33] The fact that the equity stake will decrease in value if a party is opportunistic provides an incentive for alliance partners to behave in a more trustworthy fashion.

If alliance partners can rely on trusting interpersonal relationships to govern the alliance arrangement and protect each party's interests, this provides additional protection. As uncertainty decreases and alliances are used more for positioning than for knowledge acquisition or as options, contractual agreements become more viable. The lower level of uncertainty makes it easier to write a contract that anticipates future events. Furthermore, the firm is usually attempting to tap into a specific resource at an alliance partner with a known (or at least well-estimated) value. Thus, contractual arrangements will become increasingly effective as the technology matures and as alliances are increasingly used for positioning purposes.

CONCLUSION

Alliances are central to managing emerging technologies, but alliances themselves present significant management challenges. They can be used to open windows on new technologies, create strategic options that serve as beachheads into particular technologies, or serve as a platform for expansion if the market and technology continue to show promise. As the uncertainty surrounding the technology and served markets is reduced, alliances are used to access new markets or position the firm in the new maturing industry.

As alliances have become more central to the development and commercialization of new technologies, capabilities in building and managing alliances have become increasingly important to firms pursuing emerging technologies. Companies that can increase their ability to learn about alliances and develop systems for creating and managing them will be able to move more quickly and effectively to take advantage of new opportunities. In particular, skills in knowledge sharing, evaluating the complementary of partners, creating and managing co-specialized assets, and governing alliances are important sources of advantage for any emerging technology business. As much as the technological capabilities in the laboratory, strengths in managing and developing these alliance relationships will be a key determinant of success.

CHAPTER 17

THE DESIGN OF NEW ORGANIZATIONAL FORMS

JENNIFER HERBER
Goldman, Sachs & Co.

JITENDRA V. SINGH
The Wharton School

MICHAEL USEEM
The Wharton School

Like aircraft designed for supersonic travel, organizations in the fast-paced and turbulent environment of emerging technologies require a different design to succeed. Managers and researchers have recognized the weaknesses of traditional, hierarchical organizational forms in meeting the new demands of rapid change and uncertainty, and they have begun to develop more dynamic and adaptable organizational structures tailored to this new environment. In this chapter, the authors examine emerging forms—including virtual organizations, network firms, spinouts, ambidextrous firms, front-back designs, and sense-and-respond firms.

Until the early 1990s, the stock brokerage industry was dominated by full-service firms that charged well-heeled clients hefty commissions in return for advice and trade execution. Discount brokerage emerged during the mid-1980s, providing a far more limited and far less expensive service focused primarily on trade execution. By the late 1990s, the Internet brought the revolution of no-frills online

This chapter builds upon research conducted under the aegis of the New Organizational Forms Research Initiative, directed by Jitendra Singh. We acknowledge gratefully the ongoing financial support of the Emerging Technologies Management Research Program, a research program of the Huntsman Center for Global Competition & Innovation at The Wharton School. We thank George Day, Robert Gunther, Paul Schoemaker, and Michael Tomczyk for their helpful comments.

brokerage, with the price of trades falling as low as $10. A transaction to buy or sell 5,000 shares of America Online at $100 per share would have resulted under the full-service regime in a fee in the thousands of dollars. But in the Internet era, the cost was sliced to fraction of that, as low as $10 to $20 at some online firms! The advent of web-based trading forced fee structures down to a hundredth of those that had prevailed for years at full-service houses.

This disruptive technological change put the full-service brokerage organizations at a disadvantage. Their financial consultants and stock analysts—supported by a lucrative fee structure   could not be fully sustained by the far smaller discount fee structure. None of the full-service brokers were able at first to create a significant presence in online service when that market exploded in the mid-1990s. And it came as no surprise that the initial leaders among the online firms were newcomers unburdened by any established structure, such as National Discount Brokers, Ameritrade, and E★Trade. One of the major discount firms, Charles Schwab, swiftly created an online service by building a separate arm, and even the two largest full-service brokers—Merrill Lynch and Morgan Stanley Dean Witter—moved aggressively in that direction by the late 1990s. But doing so required that established players completely rethink their organizational form and even cannibalize their full-service business. As they built their new service models, they found that new organizational forms were required to run them.[1]

One of the more hallowed management truisms is that organizations should adapt to changing environmental conditions. But successful organizations frequently have trouble responding to discontinuous, competence-destroying change such as the advent of the Internet.[2] Dominant players often fail to adapt because it means dismantling the very organizations that have led to their success. They had mastered current technologies and customer needs, but by virtue of having established that expertise and focus they had also become ill prepared to face innovative technologies and new customers. Past adaptations become inertial constraints, leading to a kind of "competency trap."[3] The organizational architectures that companies have built to propel their success can become as outmoded as feudal kingdoms in an age of democracy.[4]

Yet we have entered an era of intense experimentation with new organizational forms, as innovative technologies have created radically different opportunities for doing business. As a result, company architectures are changing, reporting relations are flattening, work designs are empowering, and markets are opening. We examine new organizational forms that are emerging in response to the discontinuous technological breaks of recent years. We seek to understand what makes these organizational forms distinctive and how they provide competitive advantage.

TOWARD NEW ORGANIZATIONAL FORMS

Management research and theory for much of this century have focused on the hierarchically structured organization and its dominant variations such as the vertically integrated company, the multidivisional firm, and the conglomerate corporation. Today, however, it is evident that the traditional "make-and-sell" hierarchical firm is giving way to a set of fresh forms. Company managers in many industries are in a period of intense experimentation, and the fast moving and globalizing markets they face are fostering greater diversity in effective forms than has been the case in the past.[5]

Experimenting managers have generally designed these new forms to capture two capabilities viewed as critical for success in environments of discontinuous technological change. The first capability is an effective balancing and exploration and exploitation.[6] When a company focuses entirely on the exploitation of its current competitive advantages, it certainly becomes better at what it is doing well, but at the same time it becomes vulnerable to abrupt changes that negate the value of what it does best. On the other hand, if a firm focuses solely on exploration of future capabilities, it risks near-term failure for lack of tangible results. A balance, then, of both building the future and exploiting the past is seen as essential. The second capability is a recombination of established competencies. Organizations that take what they already do well and create fresh blends can capitalize on existing competencies without being locked into them.[7]

THE ELEMENTS OF ORGANIZATIONAL FORMS

Distinctive organizational forms are defined by unique reconfigurations of six elements:

1. *Organizational goals* are the firm's broad objectives and performance-related outcomes ranging from market share and customer satisfaction to total shareholder return. They implicitly contain time

frames for measuring the extent to which they are achieved. A company goal, for instance, might be to establish the dominant market share in an emerging area over the next three years, much as Amazon.com has achieved in online bookselling.[8]

2. *Strategies* concern intended and emergent patterns of long-term methods for achieving goals at both the firm and business unit levels.
3. *Authority relations* include organizational architecture and reporting structures.
4. *Technologies* refer to information, communication, and production methods.
5. *Markets* include relationships with customers, suppliers, partners, and competitors.
6. *Processes* refer to dynamic links among these elements, such as recruitment, budgeting, compensation, and performance evaluation.

The point at which changes in these dimensions cease to be variations on an old form and constitute genuinely "new" forms remains an issue of debate. Steve Socolof, head of the new ventures group at Lucent Technologies, contends that a form is only new when it results in a strategic break in the industry.[9] Steve Haeckel, director of the Strategic Studies Advanced Business Institute at IBM, takes a more internal perspective, arguing that a new form is defined by a decisive shift in a firm's conception of it processes and its mindset toward change and customers.[10]

Management researchers differ as well, some insisting that organizational forms are only new if they are built on a different strategy, others arguing that new forms can be devised to better achieve the same strategy. In our view, new forms can be created by changing just one of the dimensions, but the most interesting forms are those that have reconfigured several at the same time. Not all possible combinations along the six identified dimensions of organizational form, however, are equally likely. Some recipes are inherently more conducive to superior performance and are thus more able to survive the test of the competitive marketplace.

EMERGING ORGANIZATIONAL FORMS

Traditional organizational forms are eroding and managers are building several distinctive forms alongside them or even in their place, but it remains unclear which if any of the new forms will become dominant. Indeed, the era ahead may be characterized by a proliferation of forms with no single model coming to dominate as the hierarchical form did in the past. Among the emerging forms, we see six relatively different and potentially enduring organizational models. They are not necessarily mutually exclusive, with some companies simultaneously building two or more at the same time. Nor are the boundaries between them precisely defined. Still, they are coming to represent relatively distinct responses to emerging technologies of production, communication, and distribution. The six new organizational forms are (1) virtual organization, (2) network organization, (3) spinout organization, (4) ambidextrous organization, (5) front-back organization, and (6) sense-and-respond organization.

Virtual Organization

The virtual form is an organization in which employees, suppliers, and customers are geographically dispersed but united by technology. A network of distributed organizational units and individuals act in concert to serve widely scattered customers. New information technologies have driven the rise of this form, as customers and companies come to utilize high-speed, broadband communication systems to buy and sell products and services anywhere rather than at a point of direct contact in a store or office. These technologies have also created mechanisms for inexpensively weaving together far-flung organizations and operations. The virtual organization is largely boundary-less, with tasks performed, suppliers accessed, and products delivered in hundreds if not thousands of widely strewn physical locations. Headquarters may be little more than the chief executive's home computer and an internet connection.

The virtual form minimizes asset commitments, resulting in greater flexibility, lower costs, and consequentially, faster growth. Its application and value can be well seen in the experience of Dell Computer Corporation. Founded in 1984, Dell seized on emergent technologies and information management to integrate supplier partnership, mass customization, and just-in-time manufacturing for fast and precise responsiveness to fast-evolving customer demand. It introduced virtual organizational forms across the entire value chain, from suppliers to manufacturers to customers.

The backbone of Dell's exceptional productivity, efficiency, and mass customization has been its company-wide coordination across businesses, customers, and suppliers.[11] In maintaining real-time links with its suppliers, for instance, Dell could provide the kind of detailed data that would allow them to reduce inventory, enhance speed, and improve logistics. The sharing of information with the suppliers enhanced their incentives to collaborate. And Dell's use of electronic logs rather than written forms reduced the cost for many functions ranging from order taking to quality inspection. The technology thus enabled Dell to benefit from a de facto vertical integration without the liabilities and inflexibilities of owning the supply chain.

Dell Computer's virtual linkage with customers via Internet and voice channels also permitted Dell to circumvent traditional dealership channels and thereby create a sustainable competitive advantage of lower selling cost and higher customer responsiveness. By the late 1990s, Dell had become the world's second largest computer maker, with 30,000 employees, annual revenues of $21 billion, and $30 million in daily Internet sales.

Electronic technologies have been used by other companies to push the limits of virtual relationships. In such relationships, products never appear in showrooms, customers never meet salespeople, and dollars never physically change hands. Amazon.com, CDNow, and thousands of kindred e-commerce start-ups have mastered the use of cyber catalogs in place of storefronts, credit cards instead of cash, and e-mail confirmations in place of paper receipts.

Virtual companies have also learned to exploit the unique potential of the two-way medium through which they both sell and learn. They have created more enduring and more customized relations with individual customers and they have constructed communities among customers. Amazon.com, for instance, added chat rooms for customers to exchange ideas among themselves; CDNow added an option for customers to list on its public Web site the CDs they would like to receive as gifts from friends.

Yet this flexibility of the virtual organizational form created its own new set of challenges, especially in the area of authority relations. The ties of communication technologies were strong enough that it no longer mattered if employees sat next to each other or even nearby. They could as well work from home offices miles away or even business offices a continent removed. They need not work full-time or from 9 to 5 either. But in that lessening of physical proximity and contact frequency, the traditional role of supervisors would change from overseeing work processes to job outcomes, from exercising authority over tasks to delegating responsibility for results. Supervisors would also no longer stand at the center of communication and coordination since the increasing horizontal collaboration negated the necessity of going up the organization to obtain downward cooperation in other operations. An enduring by-product has been for

bosses to become less central in giving feedback and appraising performance—and peers to become more central. Vertical organization gives way
to lateral relations.

Companies sometimes sought to compensate for the downsides of this
virtual form with a medicinal dose of physical contact. Diamond Technology Partners, for example, a consulting firm headquartered in Chicago
with principals spread across the United States, had invested heavily in communications technology so that its far-flung consultants could still work in
integrated, multidisciplinary teams despite their remote geographic locations. But it also scheduled frequent three-day "All-Hands Meetings" to
ensure a shared experience and personal familiarity that are the warp and
weave of any organization's culture. Other consulting companies require all
staff members to be back at their main offices at some periodic point to re-
mind them of their shared agendas and common ground.

The virtual organizational form brings many advantages to companies
that are themselves building and marketing emerging technologies. This
form serves as a magnet for attracting creative and energetic employees
who eschew bureaucracy and favor sovereignty. This advantage can turn to
disadvantage, however, when pushing the emerging technologies to their
next stage of development depends on a critical mass of creative people
working intensively together. The need for geographic proximity partially
explains why even the most technologically advanced industries—which
would seemingly lend themselves best to virtual forms—are often geographically concentrated, as seen in computer making in Silicon Valley and
telecommunication services in Northern Virginia.

Network Organization

The network form is based on an organized set of relationships among
autonomous or semi-autonomous work units for delivering a complete
product or service to a customer. Network forms are found both inside
companies and across sets of companies.

External Network Form.   External networks among companies can be
viewed as outsourcing in the extreme. At the core are organizations that
have chosen to concentrate on a particular competence or specific slice of
the value chain. The central organizations create symbiotic ties among a
host of legally independent entities to aggregate the necessary skills, assembly, and services.[12] They rely on other entities from suppliers and

distributors to complete the value chain in the delivery of a complete product or service.

Some external networks can be described as *federated* in that a set of loosely affiliated firms work relatively autonomously but nonetheless engage in mutual monitoring and control of one another.[13] Other external networks can be viewed more as evanescent *organizational webs* in which constellations of players coalesce around an emerging business opportunity and dissipate just as rapidly once it runs its course. Still another subspecies is the *strategic partnership* in which companies form cooperative deals, often across continents, with suppliers to achieve lowest cost manufacturing, or collaborate with research companies worldwide to acquire highest quality innovation.[14]

External networks are stitched together with a variety of methods, ranging from joint ventures and formal partnerships to franchising systems and research consortia.[15] Whatever the specific type of external network, they transform the competitive fray to one of rivalries between constellations of collaborating enterprises.

The textile industry in Prato, Italy, during the 1980s exemplifies the external network. Here, tiny firms came to specialize in a particular niche of the industry in response to customer demand for lower prices and greater variety. No single company dominated, and independent master brokers—*impannatores*—served as the customer interface, taking orders that far exceeded the capacity of any one producer. They divided and dispatched the orders to hundreds of producers over which they held no formal authority. The region's 15,000 independent firms, with an average of just five employees each, collectively produced what would ordinarily have only been available from a few massive companies. Though these miniature producers competed vigorously against one another, they also established strong cooperatives for tasks where economies of scale and joint practices proved more lucrative than allowing themselves to remain in a Hobbesian state of all against all. By pulling together a web of specialized makers, the *impannatores* created a network of firms that delivered the fabric within the customer's quality, quantity, and time requirements.[16]

The external network organizational form brings distinctive authority and market relations, relying on lateral communication instead of vertical clout to achieve coordination.[17] Firms in the Prato textile lattice independently received, executed, and dispatched work. They each managed their entire production process from product design and material purchase to spinning, dyeing, weaving, tailoring, financing, customs, and transport.

The informal network among them still ensured that their combined output met the customer's request, much like a top executive for a giant textile company would do through very different means.

Internal Network Form.   The internal network structure builds on much the same premise that undergirds the external network—aligned but loose relations among a set of operations can often beat a hierarchy of control among the operations—but here the premise is applied inside the firm. Strategic business units, microenterprises, and autonomous work teams are the building blocks, and their work is coordinated and disciplined but rarely directed by the top of the pyramid. Headquarters sets global strategy, allocates assets, and monitors results, but is otherwise little concerned with daily operations. Top executives establish a cultural esprit and common mindset across the operating units and teams, and then the upper echelon leaves it almost entirely to each of the operation to devise its own methods for making and selling.

An exemplary case is the Zurich-based ABB, Asea Brown Boveri, which has networked its many fully owned subsidiaries and business units to an extreme. This engineering and technology company employed 200,000 people in more than 100 countries during the late 1990s, and in 1998 it earned $2 billion on $31 billion in revenue. Yet its home office housed fewer than 100 managers, and virtually all of its decisions were centered in 1,300 operating units and 5,000 profit centers around the world. Described as "obsessively decentralized," the ABB pyramid is about as flat as they come, with a single layer of management between top executives and field managers. The field managers as a result have the autonomy to do what they want so long as their decisions are aligned with the firm's goals. The home office can be seen as managing an enormous portfolio of self-directed businesses, all overseen by managers incentivized to act like entrepreneurs. Headquarters provides shared services ranging from credit-agency relations to telecommunications systems and best practice dissemination. Almost everything is left to its thousands of entrepreneurial managers.

The internal network form fosters a different set of authority relations. ABB's corporate staff is too small to do more than set a broad framework for responsibility and accountability. It requires that each operating entity achieve goals that in the aggregate are what top management has promised investors. But the discretionary action for succeeding or failing to do so is vested in the thousands of profit-and-loss centers, and day-to-day decision making is thus far more bottom-up than top-down. Moreover, much of the value of gathering so many autonomous business units under a single umbrella is the opportunity for the transfer of technological innovations, best practices, and outstanding performers among them. And for this to be effective, a matrix is a predictable result, with managers reporting not up a narrowing hierarchy but across a host of reporting lines.

Both external and internal network organization forms benefit from the adaptive flexibility that comes with their built-in modularity. Whether inside a company or across a lattice of companies, units can be opened, moved, or closed, and each is far closer to its respective customers than anybody else in the operation. True, the independent companies or autonomous units are free to operate in their own self-interest, which is a potentially centrifugal force. But when those interests are effectively coordinated through the informal means common in external networks and the more formal mechanisms found in internal networks, the whole can produce far more than the sum of its parts, and far more than a monolithic pyramid.

This organizational form may be particularly useful in industries with fast-moving technological change and rapidly emergent new ways of producing and selling. When uncertainty is high, risk is great, and time is punishing, the modularity of the network form provides for quick response. The customer focus of the network form provides for nuanced response. And the local autonomy of the network form provides for a creative response.

Spin-Out Organization

The spin-out organizational form is built when companies establish fresh entities inside from new business concepts and then send them off at least partially on their own. The parent organization, sometimes resembling a holding company, serves as venture capitalist, protective incubator, and proud mentor, but the successful units are sooner or later pushed out of the nest. The parent may relinquish all ownership and control, or it may choose to retain a 20, 50, or 70 percent stake. Whatever the lingering tether, the spin-out is left largely to its own devices to sink or swim.

During the spin-out process, authority relations between the company and business unit evolve from parental control to adult independence. The goals of the spin-out will diverge from the parent's objectives once the offspring is legally separate. Still, gentle parental advice is frequently continued, and some offspring continue to make good use of the parent's accounting, legal, and investment functions.[18]

Thermo Electron Corporation and Safeguard Scientifics both exemplify the spin-out form. Headquartered in Massachusetts, Thermo Electron has long served as an "innovation incubator" for thermodynamic, medical, and technology related products for well-defined market niches. It develops, manufactures, and markets a variety of analytical and monitoring instruments for industrial process control, biomedical applications, and alternative energy systems. It is the leading producer of mammography systems for the early detection of breast cancer. Founded in 1956, the company in 1982 began "spinning out" promising technologies and services by offering minority shares in newly created subsidiaries to the public. To ensure that its managers of the spin-outs continued to produce great returns even when they could no longer be required to do so, Thermo Electron created highly leveraged incentive packages. Given a chance to behave like an entrepreneur and be rewarded for doing so, and with a proven product in hand from the incubation period, spin-out managers have often outperformed the market. So, too, has Thermo Electron, which by 1999 had grown into a $4 billion enterprise with 26,000 employees worldwide.

Safeguard Scientifics, based in Pennsylvania, is an information technology company that builds and operates a portfolio of firms operating in the areas of electronic commerce, software applications, and network infrastructures. Its first role is that of a venture capitalist, identifying new business opportunities and then funding internal start-ups to pursue them. Once Safeguard's strategic oversight and operating guidance have helped its nascent enterprises reach a point of successful takeoff, it spins-out these businesses. Its operating philosophy is one of partnership rather than rulership, and that attracts the kinds of self-directed entrepreneurs that Safeguard seeks to run its enterprises. After the spin-out, the partnership continues, typically with Safeguard as a minority owner but still a provider of management advice. Among its better known partnership firms are Novell and Cambridge Technology Partners. In 1998, Safeguard's revenue topped $2.2 billion, and an investor who had placed $29,000 in the company and subsequent stock offerings of its spin-outs since 1992 would have seen the investment grow to $217,000 by March, 1999, far exceeding an equivalent investment in the S&P 500.

Other companies do spin-outs less as a way of life but nonetheless as a significant part of their strategy. DuPont, one of America's premier chemical companies with revenue in 1998 of $27 billion, established an internal ventures group that fosters product ideas that do not fit into any of its existing business line. Lucent Technologies, the giant telecommunications equipment maker with 1998 revenue of $30 billion, created a new venture group in 1997 to foster innovative applications of its existing technologies that the business lines are not themselves pursuing. Both the DuPont and Lucent group collaborate with external venture capital firms to secure funding for their internal ventures. Both occasionally go outside their own companies to fund start-ups, and both anticipate that some of the successful enterprises inside may eventually be spun outside.

The records of Thermo Electron and Safeguard Scientifics suggest that spin-out organizational forms can help promising new technologies find a home even when managers of existing business lines are fixed on extracting most from what they already know best. Spin-outs capture ideas that otherwise would slip between the cracks or face inertial resistance.

Spin-outs can also constitute an excellent vehicle for not only developing but also commercializing expensive and risky emerging technologies. Because they become legally separate from the corporate parent, they can pursue variant growth strategies, financing objectives, and performance goals, permitting greater responsiveness to fast-changing market conditions and emerging possibilities. They can use stock-options to attract and retain talent who might otherwise exit the parent firm for the lack of real wealth incentives. And once the spin-outs are on their own in the market, the joint forces of demanding investors and aggressive competitors impose a financial discipline with an intensity rarely felt inside a large parent.

Ambidextrous Organization

If the spin-out form is designed to take a new venture out of the sometimes inhospitable environment of a large organization, the ambidextrous organizational form creates an environment in which both established and emerging businesses flourish side by side.[19] Some parts of the organization are working on incremental improvements in technologies, others are looking for breakthroughs. The ambidextrous form overcomes the "innovator's dilemma," the conundrum of listening so well to current customers that the company never anticipates radically new technologies that customers have not yet come to appreciate but will eventually demand. This organizational scheme is designed to ensure simultaneous dexterity in both continuous improvement and discontinuous innovation.[20]

With 125,000 employees and sales of $47 billion in 1998, Hewlett-Packard was concerned that the success of existing products would dampen new products because champions of the latter would not have the political clout to obtain funding and attention. The firm thus created an internal consulting group to help its autonomous business units do two things at once. As characterized by Stu Winby, its director of Strategic Change Services, the objective is to improve a business unit's sale of today's technologies, with a focus on raising volumes and lowering costs. But a concurrent objective is to organize part of the same business unit around future technologies, with an emphasis on entrepreneurship and speed to market. The latter's products sometimes compete head-on with existing products or even threaten to cannibalize them entirely, and managers of well-established product lines are predictably wary. Still, Hewlett-Packard's experience confirms that ways can be found to keep both agendas successfully working under the same roof.

The ambidextrous form can be especially useful for fostering emerging technologies without abandoning the old. Doing both at the same time runs the risk of sowing conflict, but when well orchestrated, this form helps reconcile otherwise opposed agendas. A critical feature is limiting their separation: Those responsible for traditional products are brought into active dialogue with those at the forefront of new ideas. Lateral linkages rather than segregated operations become important here for mutual stimulation. And when well incentivized to share rather than hoard knowledge, to communicate rather than isolate, both sides contribute more to the company's ultimate objectives and devote less energy to thwarting the other party.

Front-Back Organization

The front-back organizational form is organized around customers in the front, with all company functions placed at the back to serve the front.[21] The purpose is to provide customers with fast, responsive and customized solutions.

One type of front-back form is an inverted organization in which all line executives, systems, and support staff in effect work for the front-line person, allowing him or her to concentrate the company's capabilities on satisfying the customer. With the firm's systems and procedures so focused, the front-line person commands the resources to respond swiftly and precisely to evolving customer needs. The organization chart is turned upside down, with customers on top, customer-contact people next, and the rest below.[22]

The front-back form can be seen in many health maintenance organizations. They still divide medical practices into specialties such as radiology, anesthesiology, and cardiology, but many now also designate a primary care provider to coordinate the back-end functions to deliver a complete health package to the patient.

A second variant of the front-back form is a hybrid of vertical and horizontal process teams. Here, companies are divided into units with vertical reporting lines, but they also establish formal means for transcending vertical barriers when they get in the way. Sometimes front-back companies are focused on products, in other cases on geography or distribution channels. However configured, they come to resemble a "centerless corporation" in which resources are directed at whatever part is most in frontal contact with customers.[23]

The hybrid model with horizontal work that transects vertical reporting lines can be found in many management-consulting firms. Partners and associates at McKinsey & Company, Andersen Consulting, and kindred companies are organized into specialized practices, such as strategy, information, and change, but they also create temporary client teams drawn from several of the practices. Team leaders have command over the resources of the specialized practices for the duration of the engagement to ensure that their clients receive the right combination of technical expertise to solve the problems they face.

Eastman Kodak also employs a similar hybrid form. With 86,000 employees and sales of $13 billion in 1998, Kodak divides its operations four ways: by business, such as coatings, inks, and resins; by technical competencies, such as polymer technology; by functions, such as manufacturing; and by marketing region.[24] As characterized by a senior Kodak executive, the organization is a "pizza with pepperoni," with hierarchy, networks, and teams scattered across a common platform. Vertically structured teams around product groups are the most enduring, but global cross-functional process teams play an equally important role in areas ranging from new business generation to order fulfillment. Several scientific teams, for example, draw experts from throughout the company. One of the innovation teams thus unites people from business units, technical services, sales and marketing, and research and development.

Front-back organizations differ from traditional forms most starkly in their reconfigured authority relations. Health maintenance organizations, for example, realign incentives to foster cooperation instead of adversarial relationships among physicians, health workers, and medical plans. If doctors have traditionally been the unassailable authority in hospitals, they are now members of patient-centered teams whose leader may not be a physician at all. The hybrid form also transforms authority, as employees become more accountable to their peers elsewhere in the organization than to their bosses. Their performance depends as much on how well they communicate and work across boundaries as with their direct supervisors.

Sense-and-Respond Organization

The sense-and-respond organizational form is focused even more intensely on identifying emerging customer needs.[25] While the front-back form develops a distinctive relationship with customers, the sense-and-respond firm orients the entire organization around meeting ever-changing customer demands. The working premise is that unpredictable change is inevitable in the marketplace, and the challenge is to ready the organization to capitalize on whatever discontinuity confronts it.

Adaptability is among the foremost capacities of sense-and-respond firms. They tend to plan from the bottom up with few predetermined long-term plans, reacting almost daily to market movements. They occupy a middle ground between a strategy of "control your own destiny" and a strategy of "let your destiny happen to you." One variation of this form is what has been termed a "MegaStrategic Business Entity," found among giant, diversified companies that continuously change to stay with their same customers for years.[26]

Westpac Banking Corporation, an Australian-based firm with 31,000 employees and $6 billion in revenue in 1998, illustrates the sense-and-respond form. For a decade it has worked as a collection of capabilities and assets managed to adapt to customer requests. It is not especially efficient at processing, but its modularity ensures that it gathers detailed information from customers and responds with precisely what each needs. It sets as its main objective continuously responding to customers and anticipating their coming needs. Authority relations necessarily are more fluid to ensure flexible response to customer requests.[27]

For companies engaged in managing emerging technologies, the sense-and-respond form can serve to maintain a riveting focus on the market that is sometimes lost when firms begin to develop the new technologies. It helps companies see the first signs of emerging market needs that can only be met by innovative approaches, and then to remain attentive to further changes that may dictate fine-tuning or even right angle turns before the new products are finally ready for market.

CONCLUSIONS

In a global environment in which many competitive advantages can be rapidly replicated or swiftly undercut, organizational form is becoming an important source of sustainable advantage.[28] As more and more competitors restructure to achieve world-class productivity, adopt total quality standards, and transform one product after another into commodities, the traditional ways of competing may be sustained only with increasing difficulty.

But with the traditional company hierarchy giving way now to a proliferation of organizational forms, competitive advantage may be increasingly found in being the first to fashion a form that best capitalizes on emerging technologies in a way that is especially responsive to customer needs. And this advantage may be particularly pronounced for companies that use or produce new technologies since traditional organization forms have tended to inhibit innovations that threaten existing knowledge bases and production systems.[29]

For managers asking which of the six organizational forms holds greatest promise for their firm, the choice is a contingent one. As summarized in Table 17.1, the selection depends on the unique configuration of a company's goals and authority relations on the one hand, and the nature of its changing technologies and markets on the other. When a firm's technologies and markets are relatively fresh but its goals and authority relationships are not, the ambidextrous form may be most appropriate. When

Table 17.1
Organizational Forms and Changing Environments

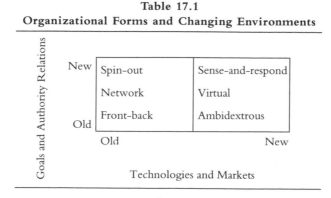

an enterprise's goals and authority relations are new but its technologies and markets are less so, the spin-out form may be most suited. When a company is facing change in both areas, the sense-and-respond form may well be more appropriate.

Organizations must look carefully at their competitive environments and internal capacities in selecting the right organizational form. The six forms described here represent distinctive models, but hybrids can be found among each and some companies have simultaneously adopted two or more at the same time. The six forms can best be seen as a starting point for thinking about a form that is uniquely tailored to meet the specific challenges facing a company. Because those challenges are so varied, we are likely to see a host of organizational forms that borrow no model intact and custom-build what will be needed to develop, manage, use, and sell emerging technologies during the years ahead.

CHAPTER 18

DESIGNING THE
CUSTOMIZED WORKPLACE

JOHN R. KIMBERLY
The Wharton School and INSEAD

HAMID BOUCHIKHI
Department of Strategy and Management, ESSEC

Across a wide range of organizations, employment relationships are being rewritten. For emerging technology firms—with their high risks, high pay-offs, dependence on knowledge, and need for flexibility—the human resource challenges are even greater. The authors explore characteristics of emerging technologies and changes in the workforce that are reshaping the relationships between employees and firms. Then they discuss an emerging new model for a "customized workplace," in which workers and employers jointly deter-mine the conditions of employment on a case-by-case basis instead of offering employees a one-size-fits-all work arrangement, or even a thinly disguised "menu" of options. These customized relationships can help companies find the talent they need to succeed by offering employees more control over their working life.

Software executive Matt Szulik had worked for five companies since the mid-1980s. This record of job shifting and lack of loyalty once might have made him suspect as a new hire in the age of the corporate man, but not today. As he started his sixth job in November 1998, his experience with multiple software start-ups instead made him a "hot hire."[1]

At the age of 40, Chris Peters had become head of Microsoft's Office Division, with stock options that made him a multimillionaire. Where he might once have used his fast-track career as a launching pad for greater conquests, he instead took a leave of absence to pursue the one goal he had not yet achieved—qualifying for the professional bowling tour.[2]

These stories, pulled from the recent headlines of the *Wall Street Journal,* illustrate the very different world of employment in the 1990s, particularly in high-technology industries.[3] Companies in these industries are developing and commercializing new and unproven technologies, which have revolutionary potential to produce extraordinary economic gains, but also high risks of failure. The rapid pace of change and the intense value of knowledge—much of which is inside the heads of employees—has rewritten the contracts and relationships between workers and organizations.

The stories of Szulik and Peters highlight a set of significant and common challenges to both firms and employees as they struggle to maintain or enhance shareholder value, earn a reasonable return on invested capital, provide a challenging and stimulating work environment, maintain reasonable flexibility in the stock of human assets, and have a rewarding, fulfilling personal life. These human resources challenges are driven by the interaction among three different but interconnected factors, as illustrated in Figure 18.1:

- The properties of emerging technologies.
- Changing character of people and what they are seeking.

Figure 18.1
Interactions Driving Human Resources Challenges

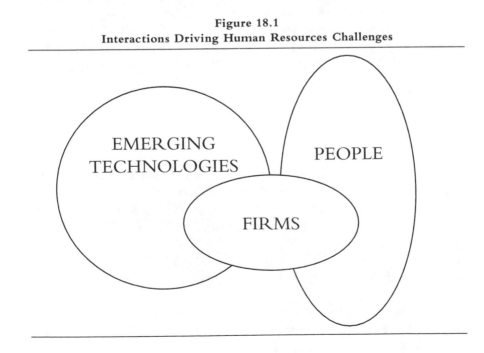

- The changing character of firms and how they are approaching their human capital.

This figure is deliberately oversimplified. In representing the interconnections in this fashion the intent is to emphasize that firms are, in a sense, caught between the attributes and dynamics associated with emerging technologies on the one hand and the changing attributes and ambitions of people (who comprise *both* their potential labor pool *and* their potential customer base) on the other. To survive and prosper, firms must be alert to changes in people—as employees and customers—as well as to new discoveries which have the potential to create new and sustainable advantage in the marketplace. Companies have shown themselves to be much better at developing and adapting to new technologies than to changes in people, and this has resulted in the kinds of problems in recruiting and retaining key employees that are illustrated by the cases of Szulik and Peters.

Firms organized to develop and commercialize emerging technologies need to redefine the mindset that tends to dominate traditional approaches to human resources and move toward a conception of the firm as a customized workplace. Companies have begun to use the real options to create flexibility in managing their investments in technologies (as discussed by William Hamilton in Chapter 12). They now need to apply a parallel approach to building flexibility into their human resource management. These customized workplaces create a setting in which the changing needs of firms and the changing expectations and aspirations of people are accommodated simultaneously.

HUMAN RESOURCES CHALLENGES OF EMERGING TECHNOLOGIES

The distinctive characteristics of emerging technologies attenuate the organizational challenges that face any firm doing business at the dawn of the twenty-first century. In particular, the high risks, potentially high returns, uncertainty, centrality of know-how, and intense competition for talent create a very different employment environment.

Unusually High Risk

One of the distinctive characteristics of emerging technologies is an unusually high level of risk and uncertainty and a low "hit rate." In biotechnology, for example, a research program can occupy dozens of scientists and technicians for several years before a major breakthrough can be achieved, if ever. In other high technology sectors, such as electronics, computers, and software, several firms work on competing offers and, usually, the one that hits the market first locks the others out. In these contexts, firms face some thorny challenges. How to attract talented experts to work in organizations where their efforts may never materialize in marketable products? How to keep people motivated and minimize the demoralizing effects of failures and setbacks? How to offer competitive compensation to scarce and much sought after people in firms where the first revenue dollar can take several years to come?

To cope with the high level of risk and uncertainty, emerging technology firms need a capacity to react quickly to both technological and market changes. Managers need to initiate and terminate projects without delay. They also need to cope effectively with the human capital implications of doing so. How can they move people across projects at minimum costs and adjustment time? How can they, for example, lay off entire teams, if necessary, without hurting morale and motivation in other teams?

Potentially Monumental Returns

While highly risky, emerging technologies also yield monumental returns when firms hit the right market with the right products. The widely diffused use of ownership stakes and stock options certainly keeps the payroll under control in cash hungry firms and motivates employees but, as the case of Microsoft's Peters shows, it also raises some unanticipated issues. How can managers of successful emerging technology firms retain, motivate, and, more importantly, stretch employees who become millionaires very early in their life?

Unpredictable Cycles of Growth

Once they find their way to the market, firms operating in emerging technologies often grow rapidly. The different phases of the development and commercialization of the technology and the firm require different human resource configuration. The start-up that initially created the technology may bear little resemblance to the more mature organization that takes it to market. In addition to the well known financial risks, rapid growth also raises challenging human capital questions. How can firms create the flexibility to handle rapid growth? How can a firm that may have operated as a small business for years suddenly recruit, train, and manage hundreds of people? How does it maintain simultaneously an entrepreneurial spirit and high quality standards?

Sophisticated, Evolving, and Appropriable Know-How

Knowledge usually is the most valuable asset of an emerging technology firm, but the knowledge base is sophisticated and evolving—and much of it is in the minds of highly mobile employees. These characteristics raise difficult, and sometimes litigious, issues around the ownership of intellectual property. What happens when a biotechnology scientist or a software engineer is hired by the competition? What is the status of the knowledge and expertise he sells to his new employer? Does it belong to him or to the former employer? How can or should the legitimate claims of engineers and scientists to ownership of their expertise and inventions be balanced with the equally legitimate ownership claims of firms that provided the resources and context for the accumulation of expertise and innovation?

The knowledge base in emerging technologies also tends to be perishable and requires permanent renewal if the firm is to keep up with a sustained pace of innovation. While managers in traditional industrial firms can easily replace old machinery and physical assets with more up-to-date ones, they have much less control over how individuals renew their knowledge base. How can managers, often not experts themselves, induce individuals to keep on top of innovation in their field of expertise?

Intense Competition for Talent

Because of the sophisticated, evolving and appropriable character of the know-how on which emerging technology firms are built, the talent pool available to them is typically very thin, a condition which leads to intense competition for the best and the brightest. This war for talent[4] is most evident in high-tech locales such as the Silicon Valley where there are many knowledge-based firms with a high proportion of start-ups.

The competition for senior executive talent has been so intense in the Valley that some companies are turning to virtual CEOs to meet the need.[5] These executives with experience in emerging technology start-ups are opting to apply their know-how to several companies simultaneously rather than working for just one. There are an estimated 300 openings for seasoned CEOs in the Valley alone, a stunning number and a stark reminder of the significance of human capital in emerging technology firms. Competitive intelligence leaks and the protection of intellectual property are serious problems in an environment in which there are severe shortages of talent and individuals are working for several firms simultaneously. The issues these individuals face, it seems, are different at least in degree, if not in kind, from those faced by consulting firms who also sell advice to several clients simultaneously.

How People Are Changing

Emerging technology firms are not alone in facing intense new workforce challenges. As these firms face issues distinctive to managing emerging technologies, they also face more general challenges created by broader changes in the people who are working for their organizations and buying their products. These changes in the population of workers and customers are also reshaping the workplace.

Later—and Earlier—Entry into the Workforce: Divergent Trends

Two opposing trends are occurring with respect to the age and level of educational attainment of persons recruited by knowledge-based firms. On the one hand, to obtain the sophisticated knowledge they need, companies are looking for workers with higher levels of education, which means they are hiring older employees. On the other hand, with severe labor shortages, many knowledge-based firms are hiring right out of high school and provide further training and development internally. In view of these divergent trends, how should a firm think about its recruiting and development efforts?

More Skeptical, Instrumental View of Work

While being employed by an organization was once seen as an essential part of personal identity, employees are increasingly taking a more detached and skeptical view. First, this generation of young professionals is more wary of promises from employers, after seeing the consequences of widespread corporate restructuring and downsizing on their parents and their parents' friends. They have heard the rhetoric of corporate commitment to employees, but many have witnessed, directly or indirectly, the reality of involuntary turnover and have thus developed a healthy skepticism of corporate claims in this regard. How, if at all, can or should firms counteract this skepticism?

Second, many young people, potential recruits into the work force, have adopted a highly instrumental view of the role of work in their lives. A recent survey reported in the *New York Times* found that nearly two-thirds of the college seniors questioned anticipated retiring by the age of 50. If this figure can be construed as a reliable predictor of actual behavior, it portends significant challenges for firms trying to attract and retain the most talented members of the workforce. Not only is work, then, viewed primarily as a means to accumulating the wealth necessary to permit people to do what they "really" want to do (at age 50), it also may be less central to the individual's sense of personal identity. How, then, do firms create the appropriate motivational context for these new recruits and how do they retain those whom they have recruited successfully? This, essentially, is Microsoft's challenge with Chris Peters.

Third, and perhaps partly in contradiction to the above, people are more sensitive to work life issues now than in the past and are less willing to invest so heavily in work that they jeopardize family relationships.[6] This change, in conjunction with that noted above, leads to an apparently paradoxical situation, one in which people in their twenties and thirties are trying to build the independent wealth to enable them to retire early but are also unwilling to commit single-mindedly to the pursuit of corporate objectives. How should firms think about their relationships with their people in the face of this apparent paradox?

Diminished Loyalty and a Generation of Nomads

There is a definite trend among firms, particularly in the United States, to move away from employment security as part of the employment contract and toward the notion of employability.[7] In making this shift, firms are changing the terms of the traditional bargain. No longer are employees assured of a job as long as they perform acceptably; now their employers promise a set of job experiences and training and development opportunities which will make them attractive in the labor market, be it internal or external. They also shift responsibility for their own career management from the firm to the employees.

This movement has had two particularly noteworthy consequences. First, there is diminished trust by the employee in the contract itself and a sort of "survivor's syndrome," a feeling of guilt with respect to those who have departed and an accompanying sense of anxiety about one's own future. The depth of these feelings is best illustrated in the common comment: "I always keep my resume updated," which speaks eloquently to the reality of the precariousness which many of those currently employed actually feel.[8] Second, there is increasing mobility of talent among firms, a phenomenon which is at its most intense in Silicon Valley and which is exquisitely illustrated by the case of Matt Szulik. If loyalty was the glue that bound talent to firms in the past, intense competition for knowledge-based assets is the lubricant which creates the nomadic context of the present. How, in the context of diminished trust and intense competition for talent do firms keep their most talented employees?

Challenging Authority and Building Reciprocal Relationships

Contemporary social life is characterized by increased questioning and rejection of traditional, one way, forms of authority. The sociologist and philosopher Francis Fukuyama has used the phrase "The Great Disruption" to characterize the crisis of social institutions, across cultures and nations, based on vertical patterns of authority and dominance.[9] The firm, a core institution of modern social life, is not spared by the increased rejection of traditional authority and the search for more horizontal and reciprocal relationships.

The ability to challenge authority and create a reciprocal relationship was one of the things that attracted recent MBA graduate Denis Harscoat to Freedikspace, a Silicon Valley Internet start-up. A fresh Rotterdam School of Management MBA graduate, Harscoat could have easily opted for a traditional and successful corporate career but he preferred, instead, to join a 6-month-old Silicon Valley start-up with an uncertain future:

> Before undertaking an MBA, I was a firefighter in Paris and thus worked in a military organization under the authority of superiors. I realized soon that I did not want to live the same situation again. (After the MBA) I did not want to go back to a context where politics are important: pleasing one's superior to get promotions instead of expressing one's feeling about the situation. Here (at Freedisk-space), we can say what we think. You can stand up and ask "Ricky, Ari, and Paul what do you think?" and everybody listens. Here we can discuss openly about strategy, compensation, and everything . . .

The work is very stimulating and we feel that we are at the leading edge of a human and technological revolution.

The founders of Freediskpace, as those of thousands of other emerging technology start-ups, are tremendously challenged by employees like Harscoat who, because they give up the privileges and rewards of a traditional career and agree to work hard for a firm that has not yet much to offer, are much more demanding, not willing to be managed in the traditional command/control fashion, claiming a role in strategic decision making, and willing to have some control of their own fate in a highly uncertain business environment.

Ownership of Their Own Tools and Reduced Dependence on Any One Firm

In contrast to the assembly line era when workers were essentially adjuncts to machines, today's high-tech knowledge workers own their own tools and are thus not dependent on the firm to provide them. Increasingly, in knowledge-based settings, the "tools" that are used are intellectual. Because they are intellectual—and hence intangible—they are difficult, if not impossible, for the firm to monitor and control in traditional ways. And they do not show up on the firm's balance sheet.

The CFO of a Silicon Valley start-up gave us a first-hand testimony about the challenges of managing autonomous people who pursue their own human capital development strategies:

Many engineers are motivated by being given ownership of their project or the software feature they are creating. Some is human nature—we try in every area of the company to delegate as much as possible and "get out of the way" so folks have a free hand to create and learn. Some is also that it appeals to the engineer's self interest—they are sometimes thinking of what "hot items" they will be able to add to their resumes. Many engineers think quite actively about their resumes, and if you've seen what they look like (unlike those of us more in general management) an engineer's resume looks like an artist's portfolio in a sense—it is a collection of their best projects.

So in practice, if "MPEG routing thru asynch transmission to a remote data host" is the hot thing for someone to have on their resume this year (just an example), then we will find that as engineers come in to the group, they will be asking to be assigned that hot task, they will be very motivated to own that task, and they will be upset if priorities change and the company decides to not do that (as in: "I was told when I came to work here that I'd get to work on MPEG routing thru asynch transmission to a remote data host!"). It really keeps managers of engineers on their toes!

The balance of power has swung in the direction of employees, as they may carry around a nontrivial portion of the firm's assets in their heads. The costs to the firm of their departure, either to the competition or to start their own firms, are potentially very high. Under these circumstances, how should the firm think about protecting its knowledge-based assets, and how can it prevent leakage or wholesale movement of these assets outside?

Connected by Fluid Networks of Opportunity

Information technology itself is changing the way potential employees interact with potential employers. People can recognize and become informed about potential entrepreneurial and employment opportunities with remarkable speed, a fact which undoubtedly influences how they approach and think about work and "careers." Denis Harscoat sought out his job at Freediskspace.com rather than answering a job posting or going for a traditional round of interviews:

> I used Freediskspace's product during my MBA studies at the Rotterdam School of Management in the Netherlands. At the time, in April 1999, the product was still under testing. Together with an Irish classmate, an engineer at Nokia, we had dreamed of this kind of product and we wrote several reports on a similar product during the MBA. My Irish friend and I were thinking about starting our own venture. But when I discovered Freediskspace.com, knew they were only three people behind the venture and that their site was becoming popular, I worked very hard to be hired. The founders hired me from the Netherlands on the basis of an e-mailed, free, consulting report about user interface and marketing strategy.

Inherent in these networks is a nascent revolution in the way "employment" is defined. Individuals are much more likely today than ever before to think of themselves as independent professionals and to define their relationships to firms in limited terms. Waves of downsizing, advances in information technology, and the explosion of contract-based temporary employment have created a cadre of workers with a defined set of competencies that they are willing to rent to firms for fixed—and usually relatively short—periods of time. These workers, many of whom would be very attractive potential full-time hires to firms in emerging technologies, are pioneering new dynamics in the labor market and are setting new ground rules for how work gets done. They are, in effect, "life entrepreneurs," engaged, as the British sociologist Anthony Giddens has suggested, in "strategic life planning."[10] How should firms respond to this new talent pool of individuals who are only interested in unconventional part-time, contract-based relationships to employment?

TOWARD THE CUSTOMIZED WORKPLACE

From the above discussion of the characteristics of emerging technologies and how people and work are evolving, it is clear that attracting and retaining a talented and independent workforce in a fast-changing technological and business environment will require a different approach to human resources.

The dominant management paradigm of the nineteenth century, one that is still alive in many industries and areas around the globe, is typically not responsive to customers *or* employees. The firm is often family owned and managed as a closed system. Customers buy whatever it makes available to them. Employees are viewed as interchangeable and replaceable, are hired and fired at will, and have little voice and choice. For them, opportunity lies in finding a paternalistic capitalist who offers some benefits to make life a little less painful and the constraints of work a little more tolerable.

A new paradigm developed over the last half of the twentieth century as customers and shareholders became more proactive and management needed to become more responsive to them. Market-driven strategies and flexible organizations have developed as a consequence. In contrast to management operating in the nineteenth-century paradigm, twentieth-century management is more open. The firm actively listens to its customers and shareholders and involves them, through different mechanisms, in a variety of decision processes. Customers are the main drivers of the firm's needs for flexibility[11] and employees are on the receiving end, being required to adjust their work schedules, tasks, vacations periods, geographic assignments, and jobs carried in light of these needs. Because they are at the receiving end, workers often complain about the constraints it puts on them.

Companies have tried to rig their old systems to respond to the new demands of the workforce. Management's efforts to respond through improvements in the "quality of work life," while highly visible, are a product of the twentieth-century paradigm and will not be sufficient for the future. Firms also have created set of financial and nonfinancial incentives such as signing bonuses, stock option plans, benefits, training, flextime, mini-sabbaticals, accelerated responsibility expansion, and a variety of exotic dining, exercise, and/or childcare facilities.[12] Interestingly, though, these solutions do not seem to have addressed the underlying issues and, in some cases, may even work against the desired outcomes. Some counterproductive consequences of stock option plans, for example, are now widely recognized: too early exercise and selling of stock or leaving the company to cash out on a stock option plan.[13] While these solutions are designed to increase flexibility, they are being developed in piecemeal fashion to respond to issues as they present themselves rather than being motivated by an overarching philosophy of connectedness between firms and their human capital.

Moreover, since most emerging technology companies in a particular locale use basically the same set of incentives to remain competitive in the war for talent, they are collectively encouraging and rewarding opportunistic and nomadic behavior among the workforce and increasing their operating costs. The high level of turnover and mobility in high technology sectors[14] suggests that the solutions to *The Human Equation*[15] lie outside traditional human resource policies and incentives.

The challenge for management in the next century, as exemplified by the questions currently being faced by firms in emerging technologies, will be to internalize fully the diverse and changing needs of individuals and to invent employee-driven flexibility. The challenge is significant, because management will have to customize the workplace to suit the needs of *both* customers *and* workers simultaneously. To do this effectively, the firm will have to apply the logics of marketing, developed for its customers, to its relationships with workers.

As long as people are effectively treated by existing human resource policies as black boxes, that is, as relatively undifferentiated responders to a given set of incentives, these will not provide viable and sustainable solutions to the challenges discussed earlier, irrespective of their cost to the firm. Employees are less fooled by the menu selection in the cafeteria and other fragmented management gimmicks—no matter how well-intentioned they may be   if they are kept away from deciding about more important issues at the intersection of work and their personal priorities. Increasingly, people are not only willing but also *expecting* to influence the decisions that affect their work and personal lives. In a sense they can be viewed as "investors" in the workplace, seeking to maximize returns for given levels of risk.[16]

Emerging New Models

What we call "the customized workplace"[17] is a managerial approach that responds to these emerging aspirations. In contrast to traditional organizational processes, where organizational structures and systems are derived from a predefined strategy, the customized workplace, as we view it, explicitly seeks to balance what matters for the firm (its strategy) with what matters for individuals (their life strategies). Employees want more control over some fundamental aspects of their work lives, including:

- What to work for.
- The content of the work.
- When and where to work.
- How to accomplish the work.
- With whom and for whom to work.
- For how long to work.
- Direction of career plan.
- Skills needed to pursue the personal career plan.

These dimensions were once set almost exclusively by the employer, in a take-it-or-leave-it package. But now, as more employees have the option to leave such packages, companies need a more dynamic model. They need to address each of these issues on a customized basis.

Although full-fledged instances of the customized workplace are yet to be invented, many business organizations around the globe already are moving toward this model. Firms in the Silicon Valley have been forced by competitive pressures to be in the vanguard. As interesting, however, are a number of firms that have moved in this direction early and without the same intensity of competitive pressure. For whatever reasons, they appear to have been among the first to recognize and respond to the sociological trends we described earlier in this chapter. Their responses point the way toward the customized workplace:

- Like many Silicon Valley start-ups, SkyStream sees attracting employees, particularly software engineers, as one of its most crucial challenges. CEO Jim Olson, a veteran of Hewlett-Packard and 3Com, has focused on creating a winning atmosphere and approaching the engineers as individuals rather than as a category. According to Olson, without a sense that the company is performing well, people are likely to leave in search of the next winner. And unless people are understood and treated as individuals, company policies are likely to be ineffective in tapping the motivations of people to invest of themselves deeply in the affairs of the firm.
- At Semco, the Brazilian company made famous by Ricardo Semler's book *Maverick,* employees are involved in deciding about the location of new facilities and the acquisition of new machinery, have substantial freedom in deciding their work schedule, and enjoy total discretion over the investment of a portion of the profits.[18]
- At Metanoiques, a French midsize company that specializes in collaborative software created by an unconventional founder, there are no employees. Every member owns an equal share of the company and acts as an independent entrepreneur with profit and loss responsibility. The company has no head office and people are free to organize their own schedules. Internal collaboration is carried out through extensive use of information and communication technology.
- The founder of CFDP, a small French insurance company, recently went so far as to sell the company to employees and partner insurance agents and dismantle the head office.[19] Through this move, he hopes to transform the organization into a "community of independent entrepreneurs" where associates are free to conduct their local business and use networks to coordinate with other members of the organization.
- Therese Rieul, the founder manager of KA-L'informatique douce, a midsize French computer and software retailer, has always refused to write formal job descriptions because she believes that individuals should be allowed to design their own jobs. She is driven by the belief that management should be concerned primarily with outcomes and leave people free to figure out the best ways to perform the tasks.[20]

As shown by these examples, creative solutions to human capital issues are being developed around the world, in firms that are young and old, high tech or low. These examples demonstrate that while customized workplaces take many forms, they represent a common philosophical attitude toward people management issues. They focus on individual needs and aspirations and use intrinsic motivators. These firms build employment relationships designed to encourage involvement, trust, and mutual commitment.

A Focus on Individual Needs and Aspirations

In the customized workplace, people are managed as individuals rather than groups. Managers who have historically managed individuals through policies designed for aggregate groups: blue collar workers, hourly workers, part timers, white collar workers, high potential executives, women, and/or minorities, will find this approach quite foreign. In the customized workplace, people are not managed as members of a given group. They need to be treated as individuals. The biggest challenge for management, then, will be to achieve sufficient predictability in a context where individual behaviors are less subject to direct control.

While individuals' needs and aspirations are viewed as disturbances in traditional human resource mindsets—or a necessary evil—they represent the foundation of the customized workplace. Balancing firms' needs for predictability and effectiveness with diverse individual needs requires a new employment contract, in which management and employees confront their strategic and life plans and seek common ground.

These companies recognize that individuals are life-entrepreneurs who do their own strategic life planning. In twentieth-century management, even in its most enlightened versions, the firm is the only strategic planning agent. Management first elaborates a corporate strategy and then creates the optimal organizational and incentive structure to motivate people to implement the firm's strategy. In a context in which individuals are not necessarily motivated to work harder for more money or social status, it is important to involve them upstream, and to give them an opportunity to influence business strategy in a sense more consistent with their own life strategies. This evolution will be difficult to accomplish given the deeply seated belief that strategy is top management's exclusive prerogative and that inviting input from lower levels would yield only mediocre ideas (and, perhaps, challenge management's competence and authority).

A Focus on Intrinsic Motivators

Extrinsic motivators such as salary and benefits can be easily copied by competitors. The customized workplace relies more upon intrinsics that are less easily imitated and are thus more likely to be effective in the recruitment and retention process. The company focuses on understanding the "deeply embedded life interests" of employees and creating employment relationships that respond to those interests.[21] Employees are likely to stay engaged to the extent that what they do is connected to these deeper interests, and the clear managerial implication, therefore, is that efforts should be made to promote such connections.

Involvement

Sharing information and responsibility for the firm's situation with employees is another ingredient of the customized workplace. Contrary to the idea widely held in management circles that people never make decisions that can hurt them, some companies have proven that sharing the burden of a difficult situation with employees can be a very effective turnaround strategy. When Bernard Martin took charge of Sulzer France, which specializes in diesel engines, he involved employees in developing creative solutions to a severe financial crisis. Instead of formulating an action plan unilaterally, he told the employees that they should not expect him to come up with a miraculous solution, that the company's fate was in their hands, and that his role was to engage a process whereby together they could develop an effective turnaround strategy. Today, this case is often cited as an exemplary turnaround story in France and Bernard Martin, now retired, is in great demand as a speaker in management circles.

The Centrality of Trust

At its core, the customized workplace is based upon trust. Individuals are willing and able to commit substantial amounts of time, resources, and personal identity to relationships based on trust. After being pushed to the background by the contractual logic of "scientific management," the centrality of trust in business life is now being rediscovered. The recent literature on interorganizational alliances and joint ventures, for example, stresses the critical role of trust for the effective maintenance of these relationships.[22]

The importance of trust is revealed most clearly in times of hardship and rapid change. Given that hardship and rapid change are the hallmark of emerging technologies, trust is central to building a successful emerging technologies organization. Only a trusting workforce can voluntarily make sacrifices or create the flexibility the firm needs to succeed.

Mutual Commitment

Trust must be built before the hard times come, and for trust to grow between two parties, reciprocity is required. People would be willing to put a part of their fate within management hands only if management also accepts to put some of its own fate within the hands of people. Reciprocity develops only when each partner in a relationship is potentially vulnerable to the decisions of the other. Managers who need to keep things and people under control will have difficulty establishing trusting relationships.

The customized workplace is not viable if it is made up of free electrons who can change their behavior or withdraw from the game at any time. Organizations that have always relied on the cooperation of autonomous and powerful individuals, like professional sports teams, have long since placed mutual commitment and accountability at the heart of the employment relationship.

Yet the balance is that this commitment cannot be the vague and open-ended "job for life" commitments large organizations made in the past. It is more like the employment relationships of professional athletes and team managers, who are bound together for an agreed number of seasons and early termination of the contract by a party entitles the other to substantial compensation. In the business arena, this sort of arrangement has been mainly reserved for the employment of very senior executives of large public corporations. In the twenty-first century, it will have to be extended to every employment relationship. When the employment relationship eventually binds the parties for a predetermined period of time and makes them accountable for opportunistic termination, people can no longer suspect the firm of treating them as disposable assets and the firm can count, in return, on their collaboration for the duration of the contract.

Because it is based on participation, power sharing, trust, negotiation, reciprocity, and mutual commitment, the customized workplace will require adult, as opposed to paternalistic relationships based on charismatic leadership. Where the old model of the organization was an adult/child relationship, adult leadership relates to individuals as adults and requires skills for listening, understanding individuals' self-identity, anticipating, mediating, compromising, trusting, and committing. The leader of the twenty-first century will not be a God but a mortal who helps other mortals to awaken the God that is in all of us.

THE EVOLVING CUSTOMIZED WORKPLACE

These changes in both the workforce and the workplace and new approaches to human resource management are not uniformly distributed. Some employees still are seeking more traditional employment relationships. Some employers still are offering these relationships. An even greater challenge for incumbent firms seeking to build new emerging technology capabilities, often the old and new systems and philosophies of employment are forced to live side-by-side in the same firm.

This evolution of employers and workers creates opportunities for mismatch. At the risk of significant oversimplification, if we think of firms as seeking either stability or change in the composition of their human capital base and of individuals as seeking either stability or change in their work lives, we arrive at the four types of cases illustrated in Figure 18.2.

Where both firms and individuals converge, in cells 1 and 4, there is a good match. *Settlers* are most productive in firms seeking stability in their workforce, and *nomads* fit in well in firms which seek change in their workforce. Where they diverge, in cells 2 and 3, there are problems. People seeking stability in firms that are seeking change are likely to be unhappy. Similarly, people seeking change in firms seeking stability are also likely to be unhappy. For firms in emerging technologies, where change and

Figure 18.2
Individuals' and Firms' Need for Stability and Change

Individuals Seeking

		Stability	Change
Firms Seeking	Stability	Settlers in settled organizations 1	Nomads in settled organizations 2
	Change	Settlers in mobile organizations 3	Nomads in mobile organizations 4

uncertainty dominate, clarity about how to create flexibility without fragmentation is a major challenge. Is this most easily achieved through seeking stability or change in the people it employs? Should they be looking for settlers or for nomads?

The logic of the customized workplace can address these challenges by developing a level of customization that is appropriate to the demands of the employees and the requirements of the firm. Part of the customization of the workplace is to determine what level of customization to offer to employees and tailor the relationship to meet those needs, even if they are more traditional settler needs.

The cliché that, "Our people are our most important asset" was often articulated by twentieth-century firms but not always accompanied by consistent management practice. Rather than putting customers or people first, we believe management will have to put shareholders, customers, and workers first. Whether the individual is acting as customer, investor, worker, spouse, parent, or community activist, he is less willing to let others make decisions for him.

Institutions that fail to take notice of this trend do so at their own peril. Much as management did not have a choice as to whether to acknowledge or ignore shareholders and customers, it will also have to cope with the demands of autonomous and proactive individuals whose collaboration and commitment can longer be taken for granted. This is increasingly the context facing firms in emerging technologies today and will be the sociological context facing firms more generally in the twenty-first century.

The characteristics of emerging technologies—with their high uncertainty, payoffs and dependence on knowledge—place firms in these industries at the forefront of rethinking relationships with workers and the designs of their organizations. High-tech companies are already engaged in this process, but often on a piecemeal basis. The more comprehensive view offered by the customized workplace could create more coherent policies and learning. Emerging technology firms can become laboratories for testing new workplace designs. Even as scientists advance the knowledge of technology in research labs, these workforce experiments—if they are carried out consciously and systematically—will produce new insights that will accelerate the convergence between changing needs and aspirations of people and firms in the new economy.

Notes

Chapter 1

1. Nanett Byrnes and Paul C. Judge, "Internet Anxiety," *Business Week* (June 28, 1999), pp. 79–88.

2. The best known statement of this viewpoint is Richard Foster, *Innovation: The Attacker's Advantage* (New York: Summit Books, 1986). Similar conclusions are reached by James M. Utterback, *Mastering the Dynamics of Innovation* (Boston: Harvard Business School Press, 1995); Alfred D. Chandler, "Organization Capabilities and Economic History of the Industrial Enterprise," *Journal of Economic Perspectives,* vol. 6 (summer 1992), pp. 79–100; and Rebecca M. Henderson and Kim B. Clark, "Architectural Innovation: The Reconfiguration of Existing Systems and the Failure of Established Firms," *Administrative Science Quarterly,* vol. 35 (March 1990), pp. 9–30.

3. David B. Yoffie, "Competing in the Age of Digital Convergence," *California Management Review,* vol. 38 (summer 1996), pp. 31–53; and Andrew Grove, *Only the Paranoid Survive* (New York: Doubleday, 1996).

4. Joseph L. Bower and Clayton M. Christensen, "Disruptive Technologies: Catching the Wave," *Harvard Business Review,* vol. 73 (January/February 1995), pp. 43–53; and Clayton M. Christensen, *The Innovator's Dilemma* (Boston: Harvard Business School Press, 1997).

5. See note 3, Grove.

6. Hugh Courtney, Jane Kirkland, and Patrick Viguerie, "Strategy under Uncertainty," *Harvard Business Review,* vol. 75 (November/December 1997), pp. 66–81.

7. Robert H. Frank, *The Winner-Take-All Society* (New York: Free Press, 1995).

8. Bernard Wysocki, Jr., "Outlook: No. 1 Can Be Runaway Even in a Tight Race," *Wall Street Journal* (June 28, 1999), p. A1.

9. Michael Tushman and Philip Anderson, "Technological Discontinuities and Organizational Environment," *Administrative Science Quarterly,* vol. 31 (1986), pp. 439–456.

10. Joseph A. Schumpeter, *The Theory of Economic Development* (Cambridge, MA: Harvard University Press, 1934).

11. Clayton G. Smith and Arnold C. Cooper, "Entry into Threatening New Industries: Challenges and Pitfalls," *Building the Strategically-Responsive Organization,* eds.

Howard Thomas, Don O'Neal, Rod White, and David Hurst (New York: Wiley, 1994).

12. Bob Davis and David Wessel. "At Dawn of Electricity, Feuds and Hype," *Wall Street Journal* (April 6, 1998), A17; Carl Shapiro and Hal Varian. "Information Rules" (Boston: Harvard Business School, 1999).

13. Paul J.H. Schoemaker, "The Quest for Optimality: A Positive Heuristic of Science?" *Behavioral and Brain Sciences* (1991), 14, pp. 205–245.

14. Eric D. Beinhocker, "Robust Adaptive Strategies," *Sloan Management Review* (spring 1999), pp. 95–106.

15. G. Christian Hill. "Technology: First Hand-Held Data Communicators Are Losers, But Makers Won't Give Up," *Wall Street Journal* (February 3, 1994), B1; Kiyonori Sakakibara, Christian Lindholm, and Antti Ainamo, "Product Development Strategies in Emerging Markets: The Case of Personal Digital Assistants," *Business Strategy Review* (winter 1995), pp. 23–38.

16. Pat Dillon, "The Next Big Thing," *Fast Company* (June/July 1998), pp. 97–110.

17. Catherine Arnst, "PDA: Premature Death Announcement," *Business Week* (September 12, 1994), pp. 88–89.

18. See note 16.

19. Centocor 1997, Annual Report No. 2.

20. Christopher Moran, "Industry Snapshot: Biotechnology," Hoover's Online, based on 1996 data.

21. Rodney Ho, "1992: Gene Therapy," *Wall Street Journal Interactive Edition* (May 24, 1999).

22. Robert Langreth and Steven Lipin, "Johnson & Johnson Is Near to Deal to Acquire Centocor for $4.9 Billion" *Wall Street Journal* (July 21, 1999), p. A1.

23. Centocor company profile.

24. See note 22.

25. "The Future Was 'Obviously Not Obvious,' " *Stanford Observer* (May/June 1994), p. 13.

26. William F. Hamilton and Graham R. Mitchell, "Managing R&D as a Strategic Option," *Research & Technology Management* (May/June 1988), pp. 15–22; Edward H. Bowman and Dileep Hurry, "Strategy through the Options Lens: An Integrated View of Resource Investments and the Incremental–Choice Process," *Academy of Management Review*, vol. 18 (1993), pp. 760–782; Rita G. McGrath, "A Real Options Logic for Initiation Technology Positioning Investments," *Academy of Management Review*, vol. 22 (1997), pp. 974–996.

27. See note 3, Grove, p. 151.

28. Michael Tushman and Charles A. O'Reilly, III, *Winning through Innovation: A Practical Guide to Leading Organizational Change and Renewal* (Boston: Harvard Business School Press, 1997).

29. Michael E. Weinstein, "Rewriting the Book on Capitalism," *New York Times* (June 5, 1999), p. B–7.

30. Adam M. Brandenburger, Barry J. Nalebuff, and Ada Brandenberger, *Coopetition* (New York: Doubleday, 1997).

Chapter 2

1. Note the similarities here with the study of 80 long-lived companies by Arie de Geus, *The Living Company* (Boston: Harvard Business School Press, 1997). Among the companies that were at least 200 years old, four traits stood out: (1) an external orientation (2) focus on core values (3) experimentation in the periphery, and (4) financial conservatism. De Geus' study goes beyond surviving technological changes, considering instead all kinds of challengers. Moreover, he focuses on overall corporate survival as opposed that of a single business unit.

2. Vincent Barabba, *Meeting of the Minds: Creating the Market-Based Enterprise* (Boston: Harvard Business School Press, 1995). The study of the potential for copiers also overlooked the huge demand for the copying of copies, beyond simply copying originals.

3. For further details, see "Encyclopedia Britannica," Harvard Business School Case, N9-396-051 (December 1995); and for an update, see "Bound for Glory?: The Venerable Encyclopedia Britannica Struggles to Survive in an Electronic Age," *Chicago Tribune Magazine* (March 1998).

4. Robin M. Hogarth and Howard Kunreuther, "Risk, Ambiguity and Insurance," *Journal of Risk and Uncertainty,* vol. 2 (1989), pp. 5–35. Also see Paul J.H. Schoemaker, "Choices Involving Uncertain Probabilities: Test of Generalized Utility Models," *Journal of Economic Behavior and Organization,* vol. 16 (1991), pp. 295–317. Other things being equal (such as expected value), people prefer the known to the unknown, especially on the gain side.

5. Chris Floyd, "Managing Technology Discontinuities for Competitive Advantage," *Prism* (Second Quarter, 1996), pp. 5–21; Clayton M. Christensen, "Exploring the Limits of the Technology S-Curve, Part I: Component Technologies," *Production and Operations Management,* vol. 1 (fall 1992).

6. For an extensive discussion of these issues in the context of the information economy, see Carl Shapiro and Hal R. Varian, *Information Rules: A Strategic Guide to the Network Economy* (Boston: Harvard Business School Press, 1999).

7. Sam Hariharan and C.K. Prahalad, "Strategic Windows in the Structuring of Industries: Compatibility Standards and Industry Evolution," *Building the Strategically-Responsive Organization,* eds. Howard Thomas, Don O'Neal, Rod White, and David Hurst (New York: Wiley, 1994), pp. 289–308.

8. Daniel Kahneman and Amos Tversky, "Prospect Theory," *Econometrica,* vol. 47 (1979), pp. 283–291; Daniel Kahneman, Jack L. Knetsch, and Richart Thaler, "Experimental Tests of the Endowment Effect and the Case Theorem," *Journal of Political Economy,* vol. 98, no. 61 (December 1990), pp. 1325–1348.

9. Clayton G. Smith and Arnold C. Cooper, "Entry into Threatening New Industries: Challenges and Pitfalls," *Building the Strategically-Responsive Organization,* eds. Howard Thomas, Don O'Neal, Rod White, and David Hurst (New York: Wiley, 1994).

10. Daniel Kahneman and Dan Lovallo, "Timid Choices and Bold Forecasts: A Cognitive Perspective on Risk Taking," *Management Science,* vol. 39 (1993), pp. 17–31.

11. J. Edward Russo and Paul J.H. Schoemaker, *Decision Traps: Ten Barriers to Brilliant Decision Making and How to Overcome Them* (New York: Simon & Schuster, 1990).

12. Janet E. Bercovitz, John de Figueiredo, and David J. Teece, "Firm Capabilities and Managerial Decision-Making: A Theory of Innovation Biases," *Technological Innovation: Oversights and Foresights,* eds. Raghu Garud, Praveen Rattan Nayyar, and Zur Baruch Shapiro (Cambridge, England: Cambridge University Press).

13. Clayton Christensen, *The Innovator's Dilemma* (Boston: Harvard Business School Press, 1997).

14. Clayton M. Christensen and Richard S. Rosenbloom, "Explaining the Attacker's Advantage: Technological Paradigms, Organizational Dynamics, and the Value Network," *Research Policy,* and Richard S. Rosenbloom and Clayton M. Christensen, "Technological Discontinuities, Organizational Capabilities, and Strategic Commitments," Working Paper 94-1 (Berkeley, CA: Consortium on Competitiveness and Cooperation, January 1994).

15. Michael Tushman and Charles A. O'Reilly, III, *Winning through Innovation: A Practical Guide to Leading Organizational Change and Renewal* (Boston: Harvard Business School Press, 1997).

16. See note 9.

17. Richard Foster, *Innovation: The Attacker's Advantage* (New York: Summit Books, 1986).

18. Gary S. Lynn, Joseph G. Morone, and Albert Paulson, "Marketing and Discontinuous Innovation: The Probe and Learn Process," *California Management Review,* vol. 38 (spring 1996), p. C-37.

19. This example is adapted from Barabba in note 2.

20. Peter Senge, "The Leader's New Work: Building Learning Organization," *Sloan Management Review* (fall 1990), pp. 7–23.

21. See note 1.

22. Richard O. Mason and Ian L. Mitroff, *Challenging Strategic Planning Assumptions* (New York: Wiley, 1981).

23. Nikil Deogun, "High-Tech Plunge: A Tough Bank Boss Takes on Computers with Real Trepidation," *Wall Street Journal* (July 25, 1996), pp. A1–2.

24. Paul J.H. Schoemaker, "When and How to Use Scenario Planning: A Heuristic Approach with Illustration," *Journal of Forecasting,* vol. 10 (1991), pp. 549–564. Or Paul J.H. Schoemaker, "Scenario Planning: A Tool for Strategic Thinking," *Sloan Management Review,* vol. 36, no. 2 (winter, 1995), pp. 25–40.

25. Gary Hamel and C.K. Prahalad, *Competing for the Future* (Boston: Harvard Business School Press, 1994).

26. Kim Warren, "Exploring Competitive Futures Using Cognitive Mapping," *Long Range Planning,* vol. 28 (1995), pp. 10–21. For a review of cognitive mapping

tools, see Josh Klayman and Paul J.H. Schoemaker, "Thinking about the Future: A Cognitive Perspective," *Journal of Forecasting,* vol. 12 (1993), pp. 161–168. For further examples, see John D.W. Morecroft and John D. Sterman, *Modeling for Learning Organizations* (Portland, OR: Productivity Press, 1994). The classic references to mental models and their cognitive functions are: Dedre Gentner and Albert L. Stevens, eds., *Mental Models* (Erlbaum, 1983); and Philip N. Johnson-Laird *Mental Models,* 2nd ed. (Cambridge, MA: Harvard University Press, 1983).

27. Gary S. Lynn, Joseph G. Morone, and Albert S. Paulson, "Marketing and Discontinuous Innovation: The Probe and Learn Process," *California Management Review,* vol. 38 (spring 1996), pp. 8–17.

28. F. Gulliver, "Post-Project Appraisals Pay," *Harvard Business Review* (March/April 1987), pp. 128–32; Steven E. Prokesh, "Unleashing the Power of Learning: An Interview with British Petroleum's John Browne," *Harvard Business Review* (September/October 1997), pp. 147–168.

29. Howard Perlmutter, "On Deep Dialog," Working Paper, Emerging Global Civilization Project (The Wharton School, University of Pennsylvania, 1999).

30. Kees van der Heijden, *The Art of Strategic Conversation* (New York: Wiley, 1998).

31. Pankaj Ghemawat, *Commitment: The Dynamic of Strategy* (New York: Free Press, 1991) and Pankaj Ghemawat and Patricio del Sol, "Commitment versus Flexibility," *California Management Review,* vol. 40, no. 4 (summer 1998), pp. 26–42.

32. Erick D. Beinhocker, "Robust Adaptive Strategies," *Sloan Management Review* (spring 1999), pp. 95–106.

33. David B. Yoffie and Michale A. Cusumano, "Building a Company on Internet Time: Lessons from Netscape," *California Management Review,* vol. 41, no. 3 (spring 1999).

34. For insights about how Silicon Valley companies succeed in fast-changing environments, see Homa Bahrami, "The Emerging Flexible Organization: Perspectives from Silicon Valley," *California Management Review,* vol. 34, no. 4 (summer 1992), pp. 33–52 or Kathleen Eisenhardt and Shona Brown, *Competing on the Edge: Strategy as Structured Chaos* (Boston: Harvard Business School Press, 1998).

35. Patrick Anslinger, Dennis Carey, Kristin Fink, and Chris Gagnon. "Equity Carve-Outs," *The McKinsey Quarterly,* vol. 1 (1997), pp. 165–172.

36. Based on a report of the findings of Rajesh Chandy and Gerard Tellis, in Jerry Useem, "Internet Defense Strategy: Cannibalizing Yourself," *Fortune* (September 6, 1999), pp. 121–134.

37. Erick Shonfeld, "Schwab Puts It All Online," *Fortune* (December 7, 1998), pp. 94–100.

Part I

1. John Hume, "Transforming Monsanto through Innovation: Faith, Hope and $2 Billion," *Valuing Corporate Innovation,* Conference Report No. 97-1 (May 2, 1997), Wharton Emerging Technologies Management Program, 10-12.

Chapter 3

1. G. Garratt. "The Early History of Radio" (London: Institute of Electrical Engineers, 1994).

2. Hugh Aitken, *The Continuous Wave: Technology and American Radio, 1900–1932* (Princeton, NJ: Princeton University Press, 1985).

3. Stephen Gould and Niles Eldredge, "Punctuated Equilibria: The Tempo and Mode of Evolution Reconsidered," *Paleobiology,* vol. 3 (1977), pp. 115–151.

4. George Basalla, *The Evolution of Technology* (Cambridge, England: Cambridge University Press, 1988), p. 141.

5. John H. Dessauer, *My Years with Xerox: The Billions Nobody Wanted* (Garden City, NY: Doubleday, 1971).

6. See note 4.

7. Clayton Christensen and Richard Rosenbloom, "Explaining the Attacker's Advantage: Technological Paradigms, Organizational Dynamics, and the Value Network," *Research Policy,* vol. 24 (1995), pp. 233–257.

8. Ron Adner and Daniel Levinthal, "Dynamics of Product and Process Innovations: A Market-Based Perspective," Unpublished Working Paper (University of Pennsylvania, 1997).

9. See, for example, Giovanni Dosi, "Technological Paradigms and Technological Trajectories," *Research Policy,* vol. 11 (1983), pp. 147–62; George Basalla, *The Evolution of Technology* (Cambridge, England: Cambridge University Press, 1988); Richard Rosenbloom and Michael Cusumano, "Technological Pioneering and Competitive Advantage: The Birth of the VCR Industry," *California Management Review* (1987), pp. 51–76.

10. See, for example, Michael Tushman and Philip Anderson, "Technological Discontinuities and Organizational Environments," *Administrative Science Quarterly,* vol. 31 (1986), pp. 439–465; and Richard D'Aveni, *Hypercompetition* (New York: Free Press, 1994).

11. Joseph A. Schumpeter, *The Theory of Economic Development* (Cambridge, MA: Harvard University Press, 1934).

12. See note 7.

13. Richard Foster, *Innovation: The Attacker's Advantage* (New York: Summit Books, 1986).

14. Nathan Rosenberg, "Technological Change in the Machine Tool Industry, 1840–1910," *Perspectives on Technology,* ed. N. Rosenberg (London: M.E. Sharpe, 1976).

15. W. Brian Arthur, "Competing Technologies, Increasing Returns, and Lock-In by Historical Events," *Economic Journal,* vol. 99 (1989), pp. 116–131; and Paul David, "Clio and the Economics of QWERTY," *American Economic Review,* vol. 75 (1985), pp. 332–336.

16. David B. Yoffie, "Competing in the Age of Digital Convergence," *California Management Review,* vol. 38 (summer 1996), pp. 31–53; and Fumio Kodama,

"Technology Fusion and the New R&D," *Harvard Business Review*, vol. 70 (1992), pp. 70–78.

17. David Teece, "Capturing Value from Technological Innovation," *The Competitive Challenge*, ed. D. Teece (New York: Harper & Row, 1987).

18. Fumio Kodama, "Technology Fusion and the New R&D," *Harvard Business Review*, vol. 70 (1992).

19. Anita McGahan, Leslie Vadasz, and David Yoffie, "Creating Value and Setting Standards: The Lessons of Consumer Electronics for Personal Digital Assistants," *Competing in the Age of Digital Convergence*, ed. D. Yoffie (Boston: Harvard Business School Press, 1997).

20. See also George Day, Chap. 12.

21. Ron Adner and Daniel Levinthal, "Dynamics of Product and Process Innovations: A Market-Based Perspective," Unpublished Working Paper (University of Pennsylvania, 1997).

22. See note 9, Rosenbloom and Cusumano.

23. Kathleen Wiegner, "Silicon Valley 1, Gallium Gulch 0," *Forbes*, vol. 141 (1988), pp. 270–272.

24. Robert Ristelhueber, "GaAS are Making a Comeback, But Profits Remain Elusive," *Electronic Business Buyer*, vol. 19 (1993), pp. 27–28.

25. Devendra Sahal, "Technological Guideposts and Innovation Avenues," *Research Policy*, vol. 14 (1985), pp. 61–82.

26. See, for example, William Abernathy, *The Productivity Dilemma* (Baltimore: Johns Hopkins University Press, 1978); Eric von Hippel, *The Sources of Innovation* (New York: Oxford University Press, 1988); M. Lambkin and G. Day, "Evolutionary Processes in Competitive Markets: Beyond the Product Life Cycle," *Journal of Marketing*, vol. 53 (1989); Dorothy Leonard-Barton, *Wellsprings of Knowledge: Building and Sustaining the Sources of Innovation* (Boston: Harvard Business School Press, 1995); Clayton M. Christensen, *The Innovator's Dilemma* (Boston: Harvard Business School Press, 1997); and Geoffrey A. Moore, *Inside the Tornado: Marketing Strategies from Silicon Valley's Cutting Edge* (New York: HarperBusiness, 1995).

27. See note 5.

28. Steven E. Prokesch, "Battling Bigness," *Harvard Business Review*, vol. 71, no. 6 (1993), p. 143.

29. See Chap. 6, note 19.

30. Joseph G. Morone, *Winning in High-Tech Markets: The Role of General Management: How Motorola, Corning, and General Electric Have Built Global Leadership through Technology* (Boston: Harvard Business School Press, 1993).

31. Clayton M. Christensen, *The Innovator's Dilemma* (Boston: Harvard Business School Press, 1997).

32. Richard Nelson and Sidney Winter, *An Evolutionary Theory of Economic Change* (Cambridge, MA: Harvard University Press, 1982); and James G. March and Herbert A. Simon, *Organizations* (New York: Wiley, 1958).

33. George S. Day and David B. Montgomery, "Diagnosing the Experience Curve," *Journal of Marketing,* vol. 47, no. 3 (1983).

34. Wesley M. Cohen and Daniel A. Levinthal, "Absorptive Capacity: A New Perspective on Learning and Innovation," *Administrative Science Quarterly,* vol. 35, no. 1 (1990), pp. 128–152.

35. Gary Hamel and C.K. Prahalad, "Corporate Imagination and Expeditionary Marketing," *Harvard Business Review,* vol. 69, no. 4 (1991), pp. 81–92.

36. M.L. Dertouzos, R.K. Lester, and R.M. Solow, "Made in America: Regaining the Productive Edge," *The MIT Commission on Industrial Productivity* (Cambridge, MA: MIT Press, 1989).

37. See, for example, note 5; R. Rosenbloom and M. Cusumano. "Ampex Corporation and Video Innovation," in *Research on Technological Innovation, Management, and Policy,* R. Rosenbloom, ed. (1985), vol. 2, pp. 113–185; Daniel A. Levinthal, "The Slow Pace of Rapid Technological Change: Gradualism and Punctuation in Technological Change," *Industrial and Corporate Change,* vol. 7 (1998), pp. 217–247; see note 14; Thomas P. Hughes. "Networks of Power: Electrification in Western Societies" (Baltimore: Johns Hopkins Press, 1983).

38. See note 17.

39. See note 5.

40. Smith and Alexander (1988).

Chapter 4

1. One of the authors (D.S.D.) lead the venture capital investment in AquaPharm as an associate at a venture capital firm and was vice president of the company from its launch to its eventual sale. He holds or shares responsibility for the managerial decisions described here.

2. M. Iansiti and J. West, "Technology Integration: Turning Great Research into Great Products," *Harvard Business Review* (May/June 1997), pp. 69–79.

3. D. Matheson, J.E. Matheson, and M.M. Menke, "Making Excellent R&D Decisions," *Research & Technology Management* (November/December 1994), vol. 37, no. 6, pp. 21–24; and M.M. Menke, "Managing R&D for Competitive Advantage," *Research & Technology Management* (November/December 1997), vol. 40, no. 6, pp. 40–42.

4. Gary Hamel and C.K. Prahalad, *Competing for the Future* (Boston: Harvard Business School Press, 1994), pp. 129–130.

5. The core competence view stands in contrast to the position view best developed by Michael Porter, *Competitive Strategy* (New York: Free Press, 1980); and Michael Porter, *Competitive Advantage* (New York: Free Press, 1985).

6. On core competencies, see Margaret Peterarf, "The Cornerstones of Competitive Advantage: A Resource Based View," *Strategic Management Journal,* vol. 14 (1993), pp. 179–191; C.K. Prahalad and G. Hamel, "The Core Competence of the Corporation," *Harvard Business Review* (May/June 1990), pp. 79–91; D.J. Collis and C.A. Montgomery, "Competing on Resources: Strategy in the 1990s," *Harvard*

Business Review (1995), pp.118–128; Kathleen Conner, "A Historical Comparison of Resource Based Theory and Five Schools of Thought within Industrial Organization: Do We Have a New Theory of the Firm?" *Journal of Management* (March 1991), pp. 121–154.

7. *Fortune,* vol. 9, no. 138 (November 9, 1998), p. 104.

8. Mark Meyers and Richard S. Rosenbloom, "Rethinking the Role of Research," *Research & Technology Management,* vol. 39, no. 3 (May/June 1996), pp. 14–18

9. George Day estimated these probabilities based on numerous studies of acquisitions, joint ventures, and new product failure rates.

10. Position and learn is most akin to the useful typology for uncertain environments, strategies, and actions under uncertainty developed by Courtney and coworkers at McKinsey of "reserving the right to play" whereas sensing and following or leading may both be shaping strategies. H. Courtney, J. Kirkland, and P. Viguerie, "Strategy under Uncertainty," *Harvard Business Review* (November/December 1997), pp. 67–79.

Chapter 5

1. S. Kobrin and E. Johnson, "We Know All About You: Personal Privacy in the Information Age," Working Paper, Wharton Forum on Electronic Commerce (The Wharton School, February 27, 1999).

2. This early history is distilled from www.cs.washington.edu/homes/lazowska /cra/networks.html, a short historical piece written by Vinton Cerf, the "founding father" of the Internet.

3. See, for example, B. Kahin, ed., *Building Information Infrastructure* (New York: McGraw-Hill, 1992), for discussions and concerns about the (at the time) coming privatization of the NSFNet backbone network.

4. Andrew Freeman, "Technology in Finance Survey," *Economist* (October 26, 1996). As of this writing, about 45 percent of U.S. households have PCs and about 1 in 4 people are online. Household penetration in Europe is significantly lower; about 1 in 20 people are online in France, and about 1 in 10 in the United Kingdom.

5. NSFNet architecture was phased out by April, 1995; no comparable usage statistics are now available. These usage numbers do not reflect total Internet bytes, but only those traversing the NSFNet backbone.

6. Glenda Korporaal, "Baby Bell's $90bn Mother of All Mergers," *Financial Review* (October 14, 1993) for an account of the merger announcement. Ian Scales, "Irreconcilable Differences?" *Communications International* (December 1994) for an account of the failure of the merger negotiations.

7. Andrew C. Barrett, "Shifting Foundations: The Regulation of Telecommunications in an Era of Change," *Federal Communications Law Journal,* vol. 46, no. 1 (December 1993).

8. Mark K. Lottor, *Internet Domain Survey* (Network Wizards, Inc., October 1994).

9. Todd Spangler, "The Net Grows Wider: Internet Services," *PC Magazine,* vol. 15, no. 20 (November 19, 1996).

10. For an interesting legal perspective on controlling pornography in different countries, see Dawn A. Edick, "Regulation of Pornography on the Internet in the United States and the United Kingdom: A Comparative Analysis," *Boston College International and Comparative Law Review,* vol. 21 (summer 1998).

11. See "The Music Industry: A Note of Fear," *Economist* (October 31, 1998), p. 67.

12. For a primer on intellectual property law and the Internet, see the several articles on this subject in *Berkeley Technology Law Journal,* vol. 13 (1998).

13. See the special issue of the *South Carolina Law Review,* vol. 49 (summer 1998), which features a number of articles discussing these electronic commerce concerns.

14. I use the term *broadband* to refer to an electronic signal (or the facilities designed to transmit that signal) carrying information substantially greater than voice, such as video or high-speed data. For the more engineering oriented, I consider ISDN to be more than voice but less than broadband. Generally, a useful if not wholly accurate benchmark would be signals of 10 MHz or above. Note that modern compression technologies may eventually permit the practical carriage of such signals across telephone lines originally designed for voice.

15. Gerald R. Faulhaber, "Public Policy in Telecommunications: The Third Revolution," *Information Economics and Policy,* vol. 7 (1994), for supporting material.

16. See note 7.

17. Even cellular telephone, once thought to be a product targeted to wealthy stockbrokers phoning in buy and sell orders from their BMWs, has achieved a market penetration substantially beyond that originally predicted. In 1995, the cellular market grew by 36 percent to 32 million subscribers (compared to 145 million landline telephone subscribers). Tom McCall, "US Cellular Market Exhibits Solid Growth," *DataQuest Interactive* (March 25, 1996). Today, it is as likely that the person using a cellular phone next to you in a traffic jam is driving a pickup truck as a BMW.

18. For an early reference (among many others), see Ronald Brauetigam and Bruce Owen, *The Regulation Game: Strategic Use of the Administrative Process* (Cambridge, MA: Ballinger Publishing, 1978).

19. See, for example, Tom Hazlett, "Duopolistic Competition in Cable Television," *Yale Journal of Regulation,* vol. 7, no. 1 (1990), pp. 65–119; and Stanford Levin and John Meisel, "Cable Television and Competition," *Telecommunications Policy,* vol. 15, no. 11 (1991), pp. 521–522.

20. "Focus on Universal Service," *Telco Competition Report* (BRP Publications, October 24, 1996).

21. Oliver Williamson, "Franchise Bidding for Natural Monopolies—In General and with Respect to CATV," *Bell Journal of Economics,* vol. 7, no. 1 (1976).

22. Alfred Kahn, *The Economics of Regulation* (New York: Wiley, 1970).

23. This is not to say that cable or broadcast firms actually *produce* their own content (although broadcasters do produce their own news shows), but rather they *control* the content, which they generally purchase from outside entertainment suppliers.

24. See, for example, M. Kapor, "Where Is the Digital Highway Really Heading? The Case for a Jeffersonian Information Policy," *Wired,* vol. 1, no. 3 (July/August 1993).

25. This is a much different issue than the current litigation by the Department of Justice of Microsoft, in which the focus is the alleged attempt by Microsoft to "leverage" its power in the OS (conduit) market to dominate the emerging Internet browser market (which itself can be viewed as an alternative conduit market).

26. United States v. Microsoft Corporation, Civil Action No. 94-1564 (1994). U.S. District Court of the District of Columbia ("Final Judgment" entered August 21, 1995).

27. Gerald R. Faulhaber and Christiaan Hogendorn, "The Market Structure of Broadband Telecommunications," Working Paper, Public Policy & Management Department (Wharton School, University of Pennsylvania, August 1998).

28. AT&T/TCI has recently proposed a rollout of broadband services to residential customers via cable modem in areas where TCI distributes cable. Many are calling for some form of regulation on this new "monopoly." It is far more likely that AT&T is simply the first entrant in this market, with others likely to follow. In fact, AT&T's strategy of serving many areas at the outset of service could be a form of preemption, as mentioned above.

29. This analysis draws on James Kaplan, "Integration, Competition, and Industry Structure in Broadband Communications" (Wharton School Advanced Study Project Paper, 1996).

30. FCC Report and Order and Notice of Proposed Rulemaking, Docket 96-99 (March 11, 1996).

Part 2

1. Ken Henriksen, "Telecommunications Infrastructure," *The Convergence of Information Technologies: New Rules of Competition* (Wharton Emerging Technologies Management Program, April 17, 1998), pp. 22–23.

Chapter 6

1. David Stipp, "Gene Chip Breakthrough," *Fortune* (March 31, 1997), pp. 56–73.

2. Andrew H. Van de Ven and R. Garud, "A Framework for Understanding the Emergence of New Industries," *Research in Technological Innovation and Policy,* vol. 4 (1989), pp. 192–225.

3. Gina Colarelli O'Connor, "Market Learning and Radical Innovation: A Cross Case Comparison of Eight Radical Innovation Projects," *Journal of Product Innovation Management,* vol. 15 (March 1998), pp. 151–166.

4. More specific guidance can be found in a companion volume in this series: George S. Day and David B. Reibstein, *Wharton on Dynamic Competitive Strategies* (New York: Wiley, 1998).

5. Everett M. Rogers, *Diffusion of Innovations* (New York: Free Press, 1983).

6. Clayton M. Christensen, *The Innovator's Dilemma* (Boston: Harvard Business School Press, 1997).

7. This belief is not well founded. For a sampling of the debate on first-mover advantages see Marvin B. Lieberman and David B. Montgomery, "First-Mover Advantages," *Strategic Management Journal,* vol. 9 (summer 1988), pp. 41–58; and Gerard J. Tellis and Peter N. Golder, "First to Market, First to Fail? Real Causes of Enduring Market Leadership," *Sloan Management Review* (winter 1996), pp. 65–75.

8. Giovanni Dosi, "Sources, Procedures and Microeconomic Effects of Innovation," *Journal of Economic Literature,* vol. 26 (September 1988), pp. 1120–1171.

9. Adapted from Geoffrey A. Moore, *Inside the Tornado* (New York: HarperCollins, 1995).

10. See note 9.

11. Gary S. Lynn, Joseph G. Morone, and Albert Paulson, "Marketing Discontinuous Innovation: The Probe and Learn Process," *California Management Review,* vol. 38 (spring 1996), p. C-37.

12. Kiyonori Sakakibara, Christian Lindholm, and Antti Ainamo, "Product Development Strategies in Emerging Markets: The Case of Personal Digital Assistants," *Business Strategy Review* (winter 1995), pp. 23–38.

13. Glenn Bacon, Sarah Beckman, David C. Mowery, and Edith Wilson, "Managing Product Definition in High-Technology Industries: A Case Study," *California Management Review,* vol. 36 (spring 1994), pp. 32–56.

14. Marjorie E. Adams, George S. Day, and Deborah Dougherty, "Enhancing New Product Development Performance: An Organizational Learning Perspective," *Journal of Product Innovation Management,* vol. 15 (September 1998), pp. 403–422.

15. See note 11.

16. Gary Hamel and C.K. Prahalad, *Competing for the Future* (Boston: Harvard Business School Press, 1994).

17. See note 11.

18. See note 11, p. 14.

19. More details on this methodology can be found in Eric von Hippel, *The Sources of Innovation* (New York, Oxford University Press, 1988). Portions of this discussion were adapted from Eric von Hippel, Stefan Thomke, and Mary Sonnack, "Creating Breakthroughs at 3M," *Harvard Business Review,* vol. 77 (September/October 1999), pp. 47–57; and Stefan Thomke and Ashok Nimgade, "Note on Lead User Research," Paper Number 9-699-014 (Cambridge, MA: Harvard Business School, October 16, 1998).

20. Dorothy Leonard and Jeffrey Rayport, "Spark Innovation through Empathic Design," *Harvard Business Review* (November/December 1997), pp. 102–113.

21. Daniel Roth, "My What Big Internet Numbers You Have," *Fortune* (March 15, 1999), pp. 114–120.

22. Useful sources on the Bass model are Vijay Mahajan, Eitan Muller, and Frank Bass, "New Product Diffusion Models in Marketing: A Review and Directions for Research," *Journal of Marketing*, vol. 54 (January 1990), pp. 1–26; Vijay Mahajan, Eitan Muller, and Frank Bass, "Diffusion of New Products: Empirical Generalizations and Managerial Uses," *Marketing Science*, vol. 14, no. 3 (1995), pp. 679–688; and Daniel Roth, "My, What Big Internet Numbers You Have," *Fortune* (March 15, 1999), pp. 114–120.

23. Glen L. Urban, Bruce D. Weinberg, and John R. Hauser, "Premarket Forecasting of Really-New Products," *Journal of Marketing*, vol. 60 (January 1996), pp. 47–60 for a description of the method; and Eric Almquist and Gordon Wyner, "Identifying the Opportunities of the Future," *Mercer Management Journal*, vol. 10 (1998), pp. 31–40 for its application to the broadband market.

24. For more information on these methods see the chapter on scenario planning in this book, by Schoemaker and Mavaddat, and Rita G. McGrath and Ian C. MacMillan, "Discovery-Driven Planning," *Harvard Business Review* (July/August 1995), pp. 4–12.

Chapter 7

1. P.A. Roussel, "Technological Maturity Proves a Valid and Important Concept," *Research Management* (January/February 1984), pp. 29–30.

2. The framework proposed here for understanding the composition of markets with limited populations is based on initial work by Shintaku. See J. Shintaku, "Technological Innovation and Product Evolution: Theoretical Model and Its Applications," *Gakushuin Economic Papers*, vol. 26 (1990), pp. 53–67.

3. M. Tushman and P. Anderson, "Technological Discontinuities and Organizational Environments," *Administrative Science Quarterly*, vol. 31 (1986), pp. 439–465.

4. E. von Hippel, *The Sources of Innovation* (New York: Oxford University Press, 1988).

5. See, for example, Rita G. McGrath, Ian C. MacMillan, and Michael L. Tushman, "The Role of Executive Team Actions in Shaping Dominant Designs: Towards the Strategic Shaping of Technological Progress," *Strategic Management Journal*, vol. 13 (1992), pp. 137–161; Ian C. MacMillan and Rita G. McGrath, "Technology Strategy," *Advances in Global High-Technology Management*, eds. M.W. Lawless and L.R. Gomez-Mejia, vol. 4 (Greenwich, CT: JAI Press, 1994), pp. 27–66; and D. Levinthal, "Adaptation on Rugged Landscapes," *Management Science*, vol. 43 (1997), pp. 934–950.

6. J. Schumpeter, *Capitalism, Socialism, and Democracy* (3rd ed.) (New York: Harper & Row, 1950).

7. C. Christensen and J. Bower, "Customer Power, Strategic Investment, and the Failure of Leading Firms," *Strategic Management Journal,* vol. 17 (1996), pp. 197–219.

8. I.C. MacMillan and R.G. McGrath, "Discover Your Products' Hidden Potential," *Harvard Business Review,* vol. 74 (1996).

9. For a discussion of conjoint analysis techniques and software to facilitate this analysis, see Green and Krieger (1996).

10. G.R. Mitchell and W.F. Hamilton, "Managing R&D as a Strategic Option," *Research & Technology Management,* vol. 27 (1988), pp. 15–22; E.H. Bowman and D. Hurry, "Strategy through the Option Lens: An Integrated View of Resource Investments and the Incremental-Choice Process," *Academy of Management Review,* vol. 18 (1993), pp. 760–782; and R.G. McGrath, "A Real Options Logic for Initiating Technology Positioning Investments," *Academy of Management Review,* vol. 22 (1997), pp. 974–996.

11. C. Christensen, "When New Technologies Cause Great Firms to Fail," *The Innovator's Dilemma* (Boston: Harvard Business School Press, 1997).

Chapter 8

1. Typesetter discussion is based on Mary Tripsas, "Unraveling the Process of Creative Destruction: Complementary Assets and Incumbent Survival in the Typesetter Industry," *Strategic Management Journal,* vol. 18, no. S1 (July 1997), pp. 119–142.

2. A.C. Cooper and D. Schendel, "Strategic Responses to Technological Threats," *Business Horizons* (1976), pp. 61–69; L.M. Tushman and P. Anderson. "Technological Discontinuities and Organizational Environments," *Administrative Science Quarterly,* 31 (1986), pp. 439–465; R.M. Henderson and K.B. Clark. "Architectural Innovation: The Reconfiguration of Existing Product Technologies and the Failure of Established Firms," *Administrative Science Quarterly,* 35 (1990), pp. 9–30.

3. L. Wallis, *Typomania* (Upton-upon-Severn, UK: Severnside Printers, 1993).

4. D. Teece, "Profiting from Technological Innovation: Implications for Integration, Collaboration, Licensing and Public Policy," *Research Policy,* vol. 15 (1986), pp. 285–305.

5. W. Mitchell, "Whether and When? Probability and Timing of Incumbent's Entry into Emerging Industrial Subfields," *Administrative Science Quarterly,* vol. 34 (1989), pp. 208–234.

6. Compugraphic correspondence (1983).

7. Carl Shapiro and H. Varian, *Information Rules* (Boston, MA: Harvard Business School Press, 1999).

8. M. Tripsas, "Adobe Systems: A Case Study in Architectural Leadership" (1998). Working paper.

9. B.A. Majumdar, *Innovations, Product Developments and Technology Transfers: An Empirical Study of Dynamic Competitive Advantage, the Case of Electronic Calculators* (Washington DC: University Press of America, 1982).

10. Clayton M. Christensen, *The Innovator's Dilemma* (Boston, MA: Harvard Business School Press, 1997).

11. A. Afuah, "How Much Do You 'Co-Opetitor' Capabilities Matter in the Face of Technological Change?" Forthcoming, *Strategic Management Journal.*

12. J.F. Reinganum, "Uncertain Innovation and the Persistence of Monopoly," *American Economic Review,* vol. 73 (1983), pp. 741–748; R.J. Gilbert and H.M. Newberry, "Preemptive Patenting and the Persistence of Monopoly," *American Economic Review,* vol. 72 (1982), pp. 514–526.

13. See note 10.

14. R. Nelson and S. Winter, *An Evolutionary Theory of Economic Change* (Cambridge, MA: Harvard University Press, 1982).

15. R.M. Henderson and K.B. Clark, "Architectural Innovation: The Reconfiguration of Existing Product Technologies and the Failure of Established Firms," *Administrative Science Quarterly,* vol. 35 (1990), pp. 9–30.

Part III

1. William E. Coyne, "Technology with Purpose," *3M Technology Platforms* (1996), p. 5.

Chapter 9

1. S.M. Davis and C. Meyer, *Blur: The Speed of Change in the Connected Economy* (Reading, MA: Addison-Wesley, 1998).

2. J.Y. Wind and J. Main, *Driving Change* (New York: Free Press, 1998).

3. See, for example, G. Hamel, "Strategy as Revolution," *Harvard Business Review,* vol. 74, no. 4 (1996); or H. Mintzberg, B.W. Ahlstrand, and J. Lampel, *Strategy Safari: A Guided Tour through the Wilds of Strategic Management* (New York: Free Press, 1998).

4. L. Downes and C. Mui, *Unleashing the Killer App* (Boston: Harvard Business School Press, 1998).

5. S.L. Brown and K.M. Eisenhardt, *Competing on the Edge: Strategy as Structured Chaos* (Boston: Harvard Business School Press, 1998).

6. H. Mintzberg, "Crafting Strategy," *Harvard Business Review,* vol. 65, no. 4 (1987), p. 72.

7. P.F. Drucker, "The Theory of Business," *Harvard Business Review,* vol. 72, no. 5 (1994), pp. 95–104.

8. K.E. Weick, "Theory Construction as Disciplined Imagination," *Academy of Management Review,* vol. 14, no. 4 (1989), pp. 516–531.

9. J.A. Byrne, "Strategic Planning is Back," *Business Week* (1996), pp. 46–52.

10. G. Hamel and C.K. Prahalad, "Competing for the Future," *Harvard Business Review,* vol. 72, no. 4 (1994), pp. 122–128.

11. E.H. Bowman, ed., "Next Steps for Corporate Strategy," *Advances in Strategic Management* (JAI Press, 1995), pp. 39–64.

12. G.T. Allison, *Essence of Decision: Explaining the Cuban Missile Crisis* (Boston: Little, Brown & Company, 1971).

13. This model further assumes that ensuing action is a steady state choice rather than a stream of partial choices, thus precluding adjustment and therefore learning.

14. H. Courtney, J. Kirkland, and P. Viguerie, "Strategy under Uncertainty," *Harvard Business Review,* vol. 75, no. 6 (1997), pp. 66–79.

15. J. Bower, *Managing the Resource Allocation Process* (1972).

16. L.J. Bourgeois, III and K.M. Eisenhardt, "Strategic Decision Processes in High Velocity Environments: Four Cases in the Microcomputer Industry," *Management Science,* vol. 34, no. 7 (1988), pp. 816–835.

17. See note 16.

18. J.W. Fredrickson and T.R. Mitchell, "Strategic Decision Processes: Comprehensiveness and Performance in an Industry with an Unstable Environment," *Academy of Management Journal,* vol. 27, no. 2 (1984), pp. 399–423.

19. K.P. Coyne and S. Subramaniam, "Bringing Discipline to Strategy," *McKinsey Quarterly,* vol. 4 (1996), pp. 14–25.

20. See note 19, p. 18.

21. See note 14.

22. K.M. Eisenhardt, "Making Fast Strategic Decisions in High-Velocity Environments," *Academy of Management Journal,* vol. 32, no. 3 (1989), pp. 543–576.

23. See note 8.

24. P.J.H. Schoemaker and J.E. Russo, *It's All in How You Frame It: Simple Steps to Make the Right Decision* (Mimeo, 1996).

25. R.L. Ackoff, "The Art and Science of Mess Management," *Interfaces,* vol. 11, no. 1, (1981), pp. 20–26.

26. Peter Williamson, "Strategy as Options on the Future," *Sloan Management Review,* vol. 40., no. 3 (spring 1999).

27. See, for example, G.S. Day, *Market Driven Strategy: Processes for Creating Value* (New York: Free Press, 1990).

28. G. Hamel and C.K. Prahalad, "Corporate Imagination and Expeditionary Marketing," *Harvard Business Review* (July/August 1991), pp. 81–92.

29. See note 3, Hamel.

30. See note 6.

31. See note 6.

32. H.A. Simon, "Strategy and Organizational Evolution," *Strategic Management Journal*, vol. 14 (special issue, winter 1993), pp. 131–142.

33. P.J.H. Schoemaker, "Scenario Planning: A Tool for Strategic Thinking," *Sloan Management Review* (winter 1995), pp. 25–40.

34. O. Gadiesh and J.L. Gilbert, "Profit Pools: A Fresh Look at Strategy," *Harvard Business Review*, vol. 76, no. 3 (1998).

35. S.G. Makridakis, *Forecasting, Planning, and Strategy for the 21st Century* (New York: Free Press, 1990), pp. 132–133.

36. See note 3.

37. S.J. Wall and S.R. Wall, *Dilemmas in Strategy-Making: The New Strategists, Creating Leaders at All Levels* (New York: Free Press, 1995).

38. "Making Strategy," *Economist* (1997). Economist 342(8006). Of course, this argument relates more to the execution of strategy, but it does not specify how often they rethink their business model, which is an important moment when the value of diversity comes into play.

39. See note 37.

40. Implementation may be faster, however, if there is broader participation in the formulation process.

41. M. Crossan and M. Sorrenti, "Making Sense of Improvisation," *Advances in Strategic Management*, vol. 14 (1997), p. 170.

42. Cited in Robert L. Glass, *Software Creativity* (Prentice Hall, ECS Professional, 1995).

43. See note 33; and personal communication.

44. S. Hart and C. Banbury, "How Strategy-Making Processes Can Make a Difference," *Strategic Management Journal*, vol. 15, no. 4 (1994), pp. 251–269.

45. See note 6.

46. "Walt Disney Imagineering: A Behind the Dreams Look at Making the Magic Real," *Imagineers* (New York: Hyperion, 1996).

Chapter 10

1. Thomas B. Rosenstiel, "Old Demons at Bay—U.S. Newspapers Face Future with New Confidence," *Los Angeles Times* (April 30, 1986).

2. Gail DeGeorge, "Knight-Ridder: Running Hard, But Staying in the Same Place," *Business Week* (February 26, 1996).

3. Foster (1986); Utterback (1995); Chandler (1992); Christensen (1998).

4. Paul J.H. Schoemaker and Cornelius A.J.M. van der Heijden, "Integrating Scenarios into Strategic Planning at Royal Dutch/Shell," *Planning Review*, vol. 20, no. 3 (1992), pp. 41–46. For additional references on scenario planning see Pierre Wack, "Scenarios: Uncharted Waters Ahead," *Harvard Business Review* (September/October 1985); Gil Ringland, *Scenario Planning* (New York: Wiley, 1998); Paul

J.H. Schoemaker, "Scenario Planning: A Tool for Strategic Thinking,"*Sloan Management Review,* vol. 36, no. 2 (winter 1995), pp. 25–40.

5. Richard Mason and Ian Mitroff, *Challenging Strategic Planning Assumptions* (New York: Wiley, 1981).

6. G. Shaw, R. Brown, and P. Bromiley, "Strategic Stories: How 3M is Rewriting Business Planning," *Harvard Business Review,* vol. 76, no. 3 (1998), p. 50.

7. Paul J.H. Schoemaker, "When and How to Use Scenario Planning: A Heuristic Approach with Illustration," *Journal of Forecasting,* vol. 10 (1991), pp. 549–564.

8. See note 7.

9. Peter Schwartz, *The Art of the Long View* (New York: Doubleday, 1991).

10. Paul J.H. Schoemaker, "Multiple Scenario Development: Its Conceptual and Behavioral Basis," *Strategic Management Journal,* vol. 14, no. 1 (1993), pp. 193–213; also see Paul J.H. Schoemaker, "How to Link Strategic Vision to Core Capabilities," *Sloan Management Review,* vol. 34, no. 1 (fall 1992), pp. 67–81.

11. Howard V. Perlmutter, "On Deep Dialog," Working Paper, Emerging Global Civilization Project (The Wharton School, University of Pennsylvania, 1999).

12. Cornelius van der Heijden, *The Art of Strategic Conversation* (New York: Wiley, 1998).

13. John Morecroft and John Sterman, *Modeling for Learning Organizations* (Productivity Press, 1994).

14. Pankaj Ghemawat, *Commitment: The Dynamic of Strategy* (New York: Free Press, 1991).

15. See note 7.

16. See note 6.

17. Table from *Economist* (September 11, 1999), p. 68.

18. Based on an Advanced Study Project by Wharton MBA students Kevin Kemmerer, Jim Obsitnik, and Tim Sheerin, conducted under the supervision of Prof. Paul J.H. Schoemaker in the Spring of 1997.

19. Eric von Hippel, Stefan Thomke, and Mary Sonnack, "Creating Breakthroughs at 3M," *Harvard Business Review,* vol. 77 (September/October 1999), pp. 47–57.

20. For a more complete list, see Paul J.H. Schoemaker, "Twenty Common Pitfalls in Scenario Planning," *Learning from the Future,* eds. L. Fahey & R. Randall (New York: Wiley, 1998), pp. 422–431. More general traps in inference and judgment are described in *Decision Traps,* eds. J. Edward Russo and Paul J.H. Schoemaker (New York: Simon & Schuster, 1989).

21. Gary Hamel, "Bringing Silicon Valley Inside," *Harvard Business Review* (September/October 1999), pp. 70–86.

22. Kathleen Eisenhardt and Shona Brown, *Competing on the Edge: Strategy as Structured Chaos* (Boston: Harvard Business School Press, 1998).

23. Arie de Geus, "Planning as Learning," *Harvard Business Review* (March, 1988).

Chapter 11

1. The term *intellectual property* is sometimes used quite broadly. It is here used narrowly, to refer to the sorts of things that are (arguably) eligible for recognition and protection under the laws of patent, copyright, trademark and trade secrets. A first bit of perspective is implicit in this usage: for example, ideas are products of the intellect but there is no legally recognized property right in ideas as such; hence it does not seem helpful to think of ideas as intellectual property.

2. W.M. Cohen, R.R. Nelson, and J. Walsh, "Protecting Their Intellectual Assets: "Appropriability Conditions and Why U.S. Manufacturing Firms Patent (or Not)" Working Paper (Pittsburgh: Carnegie Mellon University, 1999).

3. See note 2, p. 5.

4. H. Schultz and D.J. Yang, *Pour Your Heart Into It: How Starbucks Built a Company One Cup at a Time* (New York: Hyperion, 1997), pp. 76–77.

5. From a public policy standpoint, the case for strengthening protection where it is currently weak deserves more serious consideration. The different levels of innovativeness that prevail across sectors and types of innovation undoubtedly reflect to some extent the differential effectiveness of the intellectual property system—though not in the simple way that is sometimes suggested The complexity arises from the fact that enhanced protection of a given innovation generally means partial suppression of knowledge spillovers, which in turn means that the costs of further innovation are increased. See, for example, R.P. Merges and R.R. Nelson, "On the Complex Economics of Patent Scope," *Columbia Law Review,* vol. 90 (1990), pp. 839–916.

6. Other systems of categories have been used; for example, the CMU study cited above treats complexity as a separate category, while it is here treated as contributing to secrecy. Trade secret law is also usefully treated under that heading, although it is obviously a form of legal protection and often viewed as a cousin to patent law. The CMU study also explores the effects of combination of mechanisms, finding unsurprisingly that complementary assets and lead time advantages go together. See note 2, p. 7.

7. W. Barnett, "Telephone Companies," *Organizations in Industry: Strategy, Structure, and Selection,* eds. G.R. Carroll and M.T. Hannan (New York: Oxford University Press, 1995).

8. This lack of attention to utility means there is abundant work for anthologists of "wacky and wonderful" patents. For some of these, consult www.colitz.com/site /wacky/wackyold

9. Previously, protection normally extended for seventeen years from the issue date. This change was enacted pursuant to international agreements on harmonization of intellectual property law made at the Uruguay Round of GATT (1994).

10. See note 7.

11. J. Hirshleifer, "The Private and Social Value of Information and the Reward to Inventive Activity," *American Economic Review,* vol. 61 (1971), pp. 561–574.

12. R.C. Levin, R. R. Nelson, and S. G. Winter, "Appropriating the Returns from Industrial Research and Development," *Brookings Papers on Economic Activity* No. 3783–820 (1987). The surveys differed in the framing of specific questions, and they were conducted eleven years apart, so not too much should be made of the difference in results between the two.

13. See note 12, pp. 802–803.

14. See note 2, p. 13.

15. Graver Tank & Manufacturing. Co. v. Linde Air Products Co., 339 U.S. 605, 608, as quoted by Merges and Nelson, see note 5; see the latter for discussion of the doctrine.

16. R. Henderson, R.L. Orsenigo, and G.P. Pisano, "The Pharmaceutical Industry and the Revolution in Molecular Biology: Exploring the Interactions among Scientific, Institutional and Organizational Change," *Sources of Industrial Leadership: Studies of Seven Industries,* eds. D.C. Mowery and R.R. Nelson (New York: Cambridge University Press, 1999).

17. G.P. Pisano, *The Development Factory: Unlocking the Potential of Process Innovation* (Boston: Harvard Business School Press, 1997), p. 65.

18. Converted from 2.4 billion pounds at 1.6 dollars per pound.

19. See note 2, p. 24.

20. As discussed below, government institutions supplement the "natural" effects of secrecy to some extent by enforcing trade secret laws and contractual obligations relating, for example, to confidentiality or noncompete agreements.

21. D. Shapley, "Electronics Industry Takes to 'Potling' Its Products for Market," *Science,* vol. 202 (1978), pp. 848–849.

22. R.H. Hayes, G.P. Pisano, and D.M. Upton, *Strategic Operations: Competing through Capabilities* (New York: Free Press, 1996), p. 54.

23. S.G. Winter, "Knowledge and Competence as Strategic Assets," *The Competitive Challenge: Strategies for Industrial Innovation and Renewal,* ed. D.J. Teece (Cambridge, MA: Ballinger, 1987), pp. 159–184.

24. U. Zander and B. Kogut, "Knowledge and the Speed of Transfer and Imitation of Organizational Capabilities: An Empirical Test," *Organization Science,* vol. 6, no. 1 (1995), pp. 76–92.

25. J.P. Liebeskind, "Keeping Organizational Secrets: Protective Institutional Mechanisms and Their Costs," *Industrial and Corporate Change,* vol. 6 (September 1997), pp. 623–663.

26. M.J.C. Martin, *Managing Innovation and Entrepreneurship in Technology-Based Firms* (New York: Wiley, 1994), p. 45.

27. E. von Hippel, "Cooperation between Rivals: Informal Know-How Trading," *Research Policy,* vol. 16 (1987), pp. 291–302; see note 25.

28. See note 2, p. 11.

29. D.J. Teece, "Profiting from Technological Innovation," *Research Policy,* vol. 15, no. 6 (1986), pp. 285 305.

30. On this subject, see note 16, on which the present account draws.

31. This discussion of IBM draws heavily on S.W. Usselman, "IBM and Its Imitators: Organizational Capabilities and the Emergence of the International Computer Industry," *Business and Economic History,* vol. 22 (winter 1993), pp. 1–35.

32. For an example of a dismissive comment about the relevance of patents to IBM's success, see G.W. Brock, *The U.S. Computer Industry: A Study of Market Power* (Cambridge, MA: Ballinger, 1975), p. 64. Studies that make essentially no mention of patents include F.M. Fisher et al., *Folded, Spindled, and Mutilated: Economic Analysis and U.S. v. IBM* (Cambridge, MA: MIT Press, 1983); and R.T. DeLamarter, *Big Blue: IBM's Use and Abuse of Power* (New York: Dodd, Mead & Co., 1986). These two represent the *pro-* and *contra*-IBM views of the most recent antitrust case. Also see, B.G. Katz and A. Phillips, "The Computer Industry," *Government and Technical Progress: A Cross-Industry Analysis,* ed. R.R. Nelson (New York: Pergamon Press, 1982); R. Langlois, "Cognition and Capabilities: Opportunities Seized and Missed in the History of the Computer Industry," *Technological Innovation: Oversights and Foresights,* eds. R. Garud, P.R. Nayar, and Z.B. Shapira (Cambridge, MA: Cambridge University Press,1995); T.F. Bresnahan and F. Malerba, "Industrial Dynamics and the Evolution of Firms' and Nations' Competitive Capabilities in the World Computer Industry," *Sources of Industrial Leadership: Studies of Seven Industries,* eds. D.C. Mowery and R.R. Nelson (New York: Cambridge University Press, 1999); Usselman, "IBM. . . ." Although "absence of evidence is not evidence of absence," the uniform neglect of the topic by all these authors must surely tell us something about the actual importance of patents to IBM.

33. D. Teece, G. Pisano, and A. Shuen, "Dynamic Capabilities and Strategic Management," *Strategic Management Journal,* vol. 18, no. 7 (1997), pp. 509–533. Cohen et al. (see note 2, p. 23), point out that patents also may confer greater advantages when massed as patent "fences" than they do for individual innovations.

34. Lead time is referred to in the survey as the advantage of "being first to market."

35. See note 2, Table 1.

36. The innovator in such cases is playing the historical role of the entrepreneur, as that role was described in Joseph Schumpeter's classic book, *The Theory of Economic Development* (Cambridge, MA: Harvard University Press, 1934). First published in German in 1911.

37. For critical reviews of the "first mover" literature, see M.B. Lieberman and D.B. Montgomery, "First-Mover Advantages," *Strategic Management Journal,* vol. 9 (1988), pp. 41–58; and M.B. Lieberman and D.B. Montgomery "First-Mover (Dis)Advantages: Retrospective and Link with the Resource-Based View," *Strategic Management Journal,* vol. 19 (1998), pp. 1111–1125. On "hypercompetition," see R. D'Aveni, *Hypercompetition: The Dynamics of Strategic Maneuvering* (New York: Free Press, 1994).

38. This account draws primarily on R.S. Rosenbloom, *Nucor: Expanding Thin-Slab Capacity* (Boston: Harvard Business School Press, 1991) Case 9-792-023; and P. Ghemawat, *Nucor at a Crossroads* (Boston: Harvard Business School Press, 1992) Case 9-793-039.

39. *Fortune,* vol. 131, "The Fortune 500" (May 15, 1995), p. F-57.

40. A. Phillips, *Technology and Market Structure: A Case Study of the Aircraft Industry* (Lexington, MA: D.C. Health, 1971).

41. The level of the innovator's price affects how much room the followers have to attract customers by cutting price without making loses. This situation illustrates the point that an innovator may be able to protect the length and breadth of the rent stream by sacrificing some of its depth.

42. R. Makadok, "Can First-Move and Early-Mover Advantages Be Sustained in an Industry with Low Barriers to Entry/Imitation?" *Strategic Management Journal,* vol. 19 (1998), pp. 683–696.

43. S.G. Winter and G. Szulanski, "Replication as Strategy," *Organization Science* (2000).

44. A. Ryan et al., "Biotechnology Firms," *Organizations in Industry: Strategy, Structure and Selection,* eds. G.R. Carroll and M.T. Hannan (New York: Oxford University Press, 1995).

45. Quoted in Merges and Nelson, note 5. Whether the last clause adds something beyond what the word "pioneer" itself conveys seems open to doubt.

46. See note 5.

47. M. Tushman and P. Anderson, "Technological Discontinuities and Organization Environments," *Administrative Science Quarterly,* vol. 31 (1986), pp. 439–465.

Part IV

1. Raymond V. Gilmartin, "Chairman's Message," *1998 Annual Report* (Merck & Co.), p. 4.

2. Thomas Woodward, "Real Options Analysis at Merck," *Managing Uncertainty Using Scenario Planning,* Emerging Technologies Research Program conference report (November 22, 1996), p. 9.

Chapter 12

1. "Keeping All Options Open," *Economist* (August 14, 1999), p. 62; Peter Coy, "Eploiting Uncertainty," *Business Week* (June 7, 1999), pp. 118–124; Martha Amram and Nalin Kulatilaka, "Disciplined Decisions: Aligning Strategy with the Financial Markets, *Harvard Business Review* (January/February 1999), pp. 95–104.

2. Stewart Myers, "Determinants of Corporate Borrowing," *Journal of Financial Economics* (November 1977), pp. 147–176.

3. Graham R. Mitchell and William F. Hamilton, "Managing R&D as a Strategic Option," *Research Technology Management* (May/June 1988), pp. 15–22.

4. See, for example, K. Avinash Dixit and Robert S. Pindyck, "The Options Approach to Capital Investment," *Harvard Business Review* (May/June 1995), pp. 105–115; "Investment Under Uncertainty," (Princeton, NJ: Princeton University Press, 1994); Lenos Trigeorgis and Scott P. Mason, "Valuing Managerial Flexibility," *Midland Corporate Finance Journal* (spring 1987), pp. 14–21; Stewart C. Myers, "Finance Theory and Financial Strategy," *Interfaces,* vol. 14, no. 1 (January/February 1984), pp. 126–137.

5. Keith J. Leslie and Max P. Michaels, "The Real Power of Real Options," *The McKinsey Quarterly,* 3, (1997), pp. 4–22.

6. See note 1, Coy.

7. Based on discussion by Terrence W. Faulkner, "Applying 'Options Thinking' to R&D Valuation," *Research Technology Management* (May/June 1996), pp. 50–56.

8. Bruce Kogut and Nalin Kulatilaka, "Options Thinking and Platform Investments: Investing in Opportunity," *California Management Review* (winter 1994), pp. 52–71.

9. F. Peter Boer, *The Valuation of Technology:Business and Financial Issues in R&D* (New York: Wiley, 1999); Terrence W. Faulkner, "Options Pricing Theory and Strategic Thinking: Literature Review and Discussion, Version 3.3," (Eastman Kodak Company, May 1996); and Peter A. Morris, Elizabeth O. Teisberg, and A. Lawrence Kolbe, "When Choosing R&D Projects, Go With Long Shots," *Research Technology Management* (January/February 1991), pp. 35–40; Thomas Copeland and Philip T. Keenan, "Making Real Options Real," *The McKinsey Quarterly,* 3, (1998), pp. 128–141.

10. Gordon Sick, *Capital Budgeting with Real Options* Monograph Series in Finance and Economics, Stern School of Business (New York: New York University), Monograph No. 1989-3; Lenos Trigeorgis, "Real Options and Interactions with Financial Flexibility," *Financial Management* (autumn 1993), pp. 202–224.

11. For comprehensive reviews and examples, see Lenos Trigeorgis, ed., *Real Options in Capital Investments* (Westport, CT: Praeger, 1995); and Lenos Trigeorgis, *Real Options* (Cambridge, MA: MIT Press, 1996). Review: Gordon A. Sick, *The Journal of Finance,* vol. 51, no. 5 (December 1996), pp. 1974–1977.

12. See note 10.

13. Edward H. Bowman and Gary T. Moskowitz, "A Heuristics Approach to the Use of Options Analysis in Strategic Decision Making," (Reginald H. Jones Center, Wharton School, January 6, 1998); Bruce Kogut and Kulatilaka Nalin. "Capabilities as Real Options," Prepared for the conference *Risk, Managers, and Options* (Reginald H. Jones Center, Wharton School, November 18, 1997); and Timothy A. Luehrman, "Investment Opportunities as Real Options: Getting Started on the Numbers," *Harvard Business Review* (July/August 1998), pp. 51–67.

14. Elizabeth O. Teisberg, "Methods for Evaluating Capital Investment Decisions under Uncertainty," in ed. Leon Trigeorgis, Real Options in Capital Investment (Westport, CT: Praeger, 1995), pp. 31–46.

15. John F. Magee, "How to Use Decision Trees in Capital Investment," *Harvard Business Review* (September/October 1964), pp. 79–96; James E. Smith and Robert F. Nau, "Valuing Risky Projects: Option Pricing Theory and Decision Analysis," *Management Science,* vol. 41, no. 5 (May 1995), pp. 795–816.

16. Data 3.5—Release 3.5.3/32 Bit Demo. Copyright 1988–1998 Treeage Software, Inc.

17. James E. Smith and Robert F. Nau, "Valuing Risky Projects: Option Pricing Theory and Decision Analysis," *Management Science,* vol. 41, (May 1995), pp. 795–816.

Chapter 13

1. P. Romer, L. Perlman, S. Shih, M. Volkema, and D. Lessard, "Bank of America Roundtable on the Soft Revolution: Achieving Growth by Managing Intangibles," *Journal of Applied Corporate Finance,* vol. 11 (1998), pp. 2–27.

2. M.C. Jensen and W. Meckling, "The Theory of the Firm: Managerial Behavior, Agency Costs and Ownership Structure," *Journal of Financial Economics,* vol. 3 (1976), pp. 305–360.

3. S.C. Myers and N. Majluf, "Corporate Financing and Investment Decisions When Firms Have Information That Other Investors Do Not Have," *Journal of Financial Economics,* vol. 13 (1984), pp. 187–221.

4. J.E. Stiglitz and A. Weiss, "Credit Rationing in Markets with Imperfect Information," *American Economic Review,* vol. 71 (1981), pp. 393–410.

5. J. Freear and W.E. Wetzel, "Who Bankrolls High-Tech Entrepreneurs?" *Journal of Business Venturing,* vol. 5, pp. 77–89.

6. This section draws on G.W. Fenn, N. Liang, and S. Prowse, *The Economics of the Private Equity Market* (Washington DC: Board of Governors of the Federal Reserve System, 1995); P. Gompers and J. Lerner, "An Analysis of Compensation in the U.S. Venture Capital Partnership," *Journal of Financial Economics,* vol. 51 (1999), pp. 3–44; W.A. Sahlman, "The Structure and Governance of Venture-Capital Organizations," *Journal of Financial Economics,* vol. 27 (1990), pp. 473–521; M. Wright and K. Robbie, "Venture Capital and Private Equity: A Review and Synthesis," *Journal of Business Finance and Accounting,* vol. 25 (1998), pp. 521–569; and B. Zider, "How Venture Capital Works," *Harvard Business Review* (November/December 1998), pp. 131–139.

7. VentureOne, *National Venture Capital Association 1996 Annual Report* (San Francisco, CA: VentureOne, 1997).

8. For a discussion, see J. Lerner, " 'Angel' Financing and Public Policy: An Overview," *Journal of Banking and Finance,* vol. 22 (1998), pp. 773–783; and S. Prowse, "Angel Investors and the Market for Angel Investments," *Journal of Banking and Finance,* vol. 22 (1998), pp. 785–792.

9. J. Freear, S. Sohl, and W. Wetzel, "Creating New Capital Markets for Emerging Ventures," Unpublished Manuscript (Durham, NH: University of New Hampshire, 1996).

10. IPOs have been widely studied in the academic literature. An excellent survey is contained in R.G. Ibbotson and J.R. Ritter, "Initial Public Offerings," *Finance,*

Volume 9 of Handbooks in Operations Research and Management Science, eds. R.A. Jarrow, V. Maksimovic, and W.T. Ziemba (Amsterdam: North Holland, 1995).

11. B.S. Black and R.J. Gilson, "Venture Capital and the Structure of Capital Markets: Banks versus Stock Markets," *Journal of Financial Economics,* vol. 47 (1998), pp. 243–277.

12. L.A. Jeng and P.C. Wells, "The Determinants of Venture Capital Funding: Evidence Across Countries," Working Paper (Cambridge, MA: Harvard Business School, 1998).

13. D.R. Emery and J.D. Finnerty, *Corporate Financial Management* (Upper Saddle River, NJ: Prentice Hall, 1997).

14. J.R. Ritter, "The Long Run Performance of Initial Public Offerings," *Journal of Finance,* vol. 46 (1991), pp. 3–27.

15. Based on study of how 26 venture capital funds exited 442 investments from 1970–1982 in the U.S., T.A. Soja and J.E. Reyes, *Investment Benchmarks: Venture Capital* (Needham, MA: Venture Economics, 1990).

16. J.W. Petty, W.D. Bygrave, and J.M. Shulman, "Harvesting the Entrepreneurial Venture: A Time for Creating Value," *Journal of Applied Corporate Finance,* vol. 7 (1994), pp. 48–58.

Chapter 14

1. Based in part on Bradford Cornell and Alan C. Shapiro, "Financing Growth Companies," *Journal of Applied Corporate Finance* (summer 1988), pp. 6–22.

2. See Henry Grabowski and John Vernon, "A New Look at the Returns and Risks to Pharmaceutical R&D," *Management Science* (July 1990), pp. 804–821. Grabowski and Vernon examined the R&D costs and returns for 100 new drugs in the U.S. during the decade of the 1970s. R&D costs after tax, expressed in 1986 dollars, were $81 million on average. However, the top decile had R&D costs of $457 million and the second decile ran about $160 million.

3. For example, Alpha Interferon was first approved to treat a rare cancer called hairy-cell leukemia. But it has since found an $800 million worldwide market as a treatment for herpes and hepatitis, and is currently being tested against AIDS.

4. Benjamin Klein, Robert G. Crawford, and Armen Alchian, "Vertical Integration, Appropriable Rents, and the Competitive Contracting Process," *Journal of Law and Economics,* vol. 21 (1978), pp. 297–326; Bengt Holmstrom and Jean Tirole, "The Theory of the Firm," *Handbook of Industrial Organization,* eds. R. Schmalensee and R.D. Willig (New York: Elsevier Science Publishers B.V., 1989), chap. 3; and Oliver E. Williamson, "Transaction Cost Economics," *Handbook of Industrial Organization,* vol. 1, eds. R. Schmalensee and R.D. Willig (New York: Elsevier Science Publishers), chap. 3.

5. David J. Teece, "Towards an Economic Theory of the Multiproduct Firm," *Journal of Economic Behavior and Organization,* vol. 3 (1982), pp. 39–63; and *The Competitive Challenge: Strategies for Industrial Innovation and Renewal,* ed. David J. Teece (Cambridge, MA: Ballinger Publishing Co., 1987).

6. George A. Akerloff, "The Market for Lemons: Qualitative Uncertainty and the Market Mechanism," *Quarterly Journal of Economics* (August 1970), pp. 488–500.

7. William A. Sahlman, "Aspects of Financial Contracting in Venture Capital," *Journal of Applied Corporate Finance* (summer 1988), pp. 23–36 for a fuller discussion of why venture capitalists typically invest using convertible preferred instead of common stock.

8. The problem of information asymmetry in financing is discussed in S.C. Myers and N. Majluf, "Corporate Financing and Investment Decisions When Firms Have Information That Other Investors Do Not Have," *Journal of Financial Economics,* vol. 13 (1984), pp. 187–221.

9. The problem of financing growth options with debt is discussed in Stewart C. Myers, "Determinants of Corporate Borrowing," *Journal of Financial Economics* (November 1977), pp. 138–147.

10. Merton H. Miller, "Debt and Taxes," *Journal of Finance* (May 1977), pp. 261–276.

11. This estimate is from *Bridging the Gap* (Ernst & Young, 1999).

12. Michael Brennan and Eduardo Schwartz, "The Case for Convertibles," *Chase Financial Quarterly* (spring 1982), pp. 27–46.

13. See, for example, John J. McConnell and Chris J. Muscarella, "Corporate Capital Expenditure Decisions and the Market Value of the Firm," *Journal of Financial Economics* (September 1985), pp. 399–422; and Gregg A. Jarrell, Kenneth Lehn, and Wayne Marr, "Institutional Ownership, Tender Offers, and Long-Term Investments," (The Office of the Chief Economist, Securities and Exchange Commission, April 19, 1985).

14. These data come from "Biotechnology Industry Valuation: 1992 Sector Analysis," (Paine Webber Health Care Group, February 1992).

15. The use of staged financing is common in venture capital for similar reasons (see note 7).

16. The announcement of new products that never materialize, or do so only after long delays, is a common and well-recognized strategy in the software industry. Many companies announce what has come to be known as "ghostware" or "vaporware," referring to the lack of any tangible product.

17. Ashish Arora and Alfonso Gambardella, "Complementary and External Linkages: The Strategies of the Large Firms in Biotechnology," *Journal of Industrial Economics* (June 1990), pp. 361–379.

18. Stewart C. Myers, "The Capital Structure Puzzle," *Journal of Finance* (July 1984), pp. 575–592.

Part V

1. Diane Brady, "GE's Welch: 'This Is the Greatest Opportunity Yet,' " *Business Week* (June 28, 1999).

Chapter 15

1. L. Rosenkopf and M. Tushman, "The Coevolution of Community Networks and Technology: Lessons from the Flight Simulation Industry," *Industrial and Corporate Change,* vol. 7 (1998), pp. 311–346.

2. M. Tushman and L. Rosenkopf, "On the Organizational Determinants of Technological Change: Toward a Sociology of Technological Evolution," *Research in Organizational Behavior,* eds. B. Staw and L. Cummings, vol. 14 (Greenwich, CT: JAI Press, 1992), pp. 311–47.

3. M. Cusumano and R. Selby, *Microsoft Secrets: How the World's Most Powerful Software Company Creates Technology, Shapes Markets, and Manages People* (New York: Free Press, 1995).

4. L. Rosenkopf and M. Tushman, "The Co-Evolution of Technology and Organizations," *Evolutionary Dynamics of Organizations,* eds. J. Baum and J. Singh (Oxford Press, 1994).

5. W. Powell, K. Koput, and L. Smith-Doerr, "Interorganizational Collaboration and the Locus of Innovation: Networks of Learning in Biotechnology," *Administrative Science Quarterly,* vol. 41 (1996), pp. 116–145.

6. L. Rosenkopf and A. Nerkar, "On the Complexity of Technological Evolution: Exploring Coevolution within and across Hierarchical Levels in Optical Disc Technology," *Variations in Organization Science: In Honor of D.T. Campbell,* eds. J. Baum and B. McKelvey (Sage Publications, 1999).

7. G. Hamel, "Competition for Competence and Inter-Partner Learning within International Strategic Alliances," *Strategic Management Journal,* vol. 12 (1991), pp. 83–103; Y.L. Doz, "The Evolution of Cooperation in Strategic Alliances: Initial Conditions or Learning Processes?" *Strategic Management Journal,* vol. 17 (1996), pp. 55–83. Evolutionary Perspectives on Strategy Supplement; and Dyer and Singh, Chap. 9, this book.

8. J. Dyer, "Effective Interfirm Collaboration: How Transactors Minimize Transaction Costs and Maximize Transaction Value," *Strategic Management Journal,* vol. 18 (1996), pp. 535–556; and Dyer and Singh, Chap. 16, this book.

9. See note 1.

10. E. von Hippel, "Cooperation between Rivals: Informal Know-How Trading," *Research Policy,* vol. 16 (1987), pp. 291–302.

11. L. Rosenkopf and M. Faraoni, "The Coevolution of Formal and Informal Interorganizational Networks," Working Paper (University of Pennsylvania, 1996); T. Lant and A. Eisner, "The Role of Medical Professionals in Facilitating Strategic Alliances among Pharmaceutical Firms," Working Paper (New York: New York University, 1996); and L. Rosenkopf and A. Turcanu, "Strategic Participation in Cooperative Technical Networks: Emergence, Evolution and Effects of Informal Interfirm Networks," Working Paper (University of Pennsylvania, 1998).

12. J. Liebeskind, A. Oliver, L. Zucker, and M. Brewer, "Social Networks, Learning and Flexibility: Sourcing Scientific Knowledge in New Biotechnology Firms," *Organization Science,* vol. 7 (1996), pp. 428–443.

13. G.F. Davis, "Agents without Principles? The Spread of the Poison Pill through the Intercorporate Network," *Administrative Science Quarterly,* vol. 36, no. 4 (1991), pp. 583–613.

14. P. Haunschild, "Interorganizational Imitation: The Impact of Interlocks on Corporate Acquisition Activity," *Administrative Science Quarterly,* vol. 38 (1993), pp. 564–592.

15. R. Gulati and J. Westphal, "The Dark Side of Embeddedness: An Examination of the Influence of Direct and Indirect Board Interlocks and CEO/Board Relationships on Interfirm Alliances," *Administrative Science Quarterly* (1998).

16. P. Almeida and B. Kogut, "Localization of Knowledge and the Mobility of Engineers," *Management Science* (1998).

17. K.M. Eisenhardt and C.B. Schoonhoven, "Resource-Based View of Strategic Alliance Formation: Strategic and Social Effects in Entrepreneurial Firms," *Organization Science,* vol. 7, no. 2 (1996), pp. 136–150.

18. W. Boeker, "Executive Migration and Strategic Change: The Effect of Top Manager Movement on Product-Market Entry," *Administrative Science Quarterly,* vol. 42, no. 2 (1997), pp. 213–236.

19. P. Almeida and L. Rosenkopf, "Interfirm Knowledge Building by Semiconductor Startups: The Role of Alliances and Mobility," Working Paper (University of Pennsylvania, 1997).

20. See note 19.

21. T.E. Stuart and J.M. Podolny, "Local Search and the Evolution of Technological Capabilities," *Strategic Management Journal,* vol. 17 (1996), pp. 21–38. Evolutionary Perspectives on Strategy Supplement.

22. P. Lane and M. Lubatkin, "Relative Absorptive Capacity and Interorganizational Learning," *Strategic Management Journal,* vol. 19, no. 5 (1998), pp. 461–477.

23. D.-J. Kim and B. Kogut, "Technological Platforms and Diversification," *Organization Science,* vol. 7, no. 3 (1996), pp. 283–301.

24. A. Saxenian, *Regional Advantage: Culture and Competition in Silicon Valley and Route 128* (Cambridge, MA: Harvard University Press, 1994).

25. See note 16.

26. L. Rosenkopf and A. Nerkar, "Crossing Technological and Organizational Boundaries for Knowledge-Building: A Typology of Exploration Behaviors and their Effects in the Optical Disc Industry," Working Paper (University of Pennsylvania, 1998); J. Sorenson and T. Stuart, "Aging, Obsolescence and Organizational Innovation," Working Paper (University of Chicago, 1998).

27. G. Ahuja, "The Antecedents and Consequences of Innovation Search: A Longitudinal Study," Working Paper (University of Texas, Austin, 1999).

28. R. Burt, *Structural Holes: The Social Structure of Competition* (Cambridge, MA: Harvard University Press, 1992).

29. L. Freeman, "Centrality in Social Networks: Conceptual Clarification," *Social Networks,* vol. 1 (1979), pp. 215–239.

30. There are many different ways of assessing cliques (also called components). For example, strong components cliques require that each member of the clique communicate with every other member of the clique. In contrast, weak components cliques only require that each member communicate with at least one other member of the clique. Our cliques represent this second option. The nature of this hypothetical network is such that if we focused on strong components cliques, only a few pairs of firms would qualify as cliques (C and F; G and E, H and J). Later in this paper, we will look at some cellular industry networks, which are very dense, and then strong components cliques are more interesting. Thus the choice of rules defining the clique will be a function of the network under observation.

31. Recognize also that these rankings are only for 1995. Dynamic analyses of these measures on a year-by-year basis would demonstrate, for example, Qualcomm's rapid rise from virtually no network presence in 1990 to their high-status standing in 1995.

32. See note 2.

33. J. Law and M. Callon, "Engineering and Sociology in a Military Aircraft Project: A Network Analysis of Technological Change," *Social Problems,* vol. 35, no. 3 (1988), pp. 284–297.

34. T. Hughes, *Networks of Power* (Baltimore: The Johns Hopkins University Press, 1983).

35. D.F. Noble, *Forces of Production* (New York: Alfred A. Knopf, 1984).

36. See note 5.

37. T. Stuart et al., "Interorganizational Endorsements and the Performance of Entrepreneurial Ventures." *Administrative Science Quarterly* (1998).

38. T. Stuart, "Network Positions and Propensities to Collaborate: An Investigation of Strategic Alliance Formation in a High-Technology Industry," *Administrative Science Quarterly* (1999).

39. G. Ahuja, "Collaboration Networks, Structural Holes, and Innovation: A Longitudinal Study," Working Paper (University of Texas, Austin, 1998).

40. See note 21.

41. Two software packages were used for the analyses contained in this paper: (1) STRUCTURE (Columbia University, 1991); and (2) UCINET V (Analytic Technologies, 1998).

42. See note 16.

43. For further reading, see R. Burt, *Structural Holes: The Social Structure of Competition* (Cambridge, MA: Harvard University Press, 1992). Burt discusses the strategic gaming associated with managing network structure. He also introduces several more complex calculations, called "constraint" and "structural autonomy," which are other ways to assess network advantage.

Chapter 16

1. J. Hagedoorn, "Understanding the Rationale of Strategic Technology Partnering," *Strategic Management Journal,* vol. 14 (1993), pp. 371–385.

2. B. Kogut, "Joint Ventures: Theoretical and Empirical Perspectives," *Strategic Management Journal,* vol. 9 (1989), pp. 319–332; J. Bleeke and D. Ernst, *Collaborating to Compete: Using Strategic Alliances and Acquisitions in the Global Marketplace* (New York: Wiley, 1993); *Alliance Analyst* (Newcap Communications, 1998).

3. A.M. Brandenberger and B.J. Nalebuff, *Co-opetition: 1. A Revolutionary Mindset That Redefines Competition and Cooperation: 2. The Game Theory Strategy That's Changing the Game of Business* (New York: Doubleday, 1997).

4. I. Dierickx and K. Cool, "Asset Stock Accumulation and Sustainability of Competitive Advantage," *Management Science,* vol. 35, no. 12 pp. 1504–1513.

5. Kathryn Harrigan, *Strategic Flexibility* (Lexington, MA: Lexington Books, 1985); S. Balakrishnan and B. Wernerfelt, "Technical Change, Competition and Vertical Integration," *Strategic Management Journal,* vol. 7 (July/August 1986), pp. 347–359.

6. Gordon Walker and David Weber, "A Transaction Cost Approach to Make-or-Buy," *Administrative Science Quarterly,* 29 (1984); Balakrishnan and Wernerfelt, 1986.

7. See note 5, Balakrishnan.

8. P. Botticelli, D. Collis, and G. Pisano, *Intel Corporation: 1968–1997* (Boston: Harvard Business School Press, 1998), pp. 15–16.

9. See note 1, Hagedoorn, pp. 380–381.

10. J.H. Dyer and H. Singh, "The Relational View: Cooperative Strategy and Sources of Interorganizational Competitive Advantage," *Academy of Management Review,* vol. 23, no. 4 (October 1998), pp. 660–679.

11. J.G. March and H.A. Simon, *Organizations* (New York: Wiley, 1958); W.W. Powell, K.W. Koput, and L. Smith-Doerr, "Interorganizational Collaboration and the Locus of Innovation: Networks of Learning in Biotechnology," *Administrative Science Quarterly,* vol. 41 (1996), pp. 116–145; N.S. Levinson and M. Asahi, "Cross-National Alliances and Interorganizational Learning," *Organizational Dynamics,* vol. 24 (1995), pp. 51–63.

12. E. von Hippel, *The Sources of Innovation* (New York: Oxford University Press, 1988).

13. B. Kogut and U. Zander, "Knowledge of the Firm, Combinative Capabilities, and the Replication of Technology," *Organization Science,* vol. 3, no. 3 (1992), p. 386; R.M. Grant, "Toward a Knowledge-Based Theory of the Firm," *Strategic Management Journal,* vol. 17 (winter 1996), pp. 109–122.

14. See note 13.

15. R. Nelson and S. Winter, *An Evolutionary Theory of Economic Change* (Cambridge: Belknap Press, 1982); B. Kogut and U. Zander, "Knowledge of the Firm, Combinative Capabilities, and the Replication of Technology," *Organization Science,* vol. 3, no. 3 (1992), p. 386; G. Szulanski, "Exploring Internal Stickiness: Impediments to the Transfer of Best Practice within the Firm," *Strategic Management Journal,* vol. 17 (1996), pp. 27–43.

16. W.M. Cohen and D.A. Levinthal, "Absorptive Capacity: A New Perspective on Learning and Innovation," *Administrative Science Quarterly*, vol. 35 (1990), pp. 128–152.

17. G. Szulanski, "Exploring Internal Stickiness: Impediments to the Transfer of Best Practice within the Firm," *Strategic Management Journal*, vol. 17 (1996), pp. 27–43; D.C. Mowery, J.E. Oxley, and B.S. Silverman, "Strategic Alliances and Interfirm Knowledge Transfer," *Strategic Management Journal*, vol. 17 (1996), pp. 77–91.

18. K.J. Arrow, *The Limits of Organization* (New York: W.W. Norton, 1974); R. Daft and R. Lengl, "Organizational Information Requirements, Media Richness and Structural Design," *Management Science*, vol. 32, no. 5, pp. 554–571; P.V. Marsden, "Network Data and Measurement," *Annual Review of Sociology*, vol. 16, pp. 435–463; J.L. Badaraco, Jr., *The Knowledge Link* (Boston: Harvard Business School Press, 1991).

19. W. Shan and W. Hamilton, "Country-Specific Advantage and International Cooperation," *Strategic Management Journal*, vol. 12, no. 6 (September 1991), pp. 419–432.

20. Y. Doz, "The Evolution of Cooperation in Strategic Alliances: Initial Conditions or Learning Processes," *Strategic Management Journal*, vol. 17 (1996), pp. 55–83; R.M. Kanter, "Collaborative Advantage: The Art of Alliances," *Harvard Business Review* (July/August 1994), pp. 96–108.

21. A.F. Buono and J.L. Bowditch, *The Human Side of Mergers and Acquisitions* (San Francisco: Jossey-Bass, 1989).

22. See note 20.

23. O.E. Williamson, *The Economic Institutions of Capitalism* (New York: Free Press, 1985).

24. Dyer (1996a).

25. Clark and Fiyimoto (1991) and Nishiguchi (1994).

26. Asanuma (1989) and Dyer (1996).

27. A. Saxenian, *Regional Advantage* (Cambridge, MA: Harvard University Press, 1994).

28. A. Saxenian, "Regional Networks and the Resurgence of Silicon Valley," *California Management Review* (fall 1990), p. 101.

29. See note 27, p. 430.

30. B. Klein, R.G. Crawford, and A.A. Alchian, "Vertical Integration, Appropriable Rents, and the Competitive Contracting Process," *Journal of Law and Economics*, vol. 21, pp. 297–326.

31. See note 23; D.C. North, *Institutions, Institutional Change and Economic Performance* (Cambridge, England: Cambridge University Press, 1990).

32. B. Uzzi, "Social Structure and Competition in Interfirm Networks: The Paradox of Embeddedness," *Administrative Science Quarterly*, vol. 42, no. 1 (1997), pp. 35–67.

33. See note 2.

Chapter 17

1. Jerry Useem, "Internet Defense Strategy: Cannibalize Yourself," *Fortune,* vol. 140, no. 5 (1999), 121ff.

2. Michael L. Tushman and Philip Anderson, "Technological Discontinuities and Organizational Environments," *Administrative Science Quarterly,* vol. 31 (1986), pp. 439–465.

3. James G. March of Stanford University did the pioneering thinking on this concept. See James G. March, "Exploration and Exploitation in Organizational Learning," *Organization Science,* vol. 2 (1991), pp. 71–87.

4. Somewhat less frequently, established organizations are able successfully to navigate the transition. But even in these instances, it is well known that the innovating organization needs separation from the operating organization (see also chap. 2). However, the greater the separation from the operating organization, the greater the difficulty in integrating the two organizations at a later date.

5. David A. Nadler and Michael Tushman, *Competing by Design: The Power of Organizational Architecture* (New York: Oxford University Press, 1997).

6. See note 3.

7. Bruderer and Singh develop a genetic algorithm based approach to studying organizational evolution. See Erhard Bruderer and Jitendra V. Singh, "Organizational Evolution, Learning and Selection: A Genetic Algorithm Based Model," *Academy of Management Journal,* vol. 39, no. 5 (1996), pp. 1322–1349.

8. Whereas form may be correlated with strategy, the concept of strategy applies to all dimensions including goals, authority relations, technologies, and markets and means different configurations of dimensions. In addition, strategy is often considered synonymous with structure. Today researchers write about business models that include both strategy and structure. Strategies can be mapped onto organizational forms.

9. Steve Socolof, private communication. We appreciate his contributions to this project in an interview and materials he shared.

10. Stephen H. Haeckel, *Adaptive Enterprise: Creating and Leading Sense-and-Respond Organizations* (Boston: Harvard Business School Press, 1999). We also appreciate his personal contributions to this project. He was willing to share a copy of his yet unpublished manuscript, and he spent a significant amount of time in an interview.

11. Joan Magretta, *The Power of Virtual Integration: An Interview with Dell Computer's Michael Dell* (Boston: The President and Fellows of Harvard College, 1998).

12. Rolf Wigand, Arnold Picot, and Ralf Reichwald, *Information, Organization, and Management: Expanding Markets and Corporate Boundaries* (New York: Wiley, 1997).

13. Peter G.W. Keen and Michael Scott Morton, *Decision Support Systems: An Organizational Perspective* (Reading, MA: Addison-Wesley, 1978).

14. W.W. Powell, "Hybrid Organizational Arrangements: New Form or Transitional Development?" *California Management Review,* vol. 30 (1983), pp. 67–87.

15. P.S. Ring and A. Van de Ven, "Developmental Processes of Cooperative Interorganizational Relationships," *Academy of Management Review*, vol. 19 (1974), pp. 90–118.

16. Hansverner Voss, "Virtual Organizations: The Future is Now," *Strategy and Leadership* (July 1996), pp. 12–16.

17. See note 10.

18. This is similar to the "centerless corporation" proposed by Pasternack and Viscio. Bruce A. Pasternack and Albert J. Viscio, *The Centerless Corporation: A New Model for Transforming Your Organization for Growth and Prosperity* (New York: Simon & Schuster, 1998).

19. Introduced by Tushman and O'Reilly. Michael L. Tushman and Charles A. O'Reilly, *Winning through Innovation: A Practical Guide to Leading Organizational Change and Renewal* (Boston: Harvard Business School Press, 1997).

20. Clayton M. Christensen, *The Innovator's Dilemma: When New Technologies Cause Great Firms to Fail* (Boston: Harvard Business School Press, 1997).

21. Galbraith, Lawler, and Associates first introduced this form in 1993. It is an organization that divides activities between the front end, organized by customer or geography, and the back end, organized by product and technology. This form combines the features of the single business and the divisional profit-center model. Quasi profit centers exist around the front or back, and they are less autonomous than in a related diversified/ divisionalized organization. Jay R. Galbraith, Edward E. Lawler III, and Associates. *Organizing for the Future: The New Logic for Managing Complex Organizations* (San Francisco: Jossey-Bass, 1993).

22. James Brian Quinn, *Intelligent Enterprise: A Knowledge and Service Based Paradigm for Industry* (New York: Free Press, 1992).

23. See note 18.

24. Jessica Lipnack and Jeffrey Stamps, *Virtual Teams: Reaching across Space, Time, and Organizations with Technology* (New York: Wiley, 1997).

25. See note 10.

26. Kirk W.M. Tyson, *Competition in the 21st Century* (FL: St. Lucie Press, 1997).

27. Process can be front line, or execution oriented, or foresight/insight oriented, market sensing to build on trends.

28. See note 5.

29. See note 21.

Chapter 18

1. "In an Industry Where Loyalty Means Little, It Pays to Get Around," *Wall Street Journal* (December 2, 1998), p. A1.

2. "Microsoft Executives, Hitting 40, Break with the Program: Software Giant's Top Ranks Thin as Ex-Bosses Pursue Dreams, Like Bowling," *Wall Street Journal, European Edition* (June 16, 1999).

3. P. Cappelli, *The New Deal at Work* (Boston: Harvard Business School Press, 1999); C. Handy, *Beyond Certainty: The Changing Worlds of Organizations* (Boston, MA: Harvard Business School Press, 1995); and J. Rifkin and R. Heilbroner, *The End of Work: The Decline of the Global Labor Force and the Dawn of the Post-Market Era* (New York, NY: Putnam, 1995).

4. E.G. Chambers, M. Foulton, H. Handfield-Jones, S.M. Hankin, and E.G. Michaels, "The War for Talent," *The McKinsey Quarterly,* vol. 3 (1998), pp. 44–57.

5. "Silicon Valley Companies Look to 'Virtual' CEOs," *International Herald Tribune* (August 26, 1999), p. 1.

6. S. Friedman, P. Christensen, and J. DeGroot, "Work and Life: The End of the Zero-Sum Game," *Harvard Business Review,* 1998 76/6 (1999), pp. 119–129.

7. See note 3.

8. J.J. Clancy, "Is Loyalty Really Dead?" *Across the Board,* vol. 36, no. 6 (1999), pp. 14–19.

9. F. Fukuyama, *The Great Disruption* (New York: Free Press, 1999).

10. A. Giddens, *Modernity and Self-Identity* (Stanford CA: Stanford University Press, 1991).

11. Interestingly, the management literature on flexibility is concerned primarily with firm level design and tends to address people issues from this perspective only.

12. J-M. Hiltrop, "The Quest for the Best: Human Resources Practices to Attract and Retain Talent," *European Management Journal,* vol. 17, no. 4 (1999), pp. 422–430.

13. See, for example, E. Welles, "Motherhood, Apple Pie & Stock Options," *Inc. Magazine* (February 1998).

14. M.E. Brown, "When Employees Leave," *Electronic Business,* vol. 24, no. 6 (1998), p. 43.

15. J. Pfeffer, *The Human Equation* (Boston: Harvard Business School Press, 1998).

16. T.O. Davenport, *Human Capital: What It Is and Why People Invest It* (San Francisco: Jossey-Bass, 1999).

17. H. Bouchikhi and J.R. Kimberly, "The Customized Workplace," *Management 21C,* ed. S. Chowdhury (London: Financial Times Publishing, 1999).

18. R. Semler, *Maverick* (New York: Warner Books, 1993).

19. Compagnie Française de Défense et de Protection.

20. J.R. Kimberly and H. Bouchikhi, "The Dynamics of Organizational Development and Change: How the Past Shapes the Present and Constrains the Future," *Organization Science,* vol. 6, no. 1 (1995), pp. 9–18.

21. T. Butler and J. Waldroop, "Job Sculpting: The Art of Retaining Your Best People," *Harvard Business Review,* vol. 77, no. 5 (1999), pp. 144–152.

22. N. Kumar, "The Power of Trust in Manufacturer-Retailer Relationships," *Harvard Business Review* (November/December 1996), pp. 92–106.

INDEX